NORTH DEVON

MEDICAL LIBRARY

# Mould's Medical Anecdotes
## Omnibus Edition

to be returned on or before
the last date stamped below.

Dated bt
relevant
15/11/15

1998

Withdrawn

Reason:

Date:

By:

NORTH DEVON

ICAL LIBRARY

D1343338

# MOULD'S

# Medical Anecdotes

## *Omnibus Edition*

CAN BE TAKEN IN LARGE OR SMALL DOSES

Suitable for after dinner,
recommended as a tonic for the
general public,
can be taken with alcohol,
cures boredom
and all bad humours.

INGREDIENTS INCLUDE:

quacks, shrunken heads, spooks, X-rays, skeletons,
radium, saints, witch doctors and foreign bodies.

This unique patent medicine can be obtained at a very
reasonable price from the sole manufacturer, or from
any respectable bookshop.

Institute of Physics Publishing
Bristol and Philadelphia

This selection © Richard F Mould 1984, 1989, 1996

Pages 1–145 of this book were first published in *Mould's Medical Anecdotes* (1984; ISBN  0 85274 762 4); pages 147–406 were first published in *More of Mould's Medical Anecdotes* (1989; ISBN 0 85274 119 7)

All rights reserved.  No part of this publication may be reproduced, stored in a retrieval system or transmitted in any form or by any means, electronic, mechanical, photocopying, recording or otherwise, without the prior permission of the publisher.  Multiple copying is permitted in accordance with the terms of licences issued by the Copyright Licensing Agency under the terms of its agreement with the Committee of Vice-Chancellors and Principals.

*British Library Cataloguing-in-Publication Data*

A catalogue record for this book is available from the British Library

ISBN 0 7503 0390 5

*Library of Congress Cataloging-in-Publication Data are available*

Published by Institute of Physics Publishing, wholly owned by The Institute of Physics, London
Institute of Physics Publishing, Techno House, Redcliffe Way, Bristol BS1 6NX, UK
US Editorial Office:  Institute of Physics Publishing, The Public Ledger Building, Suite 1035, 150 South Independence Mall West, Philadelphia, PA 19106, USA

Printed in Great Britain by J W Arrowsmith Ltd, Bristol

To Stuart ...

... and Imogen

with love from Grandad

Dick Mould was a hospital physicist and cancer statistician at the Royal Marsden and Westminster Hospitals in London for 30 years. He is now 'retired' from the British National Health Service and travels the world full time as a consultant giving courses on basic medical statistics (mainly to radiation oncology residents) in countries including the USA, Germany, France, Japan, Austria, Belgium and Switzerland.

In addition, he acts as a consultant to the World Health Organization on matters relating to the Chernobyl accident and also lectures on the history of x-rays, Marie Curie, and Chernobyl. He has written some 80 scientific papers and has written or edited some 40 books.

His *Chernobyl — The Real Story* (published by Pergamon Press) is translated into Japanese and Russian, and his *Introductory Medical Statistics* (Institute of Physics Publishing) is translated into Japanese. His most recent book is *A Century of X-rays & Radioactivity in Medicine with Emphasis on Photographic Records of the Early Years* (Institute of Physics Publishing). The only type of book he has not yet written is a novel, but there is still time!

# Preface: *Mould's Medical Anecdotes Omnibus Edition*

I was delighted when my publishers planned to reprint a combined volume of *Mould's Medical Anecdotes* (1984) and *More of Mould's Medical Anecdotes* (1989), which appear respectively on pages 1–145 and 147–406, and when they also gave me the opportunity to include 56 more pages of anecdotes.

As before, the new material covers a wide range of topics from aphrodisiacs at a Texan Fiesta, Chinese gods and iron balls, a small additional dose of alcohol, America's first physiologist, some organization charts for readers to correlate with their own departmental environment or to plan international clinical trial organograms, to four longer offerings.

The American journalist's interview with Pierre Curie in 1904 is followed up by 1995 information on autopsy results on Pierre and Marie with a comment on the FBI. The 12-year follow-up on my earlier Moscow Case History refers not only to the KGB but also to chandeliers and a British Airways pilot. The medical records of Stalin and some of his cronies in the Politburo have never been published before in English and were translated (by the physician's grandson) from a journal written in the Georgian language by one of Stalin's doctors. Microbiological studies on Mayan archaeological artefacts lead to the Turin Shroud and the high probability that its carbon dating was incorrect and that the Shroud is in fact several hundred years older than this would suggest (1260–1390).

This additional material was gathered from a variety of visits I have undertaken in the 1990s, either when giving *Basic Medical Statistics Courses with Emphasis on Cancer*, when working as a consultant to the World Health Organization, or when attending conferences. It was during the last that I met in Atlanta the grandson of Stalin's physician, and the Mayan archaeological and microbiological expert in San Antonio. Texas was also where I partook of bull's testicles and tequila containing the famous 'worm'. Shanghai and Hong Kong were the source of the Chinese anecdotes, Paris for those concerning the Curies and a snake farm outside Ho Chi Minh City (Saigon) towards the Mekong Delta for the cobra tonic (which was not tested) and where the local pharmacist became very affronted when I started to laugh. However, I escaped in one piece without being bitten! The Japanese anecdotes were obtained during visits to Hiroshima, Nagasaki, Osaka and Nara.

I would also like to record my thanks to the many people who continue to send me amusing items for these books, but in particular I am grateful to Dr Frank Yeung, Dr Kimberly Hart, Dr Michael Selva, Dr Andrée Dutreix, Dr Leoncio Garza-Valdes, Reverend Frederick Brinkmann CSSR, Dr John Freeman, Dr Larry Doss, Professor Jean-Marc Cosset, Professor Toshihiko Inoue, Professor Yutuka Hirokawa, Dr Takehiro Inoue, Mr Shunji Ichikawa and Dr Nicholas Kipshidze. I would also like to express my thanks for their valuable help to Ms Pamela Whichard, Ms Sharon Toop and Mr Jim Revill of Institute of Physics Publishing.

Once again, I hope that there is a little of something for everybody and I have included an extensive subject index so that particular topics of interest can be located easily. This time I have dedicated this volume to my grandchildren, although from their photographs it can be seen that it could be some time yet before I find out if they have inherited their grandfather's sense of medical humour. Happy reading!

Richard F Mould
London, May 1996

# *Preface: More of Mould's Medical Anecdotes*

With my publishers now asking for a second volume of *Anecdotes*, I obviously got the recipe correct in the first volume with its ingredients of quacks, skeletons, x-rays, witch doctors, saints and foreign bodies of various descriptions. A repeat prescription is therefore offered for those who were kind enough, or foolhardy enough, to part with real money on the promise of a few laughs.

This collection has, to some extent, been easier to assemble, since many of my friends are now aware that I am an anecdotal squirrel. In consequence, material has been arriving in a steady stream from many countries. I would hasten, though, to apologise to some contributors for the exclusion of their favourite material, because I have felt it paramount to observe a certain amount of decorum with this anthology—most of the time!

As before, parts have been gathered whilst I have been abroad on non-trivial matters for WHO, IAEA and other bodies, and consequently the reader will find in these pages mention of events either experienced first hand (such as the case of the Tiger's Penis near the Golden Triangle) or merely reported in local newspapers and books in Cairo, Manila, Bangkok, Chernobyl and Chiang Mai. The spectrum of contents includes, as before, the historical, the unusual, the unexpected, the humorous and the mysterious from sources such as medical journals, advertisements, comic monologues and newspapers.

In addition, I have been able to incorporate material proffered during interviews when the first volume of *Anecdotes* was published. In particular, I must mention Monica O'Hara, who interviewed me for the *Liverpool Echo* and who turned out to be an expert graphologist as well as a journalist. We collaborated with my friend the late Harold Winter to produce the results of the graphology 'blind' trial of Wellington and Nelson.

Illustrations again loom large and I am most grateful to the various picture libraries such as those of the Wellcome Library for the History of Medicine and the Mansell Collection. I hope, therefore, that there will be something for everybody in this volume and that it will raise a laugh or two when readers have a few minutes to relax from the daily grind.

**Richard F Mould**
London, February 1989

# Preface: *Mould's Medical Anecdotes*

I first became attracted to the idea of producing a book of anecdotes during my schooldays some 30 years ago. Imagine my surprise when I later found an 1866 book of anecdotes compiled by one of my ancestors, another Richard Mould (1806–1871), four generations earlier. This book was entitled *Literary Pearls Strung at Random: Historical and Biographical Anecdotes, Aphorisms, Wit and Humour in Prose and Verse.* We obviously share a sense of humour as he had published one epitaph that I had also thought of choosing! This concerns Frederick, Prince of Wales, father of King George III:

Here lies Fred,
Gone down among the dead.
Had it been his Father
We had much rather;

ix

Had it been his Mother
Better than any other;
Had it been his sister,
Few would have missed her;
Had it been the whole generation,
Ten times better for the nation;
But since it is only Fred,
There is nothing more to be said.

However, the impetus to compile this current collection was not so much a wish to follow in family footsteps, but more a response to comments from students on a collection of illustrations used in lectures to keep the audience awake. When in reply to a request for questions I more than once received comments such as 'please tell us some more anecdotes' or 'have you written a book of anecdotes?', a compilation began to seem like a good idea, and gathering these articles together provided light relief while writing books of a more serious nature.

The resultant medical recipe, described on the jacket as a quack medicine, is offered to all those who enjoy reading books of 'snippets' and viewing interesting and unusual illustrations. Particular emphasis has been placed on photographs and the contents cover a spectrum of the historical, the unusual, the unexpected, the humorous and the mysterious. Medical anecdotes have been extracted from many sources including newspaper cuttings, epitaphs, quotations, medical journals, advertisements, comic monologues and case histories, and the subject matter ranges from evolution through animals, politicians, punk rockers and royalty to mummification and death, from medical schools to surgeons, from quacks and witch doctors to saints and from lead poisoning to radioactivity. The compilation also has an international flavour including articles relating to places as far apart as Egypt and Colombia, Russia and the Philippines, and Thailand and Ireland, but asking overseas friends for their favourite medical anecdote did not always provide a publishable reply!

Items have been selected from many medical journals including *British Medical Journal, The Lancet, New England Journal of Medicine* and *Journal of the American Medical Association.* The Wellcome Library for the History of Medicine provided several of the illustrations and there are also reproductions from the Tate and National Portrait Galleries in London and the Tretyakov Gallery and the Kremlin Faceted Chamber in Moscow. Hopefully, there is something for everyone, particularly those interested in medicine.

**Richard F Mould**
London, August 1983

# Acknowledgments: Mould's Medical Anecdotes

I should like to acknowledge with thanks the loan of the nineteenth century book from Mr R J Lewis for the article 'Why are the heads of men hairy?'; Mrs M Ion of the Local Studies Library, Shrewsbury for drawing my attention to the article on the 'Irish export trade in dead bodies'; and Professor Romsai Suwanik and Miss Sunantha for the photographs of the Wat Po Traditional School of Medicine.

I am most grateful for all the generous assistance provided by Mr Neville Hankins, Sponsoring Editor of Adam Hilger Ltd. In addition, I would like to express my thanks to Mrs Janet McKernan of the Production Department and my son, Timothy. The cartoons of the patient, nurse and doctor were devised by myself and Miss Ann Jeffries— who was the artist responsible for making these three persons come *alive*! The patient first appeared on page 32 of my 1976 textbook on *Introductory Medical Statistics* and is obviously still alive and well, although minus a leg at the end of *Medical Anecdotes*!

The Compiler and Adam Hilger Ltd (Institute of Physics Publishing) gratefully acknowledge permission to reproduce copyright material. Every effort has been made to trace copyright ownership and to give credit to copyright owners, but if, inadvertently, any mistake or omission has occurred, full apologies are herewith tendered.

George Allen & Unwin (Publishers) Ltd
American Psychological Association
Banco de la República, Bogotá, Colombia
Blackwell Scientific Publications Ltd
Articles first published in *British Medical Journal* and reproduced with the
    kind permission of the Authors and Editor
The Canadian Medical Association
Cuthbert Andrews Ltd
*Daily Telegraph*
Professor Vincent J Derbes
Dr W C Ellerbroek

Professor Harold Ellis
EMI Music Publishing Ltd, 138–140 Charing Cross Road, London WC2
*Evening Standard*
Dr C T Fitts
Fontana Paperbacks
ILFORD X-Ray Focus, by kind permission of ILFORD Limited
IPC Business Press Ltd, now renamed Business Press International Ltd
*Journal of the American Medical Association* © American Medical
    Association
*The Journal of X-ray Technology*
Dr Ole Didrik Laerum
Jennifer Laing
*The Lancet*
F Filce Leek
*The Malay Mail*, Kuala Lumpur
*Medical History*
Contribution by Mrs J L Monson, Oxford and reprinted with permission
    from *Reader's Digest*, London. ©1982 Reader's Digest
The National Portrait Gallery, London
*New England Journal of Medicine*
*New Scientist*
*The News and Courier*; *The Evening Post*, Charleston, SC
*The Observer*
Orbis Publishing Ltd
Oxford University Press
Pitman Books Ltd, London
*The Practitioner*
Dr William M Rambo
Reuters
George H Scherr, Publisher, *Journal of Irreproducible Results*
The Science Museum, London
Dr Karl Skullerud
Mrs E J Snell for and on behalf of the *Biometrika* Trustees
Dr Oscar Sugar
The Tate Gallery, London
*The Times*
The Tretyakov Gallery, Moscow
USSR State Committee for the Utilisation of Atomic Energy (KGAE) for
    permission to reproduce illustrations of Ivan the Terrible
Wellcome Institute Library, London
Westminster Medical School, London
*World Medicine*

# Contents

xiv

xix

# King's Evil

The King's Evil, or Queen's Evil when a Queen is on the throne, is tuberculous lymphadenopathy of the neck, commonly known as scrofula, and is a disease of the lymph glands of the neck. The illustration from Queen Mary's Manual, 1553–4, shows the Queen touching a scrofulous boy with both hands directly applied to the affected part of the neck. This has also been described in a letter written in 1556 which stated that after the Queen had pressed with her hands on the spot where the sore was '. . . she then made the sick people come to her again and taking a gold coin called an angel, she touched the place where the evil showed itself, signed it with the Cross and passed a ribbon through the hole which had been pierced in [the coin], placing one of them round the neck of each

of the patients and making them promise never to part with that coin, save in case of extreme need'.

A much earlier record of the cure of scrofula is attributed to the English King, Edward the Confessor who died in 1066. The illustration[1] is taken from a manuscript on the life of this king, written in 1139, and shows Edward touching a scrofulous woman.

Shakespeare also refers to the scrofula in *Macbeth*, act IV, scene 3 as follows.

> ... 'Tis called the evil:
> A most miraculous work in this good king,
> Which often, since my here-remain in England,
> I have seen him do. How he solicits heaven
> Himself best knows; but strangely-visited people,
> All swol'n and ulcerous, pitiful to the eye,
> The mere despair of surgery, he cures;
> Hanging a golden stamp about their necks
> Put on with holy prayers; and 'tis spoken,
> To the succeeding royalty he leaves
> The healing benediction.

REFERENCE

Crawfurd R 1911 *The King's Evil* (Oxford: Clarendon)

# A Royal Malady

History abounds with stories of mad kings and princes. One concerns King George III of Great Britain and Ireland who reigned from 1760, when he was 22 years of age, until his death in 1820 and who was on the English throne when America won its war of independence. His illness commenced when he was 50 and he had consultations with no fewer than seven physicians but there was no agreement on a diagnosis. One suggestion was *flying gout* which was supposed to have flown from the King's legs to his head and stayed there! He suffered from weakness and pain in the limbs, rarely slept and talked faster and faster becoming increasingly incoherent. In all, he had four attacks of this unknown illness, the last in 1811 when he was permanently disabled by blindness, senility and mental disturbance. At this point a Regent was appointed to rule, his son, the future King George IV.

The cause of this royal insanity remained a mystery until an article in *British Medical Journal* in 1966[1] claimed that the King's illness was due to attacks of hereditary porphyria—a rare disorder which includes mental disturbance, the development of skin lesions as a result of exposure to sunlight, and the passage of dark red urine caused by the presence of pigments called porphyrins. This is now widely accepted but some doubts still exist as to the genealogical evidence[2] which, it is suggested, shows that the porphyria gene passed directly down the royal line through nine generations from Mary Queen of Scots to King George IV[3].

From time to time this subject of porphyria and George III appears in the medical literature[4,5] but the most recent mention was in the pages of *New Scientist*[6] beginning 'Vampires and werewolves need fear the wooden stake or the silver bullet no longer. It's official! They're probably suffering from iron-deficiency porphyria'. It continues 'iron-free porphyrins build up in the skin which becomes electronically excited by light. The excitation is passed onto any oxygen present in the skin cells. This now highly-reactive excited oxygen then attacks the cells causing disfigurement. The sufferer quickly learns to avoid daylight, venturing out only at night. Now does that ring any bells?'. One tongue-in-cheek solution proposed for temporary relief of the illness was 'drinking blood!' and to emphasise werewolf characteristics, some of the chronic side effects of porphyria were listed as 'a tightening of the gums leading to the impression of large protruding teeth, a hirsute appearance and an animal nature'.

*King George III* from the studio of W Beechy.
(Courtesy: National Portrait Gallery, London.)

4

REFERENCES

1  Macalpine I and Hunter R 1966 *Br. Med. J.* **1** 65
2  Dean G 1971 *The Porphyrias* (London: Pitman Medical) 2nd edn
3  Macalpine I and Hunter R 1969 *George III and the Mad-Business* (London: Penguin)
4  Witts L J 1972 *Br. Med. J.* **4** 479
5  Illis L S 1964 *Proc. R. Soc. Med.* **57** 23
6  Milgrom L 1982 *New Sci.* **96** 244

(Courtesy: Wellcome Institute library, London.)

# The King is Dead

A book on the deaths of the rulers of England[1] was reviewed in *The Lancet*[2] and its author congratulated on his courage in diagnosing the causes of death on what, in many cases, is very slender and unreliable testimony. The mortality statistics for the rulers from the Norman Conquest to the death of King Edward VII, including Richard Cromwell and Lady Jane Grey are outlined below.

> No fewer than eight died violent deaths, either in battle or at the hands of the murderer or headsman. Acute infections and cardiovascular–renal disease have each accounted for six; syphilis, congenital or acquired, and 'dysentery', which covers a multitude of ignorances, have each disposed of four. 'Senile decay', a diagnosis no longer accepted as a certified cause of death, is mentioned in three cases and implied in one other. Stephen may have had an appendix abscess, Edward III gonorrhoea, and Richard II anorexia nervosa, but Henry I's 'surfeit of lampreys' was probably ptomaine poisoning. George IV and William IV, pathologically speaking, had much in common; both had hepatic cirrhosis, pericarditis and probably pneumonia. Those who think harshly of James I will do well to bear in mind that he had Bright's disease, enlarged tonsils, renal calculi, jaundice, haemorrhoids, dental caries and pyorrhoea, and arthritis—surely enough to sour any man.

REFERENCES

1   Yearsley M 1935 *An Account of the Deaths of the Rulers of England* (London: Unicorn)
2   1935 *The Lancet* **2** 1471

# Queen Anne's Oculist

Queen Anne reigned in England from 1702 to 1714 and gave birth to 17 children, none of whom survived her. She died grossly overweight at 20 stone due to overeating in the belief that this would produce healthy children. In addition, she had weak eyes and it is reported[1] that her favourite oculist was William Read, a tailor who, 'having failed as a mender of garments, set up as a

This illustration of *Sir William Read* appeared in an article entitled *Some Notable Quacks*[2]. Read died at Rochester in 1715.

mender of eyes'. Another of Queen Anne's oculists was 'Dr' Grant who started life as a tinker, tried preaching, and drifted into his medical specialty after failing at these two ventures. The following is an epigram of the time.

> Her majesty sure was in a surprise
> Or else was very shortsighted,
> When a tinker was sworn to look after her eyes
> And the Mountebank Read was knighted.

REFERENCES

1   Haggard H W 1929 *Devils, Drugs and Doctors* (New York: Harper and Row)
2   1911 *Br. Med. J.* **1** 1264

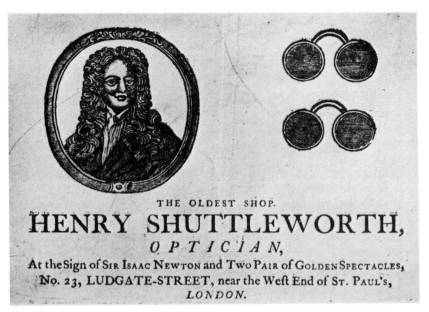

THE OLDEST SHOP.

# HENRY SHUTTLEWORTH,

*OPTICIAN,*

At the Sign of SIR ISAAC NEWTON and TWO PAIR of GOLDEN SPECTACLES,

No. 23, LUDGATE-STREET, near the West End of ST. PAUL'S,

*LONDON.*

(Courtesy: Wellcome Institute library, London.)

# *Ivan the Terrible's Skeleton*

In Cathedral Square within the Moscow Kremlin is the Archangel Cathedral, the burial place of many of the Grand Dukes of Muscovy and the Tsars including Ivan I, known as *Ivan the Moneybags*, and Ivan IV, better known as *Ivan the Terrible*, who among other cruelties murdered his son and heir. He became the first Tsar of Russia at the age of 17 in 1547 having been Grand Duke of Muscovy since the age of three. By this time he was already well known for his murderous rages, drunken acts of cruelty and sexual passions. Yet as a ruler his character was full of contradictions; he brought about much needed land and church reforms, consolidated Russia's borders to the East, and opened up diplomatic relationships with England. He died in 1584 after a continuous final twenty-year reign of terror.

In 1953 a special commission from the Soviet Ministry of Culture opened the

These two paintings from Moscow give contrasting images of Ivan IV. The left-hand picture, from the Tretyakov Gallery, shows the more characteristic *Ivan the Terrible*. The right-hand painting shows a saint-like representation of the Tsar, which is part of a mural on a window bevelling in the Faceted Chamber of the Kremlin. This chamber was the main Throne Room of the Muscovite Princes; dating from the late fifteenth century, it is one of the oldest surviving structures in the Kremlin.

sarcophagus of Ivan IV; the scene and following studies on the skeleton are described in the book by Smyth and are reproduced here.

The body lay with its arms crossed over its breast, and was clad in the dusty remnants of a monk's habit, on which were embroidered the texts of prayers. A crucifixion scene, worked in coloured silk, adorned a cloth over the torso, but as if to set off this evidence of belated piety, a drinking goblet of dark blue glass enamelled in yellow stood to the left of the head.

Adhering to the brittle skull bones were a few traces of hair from the eyebrows and beard. The Tsar had been a fairly tall man of around six feet in height; according to contemporary reports he had been very strong in his youth, but

9

towards the end of his life had put on weight until he eventually tipped the scales at 210 pounds.

Carefully stripped of its robes, the skeleton in the tomb told a clear tale of suffering to the scientists. Quite early in his life, Ivan's cartilages and ligaments had begun to ossify, or harden, and the joints of the long bones all showed traces of inflammation leading to the conclusion that he had suffered from poly-arthritis. Almost the whole skeleton showed signs of the torsion which accompanies this disease.

Chemical analysis revealed the presence of arsenic and mercury in the body; the arsenic content was judged to be normal, but the mercury count was very high. The scientists concluded that a quicksilver-based ointment had been used regularly in an attempt to alleviate the pain in his limbs.

To add to the misery of his bone disease Tsar Ivan had undergone an extremely painful and very rare experience during his fifties—at the peak of his 'reign of terror'. Judging by the state of his teeth, his adult, or secondary incisors, canines and premolars had only come through the gums at this late time, and the process must have been an agonising one.

All this did not excuse the behaviour of Ivan the Terrible, but it did go a long way towards explaining it. As a youth he had a tendency to be cruel, and the perpetual pain of his adult life, coupled with his heavy drinking in an effort to alleviate it, could only have led to a savage warping of his already embittered character.

REFERENCE

Smyth F 1980 *Cause of Death* (London: Orbis)

# Napoleon III's Bladder Stone

When Napoleon and death are mentioned together it is usually to consider the possibility that the first Emperor Napoleon was murdered in exile on the island of St Helena and that he died from accidental or homicidal arsenic poisoning. Opinion as to the technique of administration varies, and recently[1] a theory has been proposed that the arsenic contained in the wallpaper was the culprit. Earlier[2], the blame has been mentioned in relation to an English secret agent, rat poison and cooking pots made of metal containing arsenic.

However, a later Napoleon, son of one of the first Emperor's brothers became Emperor Napoleon III of France in a coup d'état in 1852. It is his illness over a period of nearly 20 years and eventual death in 1873 which are described here, based on reports published by Ellis[3].

Napoleon's first symptoms of a bladder stone appeared in 1856 with occasional spasms of pain but in 1864 there was a severe epsiode of haematuria (blood in the urine), and further attacks became more and more common. By 1869 his urine was never free from pus and it was difficult to relieve his dysuria (difficulty in passing urine). A report from his doctors was drawn up in 1870, a few days before the outbreak of the Franco-Prussian War, but was suppressed by Napoleon III, partly from reluctance to submit himself to surgery and partly because of the political implications of a sick man being at the head of a country about to go to war. The shakings of a carriage journey or of riding on horseback caused much pain; six weeks after the commencement of the war his condition was pitiful. His eyelids were puffy, his pale face was disguised with rouge and there were attacks of shivering. At the battle of Sedan he exposed himself to the enemy's fire, but his bravery may well have been a death wish to escape the tortures of his stone. However, at the end of the day, the white flag appeared over the city of Sedan and the following message was dispatched to the Emperor William of Prussia: 'Sire my brother, being unable to die at the head of my troops, nothing remains but to surrender my sword into the hands of your majesty'. The fate of France was sealed and Napoleon III became a prisoner of war of the Prussians.

Napoleon III went into exile in England, at Chislehurst, his illness got progressively worse and in 1872 he was seen by some leading London physicians and surgeons who advised exploration of the bladder—but this was refused.

The lying in state of Napolean III, a contemporary drawing from the Illustrated London News of January 25, 1873.

Eventually, though, definitive treatment was agreed and on December 26 the Emperor was anaesthetised by 'the most experienced chloroformist of the day' and a stone, which was certainly not less than the size of a date, was found. Seven days later an operation was performed to crush the stone and remove as much of the debris as possible, but this was not totally successful and a second operation had to be performed on January 6. Three days later, Napoleon III lapsed into a coma and died.

REFERENCES

1  Jones D 1982 *New Sci.* **96** 101
2  Lister J 1965 *New Engl. J. Med.* **272** 419
3  Ellis H 1969 *A History of Bladder Stone* (Oxford: Blackwell Scientific)

# One of the Favourite Jokes of Suicidal Patients

What are the things between elephants' toes?
*Slow natives*

REFERENCE

Spiegel D et al 1969 *J. Consult. Clin. Psychol.* **33** 504

# A King as Dentist

King James IV of Scotland, who was born in 1473, was unusual in that he actually paid his subjects for the privilege of allowing him to act as surgeon or dentist. For example, in 1503 he purchased 'equipment' to take out teeth and was himself the patient when he paid the barber 14 shillings to extract one of his teeth. In 1511 it is recorded that James paid compensation to a dental patient 'because the King pullit furth his teth' and in the same year, strangest of all, James IV actually operated on one of his own barber-surgeons, extracting two of his teeth.

James IV's medical practice included blood-letting, for which the fee, 28 shillings, was twice that received for dentistry, and in 1497 he had instituted measures for the isolation of those suffering from syphilis, the 'strange seiknes of Nappilis', which, following its dramatic appearance at Naples, had spread to Scotland, where it was known as grandgore or grantgore. Treatment was strongly discouraged, probably because the cure at that time, was worse than the disease.

REFERENCE

Guthrie D 1963 *Janus in the Doorway* (London: Pitman Medical)

*Dentistry 1523.* A quack extracting teeth while his assistant picks the patient's pocket; from an engraving by Luceis van Leiden. (Courtesy: Wellcome Institute library, London.)

14

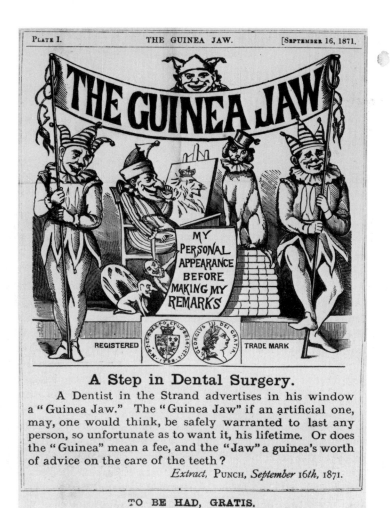

(Courtesy: Wellcome Institute library, London.)

*Donald Robertson*
*Born 1st January 1785*
*Died 4th June 1848*
*Aged 65 years*
*He was a peacable quiet man, and, to all appearance a sincere Christian. His death*
*was very much regretted—which was caused by the stupidity of Lawrence Tulloch*
*of Clotherton who sold him nitre instead of Epsom Salts by which he was killed in*
*the space of three hours after taking a dose of it.*

Gravestone inscription, Cross Kirk, Shetland

# Oliver Cromwell's Head

After the beheading of King Charles I, Oliver Cromwell became Lord Protector. When he died in 1658 after a three week illness, termed by his doctors *bastard tertian ague*, there was a hurried autopsy and embalming and, after a lying in state, he was buried in Westminster Abbey—although it was a wooden effigy used for the two public occasions! The actual date of the secret burial is unknown.

Two years later, his body and that of one of his generals, Henry Ireton, and the body of John Bradshaw, the President of the tribunal which sent Charles I to the scaffold, were exhumed and hung on a triple gallows at Tyburn. When they were taken down, they were decapitated and the heads fixed on spikes on the roof of Westminster Hall. Cromwell's head remained there for some 24 years. Since then, stories have abounded as to the last resting place of Cromwell's head[1,2,3].

The most recent review of the evidence[3] suggests that the head fell off the pike one stormy evening in 1685 and was hidden in a soldier's home for many years. Following this, it was located in a private museum in London from 1710 to 1738 and then in 1775 it reappeared in the possession of an impoverished and alcoholic actor, who offered to sell it to Sidney Sussex College, Cambridge, where Cromwell spent one year. This offer was declined and it was sold to a London jeweller, resold, and then it found its way to an exhibition in Bond Street in 1799. In 1814 it was sold to a Mr Josiah Henry Wilkinson and remained with his descendants until the mid-nineteenth century, and became known as the Wilkinson head of Oliver Cromwell.

It has now been concluded beyond all reasonable doubt that this head is in fact the real head of Oliver Cromwell and is not a forgery. Some of the evidence leading to this conclusion is fascinating as the observations of Dickson Wright[4] show.

> The head lies in the solidly made antique box, which has contained it for 200 years, and rests on a red satin cloth. I was permitted to remove the cranial vault and saw the falx, tentorium and other dural membranes all intact. The head is embalmed and very shrivelled, and the first thing I looked for was the historical wart, which Cromwell insisted on his portrait painters putting in, and there, in the proper place, was the depression from which it has been chipped. Then I looked for the little beard, that one sees in all his pictures, and it was there, as was his moustache. The long curly head of hair was gone, but it was easy to see that Cromwell was not bald when he died, but had a complete head of reddish hair.

16

*The Wilkinson head of Oliver Cromwell.* Full face showing
the wart cavity above the right eyebrow, the moustache,
the cincture of the skull-cap opened at the autopsy and the
protruding point of the pole's iron-tipped spike.
(Courtesy: Wellcome Institute library, London.)

Then I looked for the marks of the axe on the back of his neck, and the marks of
six blows of the axe were plainly to be seen. Now there is in existence an eyewitness's description of the execution, which says that it took eight blows of the axe
to sever the head because of the embalming cloths wrapped round the neck. The
nose is flattened and pushed to one side; one would expect this to occur while the
head was being chopped off.

The vault of the skull has been sawn off and prised open with a chisel, and then
the scalp has been sewn back. The iron spike goes through, under the mandibular
arch and protrudes from the vault of the skull, and the piece of oak pole attached to
the spike is very worm-eaten and some of the worm-holes in the flesh are
continuous with those in the pole, showing that the head and the pole are of the
same age.

The teeth are nearly all gone and the lips are broken, no less than one would
expect after all the vicissitudes endured by the poor man. I must say I felt queerly
moved while holding the head and thinking that from these lips came the historical
utterances of Cromwell.

The end of the story comes after the death of Canon Horace Wilkinson who
left the head to Sidney Sussex College, Cambridge. In 1960, the college decided
to give it a proper burial, the head was interred in a secret place, and a plaque at
the entrance of the college chapel reads: 'Near to this place was buried on March

25, 1960, the head of Oliver Cromwell, Lord Protector of the Commonwealth of England, Scotland and Ireland, Fellow Commoner of this College 1616–1617'.

REFERENCES

1   Pearson K and Morant G M 1935 *Biometrika* **26** 1
2   1935 *Bri. Med. J.* **1** 775
3   Bruce-Chwatt L 1982 *World Med.* 48 23 January
4   Dickson Wright A 1937 *St. Mary's Hospital Gazette* **67** 9

# Head Shrinking

The practice of shrinking a decapitated head to the size of a fist or large orange appears first in the literature among the Spanish accounts dating from the 16th century expeditions into South America. Although head *hunting* has been practised in many parts of the world, head *shrinking* appears to be confined to the 25 000 square miles between latitudes 2°–5° and longitudes 76°–79° in South America. This dense tropical jungle, supplied by the Tigre, Pastaza, Morona, Marañon, and Santiago rivers, forms part of Peru, Colombia, Ecuador, and Brazil. The gold-bearing waters of the Santiago attracted the Portuguese and Spaniards, who finally established cities in this land of the Jivaro or Jebero (hé vä Rô) Indians. The repeated cruelties of the conquerors resulted in revolts, which, by 1599, destroyed the Spanish rule, and thereafter made the governors despair of subjugating these savages—and to ask for help from the missionaries. For 150 years, these devoted churchmen were sent to preach and explore—and were usually murdered by the very savages they came to humanise. The final insurrection occurred when the Jivaros were required to bring gold tributes for the coronation of Phillip III. The rebels overcame the garrison, bound the governor, and, having melted down the gold they had brought, they poured the molten gold inside his mouth 'and his bowels burst within him'. As late as the mid 1920s, the Jivaros were still attacking Peruvian garrisons on the fringes of their lands. Most of their fighting, however, was intertribal, carried out for plunder and for revenge. Among the items prized were the women of the enemy, and it is remarkable how readily they accepted and were accepted by the very groups which slaughtered their husbands.

18

The origin of the practice of making the heads of the slain Indians smaller is unknown. Cummings, who was adopted into a tribe and married several native women, believed the primary reason was for the automatic transferral to the taker of the head, of the spiritual, mental, and moral qualities of the former head owner. Certainly it was not just for the collecting of a trophy, as appears to have been the case among the 'wild men' of the South Seas. In themselves, the heads were not so important as the use made of them, to make an object called 'tsantsa' (Sp. *chantcha*, Fr. *tzantza*), with ritual and religious significance. The ceremonies accompanying and following the head-shrinking process appear to be of much physical and psychological consequence.

The preparation of the tsantsa is variously described as taking 20 hours, weeks or months. There are many descriptions; many are obviously not at first hand. Domville-Fife was unable to persuade any tribe with which he came in contact to divulge the process. Von Hagen was unable to witness the shrinking of a human

*Two shrunken heads* with suspensory cords, lip threads, long straight hair and some facial ornamentation. Facial hair is absent. The approximate diameter of each head is 5 cm and the hair length is 36 cm (right) and 56 cm (left).

19

head, but cooperative natives illustrated the process with the head of a sloth. Flornoy combined glimpses of the early parts of human head shrinking with observations of the technique as practised for his benefit with a monkey head; later, when he was the only white man with a subgroup of Jivaros called the Huambizas, he was able to witness the final rites with a human head. Cummings included descriptions of head shrinking in his account of his life (and almost death!) among the Jivaros, implying that he was present at the procedure. Up de Graff accompanied a war party and hence considered his account to be an authentic description of a process which had baffled many commentators who had dealt with the subject. His story does agree in part with the description of Stirling, and of the more recent travellers who, like Flornoy, have taken motion pictures of the shrinking of animal heads.

Until recent years, the chief weapons of the Jivaros have been blowguns with darts to paralyse small animals, knives and lances, usually of hard palm wood (chonta). After the enemy was felled (sometimes while not yet completely lifeless), the skin of the neck was cut low on the chest, much like the modern autopsy incision, and the lower neck disarticulated with a sharp blade. A strip of bark was then passed through the mouth and down the oesophagus to the cut-off end of the neck. This permitted the head to be bound to the waist of the warrior, as the war party travelled toward home. When sufficient distance was put between the fighters and the scene of the battle, camp was made, usually along one of the plentiful streams, and the head shrinking began.

An earthenware pot, never before used, and brought along for this specific purpose, was filled with water and put on a fire to boil. The skin of the head was slit up the back of the neck to the vertex, and peeled forward, using sharp knives, originally of bamboo or chonta, and later, of metal, to work between skull and skin. The lower jaw and skull were removed and discarded, and the eyelids sewn together (from the inside, according to Cotlow). Slivers of chonta palm wood were put through the lips, and bound together with fibre, to hold the mouth closed. The 'head', now a preparation of skin, scalp and hair, was put into the boiling water with a vine whose juice is considered to have the property of preventing the hair from falling out. After boiling for two hours, the partly shrunken 'head' was cooled, and the pot thrown away, never to be used again. The slit up the back of the skin was sewed, so that there was again the semblance of a human head.

In the meantime, rounded stones had been collected and heated in the fire. These were picked up with a pair of sticks and dropped into the opening where the neck was cut off. The head was rotated with one hand, while another hot stone, held in a leaf, was used to smooth out the skin and features from the outside. Successively smaller stones were put into the head sac as it shrank, and finally heated sand was poured in, while the head was rotated to permit all parts to shrink evenly. Some eyelashes and eyebrows were plucked to keep these hairy parts in proportion to the head size. The downy hairs of the face, now disproportionately long, were singed off with a flaming stick.

20

(Courtesy: Wellcome Institute library, London.)

The crown of the head was pierced, and a string passed from outside to inside, where it was looped around a short stick, so the head could be suspended on a rack about 3 feet above the fire, where it was left to heat and smoke overnight. This process made the head darker and very hard, and the oils from the skin were buffed with a cloth, much as animal leather is polished. A 'needle' and string were used to make a purse-string around the neck skin, to close it off. Cummings described a variant process in which the head stayed in the cooled, boiled preservative overnight, in a liquid so rank that insects which fell in immediately died.

The massaging of the outside with stones was accompanied or followed by manipulation of the nose and face, to make the face resemble its original state, according to some authors, or to distort it deliberately, according to others.

21

When the head was finished, the chonta wood lip sticks were removed and replaced with coloured threads. These were knotted and allowed to dangle in front about as long as the hair behind, which was often up to half a metre in length. Karsten speaks of the Indian belief that the hair was the seat of the soul or vital power, and of the practice of Jivaro men to let their hair grow longer than that of the women. Lip sewing was done in the belief that this would prevent the spirit of the victim from cursing the slayer, and the smoking was done in the smoke of cooking fires to feed and preserve the spirit of the tsantsa, according to Cummings.

REFERENCE

Sugar O 1971 *J. Am. Med. Assoc.* **216** 117

# Why are the Heads of Men Hairy?

A small book (7 × 12 cm²) printed in London in the nineteenth century for *The Booksellers* purported to contain 'The works of Aristotle, the famous philosopher' and included 'His Book of Problems'. The question and answer below is one of these problems.

> Q. Why are the heads of men hairy? A. The hair is the ornament [sic] of the head, and the brain is purged of gross humours by the growing of the hair, from the highest to the lowest, which pass through the pores of the exterior flesh, becomes dry, and converted into hair. This appears to be the case from the circumstance that in all man's body there is nothing drier than the hair, for it is drier than the bones; and it is well known that some beasts are nourished with bones, as dogs, but they cannot digest feathers or hair, but void them undigested, being too hot for nourishment. It is answered that the brain is purged in three different ways: of superfluous watery humours by the eyes, of choler by the nose, and of phlegm by the hair.

*A good head over a pair of wooden shoes is a great deal better than a wooden head belonging to an owner whose feet are shod in calf-skin.*
OLIVER WENDELL HOLMES

22

# Holes in Their Heads

Trephining, or trepanning, is a surgical operation of quite considerable antiquity. It entails cutting a piece of bone out of the skull to expose the brain. Surgeons carry out the operation for a variety of reasons, mostly in order to effect delicate brain surgery, such as the removal of a tumour. But it is a constant source of mystery why such a hazardous operation should have been carried out in ancient times and, considering the medical resources then available, how it was done. In many societies headaches and other disorders were thought to be induced by evil spirits entering the brain, and therefore a hole had to be made to let them out. In some cases it was thought necessary to reduce pressure on the inside of the skull, while in other instances more magical reasons may have prevailed, since the removed roundels were used as charms.

It appears that in the remote past a remarkable number of trephined people survived their ordeal. Numerous skulls have been found in which the surgical incisions had healed, suggesting that the victims had lived to a ripe old age. Some of the earliest date to the Neolithic period. A good example was found at Crichel Down in Dorset in 1938, where a Beaker Period burial in a round barrow had a skull broken by a round hole. The removed roundel had been placed with the body and had presumably been carried round as an amulet in life.

One of the most interesting groups of trephined skulls has come from Anglo-Saxon cemeteries in East Anglia, datable to the sixth century AD. Four skulls were found, each cut in a distinctive manner. The expert who examined them pronounced that the people had been operated on by the same surgeon, for the method of making the incisions was unmistakable—all being elliptical instead of the usual circle or square. All had been made with the same instrument— probably a bronze chisel, used to gouge slivers of bone from the ever-deepening channel. The surgeon was no butcher; all four operations had apparently been completely successful, and did not result in later infection, as was usually the case. To have achieved this result the 'master of the Gliding Gouge', as he has been called, would have had to be extremely skilful, and prepared to deal with profuse bleeding and potential danger. How he managed such a feat is still a mystery.

REFERENCE

Laing J 1979 *Britain's Mysterious Past* (Newton Abbot: David and Charles) pp 25–6

# The 'Esau Lady'

(Courtesy: Wellcome Institute library, London.)

Bearded ladies have at all times been exhibited before the public, but the above is certainly a most striking instance. Miss Annie Jones-Elliot, called the 'Esau Lady', was born in 1865, at Marion, Smith County, Virginia. Her parents, or sisters and brothers, showed no sign of more than an ordinary supply of hair. At her birth she was already possessed of a good-size moustache, much to the distress of her parents. But when two years old, she proved a godsend to her family—her beard had by then fully developed, and she was exhibited in public with great success. She was a woman of fine physique, her skin smooth and beautiful, and she had all the accomplishments of one of the fair sex. Her hair and beard were dark brown.

# Traffic in Human Hair

In the *British Medical Journal* of August 27th, 1910 reference was made to the brisk business done in the sale of human hair which is used to supply the natural lack of what has been called woman's finest ornament. In *Chambers's Journal* for September 1st we find some details as to the hair market which may supplement the information we were able to give as to the source of the commodity. The greater bulk, says our contemporary, comes from the south-eastern corner of Bohemia. In this region the human hair market is a familiar sight, and the preparation of the material is quite a large industry. The native supply is supplemented by cargoes from China, but these combings are restricted to a certain class of article. The importation from China is very large, the hair arriving in bales packed in straw, averaging about one hundred and thirty pounds in weight. The Chinese hair, as is well known, is intensely black, and as such is of no value. Upon its arrival at the factory, therefore, the first process is to change its characteristic colour. This is accomplished by bleaching it in a solution of hydrogen peroxide. Afterwards the hair is sorted according to its length and grade, dyed and finally finished in any shade desired by the customer. We are unable to say whether the trade is a profitable one, but we note that an individual—not apparently a member of the medical profession—describing himself as an 'obesity specialist' recently attributed his bankruptcy partly to his having devoted his attention to the importation and sale of human hair, and to 'intellectual pursuits'. Bald men often ascribe the loss of their hair to the stress of mental work, the nutritive material intended for it being diverted to the brain—a theory which is received by the uncharitable as having much the same foundation as Falstaff's attribution of his huskiness of voice to singing of anthems. It is conceivable that bankruptcy may lead to baldness, but the case referred to would seem to show that the baldness—of other people—may lead to bankruptcy.

REFERENCE

1910 *Br. Med. J.* **2** 641

# The Brothers Tocci

The Tocci brothers were born in October 1877 at Locana, in the province of Turin, Italy. The illustration is from an advertisement when they were placed on display. They each had a well formed head, perfect arms and a perfect thorax to the sixth rib; they had a common abdomen, a single anus, two legs, two sacra,

(Courtesy: Wellcome Institute library, London.)

two vertebral columns, one penis but three buttocks. The right boy was christened Giovanni-Batista, and the left, Giacomo. Each individual had power over the corresponding leg on his side, but not over the other one. Walking was therefore impossible. All their sensations and emotions were distinctly individual and independent.

# Ear Surgery

A Chicago surgeon recently devised a procedure by which he hoped to transplant a healthy human ear, presumably to take the place of one destroyed by disease. But no opportunity of performing the operation presenting itself, he advertised for two subjects who should be willing to have their ears amputated for and in consideration of the sum of 300 dollars (£60 in 1902). Two persons at once came forward. But then a difficulty presented itself—the law. Inquiring of legal oracles, the surgeon was told by a judge that if he removed a healthy ear from a human head, even with the consent of the person to whom the ear and the head belonged, he would be guilty of 'mayhem'. As everyone may not know what 'mayhem' is, we quote the definition thereof given in *Blackstone's Commentaries*. In that classical work it is said: 'A man's limbs, by which for the present we only understand those members which may be useful to him in fight, and the loss of which alone amounts to *mayhem* by the Common Law, are also the gift of the wise Creator to enable him to protect himself from external injuries in a state of nature.' Another legal authority, a State Attorney, informed the surgeon that the removal of one or more ears in the manner and under the conditions aforesaid did not constitute any crime known to the law. If Blackstone's definition quoted above is to be taken as supplying the solution of this legal problem, we have no hesitation in saying that we agree with the minor prophet rather than with the judge. An ear may conceivably be useful to a man's opponent in a fight as supplying a convenient means of steadying the head for punishment, but to himself, especially if it is prominent, it scarcely affords protection against external injuries either in a state of nature or otherwise. We unfortunately do not know how the matter ended. Probably the surgeon decided that it would be better for his own comfort not to supply the lawyers with so interesting a case for argument.

REFERENCE

1902 *Br. Med. J.* **1** 161

# The Safety Tomb

In *The Lancet* of 1935 a relic of body snatching days was shown. This was a small leaflet, dated 1827, and printed on both sides (one side is shown here) which was circulated by the Rev J Scholefield of Manchester, who died in 1855. It was in

The REV. J. SCHOLEFIELD solicits the attention of such as have to perform the mournful tribute of respect to their deceased, to his

### SAFETY TOMBS,

of which the adjoining sketch will give an idea, and which are unequalled by any in the town, or indeed in the kingdom. Their security has now been proved, and they continue to increase in the estimation of the public.

A large bell is fixed and connected with a spring; so that if any one approach within a certain distance, it will ring for a length of time, and thus give an alarm to the neighbourhood and notice of DANGER to any person who may have the TEMERITY to ATTEMPT an ENTRANCE after the VAULT is SECURED. Corpses are frequently put into the vault for a certain time for security, and afterwards removed into their own graves. The Keys are placed in the possession of the Proprietor every night.

### DUES FOR THE SAFETY TOMB.

| | £ | s | d |
|---|---|---|---|
| Still born Child | £0 | 1 | 6 |
| A Child under one year old | 0 | 5 | 0 |
| A Child under five years | 0 | 6 | 0 |
| Above five years and not requiring a Bier | 0 | 7 | 0 |
| Large Corpse | 0 | 8 | 0 |

1827 and 1828 that the infamous body snatchers Burke and Hare committed their crimes, and in 1828 a Parliamentary Committee was appointed 'to inquire into the study of anatomy as practised in the United Kingdom and into the best method of obtaining bodies for the purpose'.

REFERENCE

1935 *The Lancet* **1** 719

# Irish Export Trade in Dead Bodies

In the December 29, 1824 issue of the *Eddowes Salopian Journal* the following short report was printed under the caption *Irish Export Trade in Dead Bodies*.

In our journal of the 8th inst. we recorded the proceedings of two inquests held in this town (Shrewsbury) on the preceding day, on the bodies of a male and a female unknown, which had been packed in separate boxes, and forwarded by coach from Holyhead, directed to 'Messrs Smith and Co. to be left at Aspin's Warehouse, Morgan's Lane, Tooley Street, Southwark, London' with a packet ticket nailed on each box, stating that they were 'From Thomas Williams's Steam Packet Parcel Office, 53, Lower Sackville Street, Dublin'. The paragraph was copied from our Journal into the London Papers and thence into the Dublin Papers, where it attracted (as was very natural) Mr Williams's particular attention. *The Dublin Morning Post* says, 'In consequence of the exertions made by Mr Williams to discover the person who sent them to this house, a Mr George Pearson, surgical student of St Bartholomew's Hospital, London, was taken into custody in Stephen Street yesterday. He has been fairly identified as the person who came to Mr Williams, under the name of Smith, respecting the conveyance of the boxes which were afterwards accompanied by a note signed 'J Smith'. Mr Pearson has admitted that he was the person who sent them. He is fully committed to take his trial for the offence'.

In the earlier, December 8, 1824 issue of the Journal under the caption *Inquests*, the discovery of the corpses was described. The book-keeper at the Lion Coach Office directed the boxes to be weighed preparatory to their despatch to London; the cord round one of the boxes broke; the box fell to the ground; the end burst out; the grey head of a dead man made its appearance. The body of an elderly woman was then found in the other box. The Journal went on to record their opinion that the bodies were intended for dissection and that 'we trust the parties will think it necessary for the sake of decency, to pack their *treasures* a little more carefully'.

# Skeletons for Sale

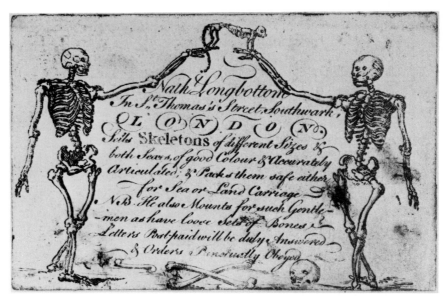

(Courtesy: Wellcome Institute library, London.)

# Kuda Bax and the Indian Fire Walk

A fire-walk demonstration by Kuda Bax organised by Mr Harry Price, the honorary secretary of the University of London Council for Psychical Investigation, took place last week in the grounds of a house at Carshalton through the courtesy of Mr Alex L Dribbel. Two trenches about 11 ft long, 6 ft wide, and 9 in deep had been dug in which had been kindled a charcoal fire. Seven tons of oak logs, one ton of fire-wood, ten gallons of paraffin, and fifty newspapers were used in preparing the fire. On the top was spread half a ton of forest-burnt oak charcoal. The fire was ignited at 8 AM in the morning of September 17th. The walk took place between 3 and 4 PM on the same day when intense heat was being given out from the fire. We are indebted to Prof C A Pannett, FRCS, for the following account of the proceedings:

Kuda Bax is the physically slight Indian type. He claims that his immunity from burning is due to an act of faith, to the influence of his mind over his bodily reactions. He is temperamental and took a dislike to one pit over which he could not be induced to walk, saying that he felt there was something wrong and unfavourable about it. The explanation of this attitude was perhaps revealed later; he had twice successfully walked over the one pit and had twice walked to the brink of it again to make a third attempt when his courage failed him. He said something had gone out from him and that the physicist who had taken the temperature of the glowing charcoal had desecrated the pit. He did not actually use that word, but said his was a religious act of faith and that it had been rendered impossible by the thrusting of the instrument (a thermocouple) into the burning embers. Up to this time the physicist had occupied himself with the pit from which Kuda Bax had already shown an aversion. When the thermocouple was placed in the fire it registered a temperature of 800°C, but when it merely made rapid contacts in imitation of footsteps it only registered, I believe, the temperature of rather warm water. On the other hand two English volunteers certainly blistered their feet in making attempts at the walk.

There appeared to be nothing abnormal about the soles of Kuda Bax's feet. The skin was soft and not at all callosed. It seemed to be very dry, much drier than the feet of the two Englishmen who made attempts to walk on the glowing embers. A skin thermometer showed the temperature of his feet to be 93.2°F. After this they were washed and dried. After an interval of about 20 minutes, during which Kuda Bax loitered about the lawn, he walked the length of the pit which took, I think, four steps. He walked lightly but not with undue haste. He did it twice. Almost immediately after the second attempt (10 seconds perhaps having elapsed) the temperature of the soles of his feet was again taken and found to be 93°F. On inspection no sign of burning of the skin could be made out unless it were that certain areas appeared a little whitened, resembling somewhat epidermis which has been scorched but not enough to cause blistering. The varying pigmentation and texture of the skin of the sole, however, made it difficult to be sure even of this. There was no sign of blistering nor any hyperæmia, nor did there seem to be any discomfort.

One of the Englishmen, Mr Digby Moynagh, who made an attempt had tried to do the same thing ten days before. The soles of his feet had numerous healing blisters on them. He acquired some more at the second attempt. His skin was moister than the Indian's and this may have had some influence, because afterwards a small piece of charcoal was found stuck to the skin, and on its removal a blister was found underneath. It is possible that moisture made the

31

charcoal adhere and perhaps prolonged the period of contact. The other Englishman, Mr Maurice Cheepen, who tried the walk was also blistered.

Before Kuda Bax made his attempt, a piece of zinc oxide plaster, 5/8 in square, was stuck to the arch of the sole of his right foot. At the end of the demonstration this piece of plaster showed no signs of burning, except that the fluff from the cotton fabric at the cut edge exhibited a very slight amount of scorching. The surface of the ignited charcoal was so irregular that contact must have been made with all parts of the sole.

No other facts were observed.

REFERENCE

1935 *The Lancet* **2** 750


# *Letter on Corpulence*


In the 1953 issue of *British Medical Journal* an editorial discussed the problems of obesity and posed various questions including 'Why do so many slim women become obstinately fat after childbirth?' They concluded the editorial with quotes from William Banting's *Letter on Corpulence*, written 90 years ago. Banting weighed 202 lb when he began dieting and a year later weighed 156 lb. His girth had meanwhile lessened by 12¼ inches (31 cm). The outline of the Banting diet is reproduced below.

> For breakfast, four or five ounces of beef, mutton, kidneys, boiled fish, bacon, or cold meat of any kind except pork, a large cup of tea (without milk or sugar), a little biscuit, and one ounce of dry toast. For dinner, five or six ounces of any fish but salmon, any meat except pork, any vegetable except potatoes, one ounce of dry toast, fruit out of pudding, game or poultry; two or three glasses of claret, sherry, or Madeira, but neither champagne, port, nor beer. For tea, two or three ounces of fruit, a rusk, and a cup of tea without milk or sugar. For supper, three or four ounces of meat or fish (as above), and a glass or two of claret and, if required, a tumbler of grog for a 'nightcap', or another glass or two of claret or sherry. (Soak the rusk in spirits, if you please.) In fact, avoid sugar-containing foods.

REFERENCE

*Br. Med. J.* **2** 1418

# Spinach as a Source of Iron

In the year that Popeye became once again a major movie star it is salutary to recall that his claims for spinach are spurious. Popeye's superhuman strength for deeds of derring-do comes from consuming a can of the stuff. The discovery that spinach was as valuable a source of iron as red meat was made in the 1890s, and it proved a useful propaganda weapon for the meatless days of the second world war. A statue of Popeye in Crystal City, Texas, commemorates the fact that single-handedly he raised the consumption of spinach by 33%. America was 'strong to finish 'cos they ate their spinach' and duly defeated the Hun.

*Popeye* . . . would have done better to eat the cans.

Unfortunately, the propaganda was fraudulent; German chemists reinvestigating the iron content of spinach had shown in the 1930s that the orginal workers had put the decimal point in the wrong place and made a tenfold overestimate of its value. Spinach is no better for you than cabbage, Brussels sprouts, or broccoli. For a source of iron Popeye would have been better off chewing the cans.

REFERENCE

Hamblin T J 1981 *Br. Med. J.* **1** 1671

# The Human Dustbin

The following case history appeared in the January 29, 1983 issue of *The Malay Mail* under the caption *The Human Dustbin: Garbage in Manuel's stomach.*

Little Manuel Gimenez's parents just couldn't get him to eat. But, strangely enough, he never seemed to lose weight.

Then Manuel, 3, began to complain of pains in his stomach. His parents took him to doctor after doctor, but nobody seemed to know what was wrong with the toddler.

After 18 months, nine doctors and three hospital visits, an X-ray revealed Manuel's amazing secret.

He was a human dustbin.

Surgeon Perdo Fernandez rushed the boy to the operating theatre and removed: thirty pebbles, two collar studs, two buttons, two coins, one screw and two plastic gambling chips.

The total weight was 6 lb 3 oz, three times what a grown up has in his stomach after eating a heavy meal. Dr Fernandez said: 'How he lived with that lot in his stomach, I just don't know. Any normal adult would have died with half the amount.'

*Manuel Gimenez . . . the human dustbin.*

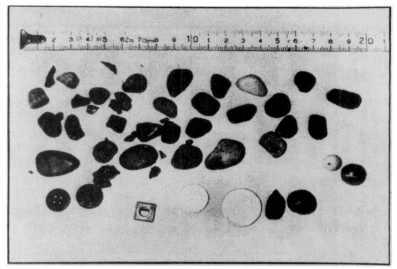

*The strange collection* inside Manuel's stomach—30 pebbles, two collar studs, two buttons, two coins, one screw and two plastic gambling chips.

The doctor, who is head surgeon of the San Jorge Hospital in Huesca, Northern Spain, went on: 'When I gave the boy a full stomach X-ray, I found all sorts of strange blurs. I decided to operate immediately. When we opened his stomach up, I just couldn't believe my eyes. There was object after object, and most of them had obviously been there for months.'

But nobody had ever seen little Manuel doing his swallowing act. His mother, 31 year old Mercedes Gimenez, explained: 'We knew he was ill because he rarely ate anything. He never lost weight. How could he with all that inside him? But Manuel was never entirely well. We kept taking him to all these doctors, who said it was something he had eaten. We never saw him swallowing these things, although he once brought up a small pebble. We fed him mainly on soup and peas. He wouldn't touch meat or fish.

'Now, thanks to God and the good doctor, Manuel is fine.'

And Dr Fernandez tried a little gentle shock treatment to stop the toddler from getting up to his old tricks. He said: 'I showed him what we had taken out of his stomach. Then I told him he would die if he ate one more pebble. He was very impressed.'

And, after the operation, Manuel celebrated by tucking in [to] a chicken chop.

*The majority of diseases are cured by nature.*

ISAAC JUDAEUS

# Self-Induced Excess Iron Therapy

Two cases of this unusual type of therapy were reported in the journal of the Westminster Medical School students' and the 'therapeutic iron' is still on display in the Medical School Museum. The cases were observed by Rosin whilst working in Rhodesia (Zimbabwe).

Case number 1 was a 28 year old male who in 1964 and 1965 was admitted to hospital on seven occasions having swallowed foreign bodies whilst a prisoner in jail. These varied as follows: nails, screws, broken razor blades, nails, nails and razor blades, nails and screws, nails and razor blades. This was the previous case history up to December 1965 when the patient acquired a taste for 18 bedsprings. This led to the Governor of Her Majesty's Prison receiving a letter instructing him 'to withold all types of edible foreign materials from the prisoner'. This proved impossible and a further 14 bedsprings were eaten in January 1966.

After serving sentence in Rhodesia, he was to be repatriated to the Republic of South Africa where he is a national. There he is wanted for gun-running charges which carry a penalty of 25 years hard labour.

He hoped by repeated swallowings of foreign bodies that he would be certified as insane and therefore unfit to face the charges in South Africa.

The patient appeared in no way to be unbalanced in mind and he described his method of swallowing the springs as follows 'I wrap each article in toilet paper, then throw my head back, press the handle of a knife to the back of my throat and slide the spring along the knife.'

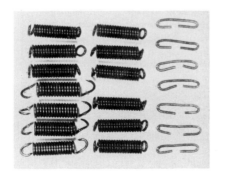 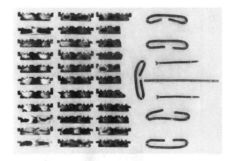

*Examples of ironmongery* removed from case number 1.

With·all this ingestion of iron, he still failed to right his iron deficiency anaemia which was discovered during his first admission.

Case number 2 was a 26 year old cellmate of the previous gentleman and was admitted to hospital in January 1966 with a history of having swallowed nine bedsprings to keep his friend company. Unlike the first case when nothing was passed per rectum and all iron had to be removed by operating on the stomach, number 2 passed three springs per rectum and also received a laparotomy when two of the springs were found hooked inside the oesophageal opening. Both cases had uneventful recoveries.

REFERENCE

Rosin R D 1966 *The Broadway Clinical Supplement* 6 September

# Foreign Bodies

Most medical school museums have a collection of foreign bodies which have been introduced by one means or another into the rectum and, for example, in Westminster's museum there is an oven door handle with a label attached to it stating that 'an unmarried man of 42 admitted that the introduction of the

handle was intended to produce an erotic sensation and that he denied ever having done this before'. There was also the case of the 54 year old company director who attended the Westminster casualty department with the story that he had sat on a rubber truncheon which disappeared up his back passage and could not be retrieved. He was somewhat reticent about details and further enquiries were not pressed.

A remarkable variety of objects have been recovered from rectums and in an ancient textbook by Keith and Treves, entitled *Surgical Anatomy*, the authors recall removing objects such as a glass tumbler, a silver matchbox, a deer horn and an umbrella handle. A more recent publication by Weston-Davies on anal fixation lists even more bizarre rectal findings: a bottle of Heinz tomato ketchup (explanation: lost door keys, climbing through pantry window, losing foothold) with a condom attached to the end with the lid; an economy size can of Heinz baked beans (a nurse was heard to remark 'fancy eating them in the tin'). This article also told the story of a retired squadron leader who for many years had used a Bofors anti-aircraft shell, vintage 1945, to replace prolapsed piles until one day it disappeared from view. The casualty surgeon wisely sent for the bomb squad as well as for the anaesthetist.

REFERENCE

Weston-Davies W 1983 *World Med.* 28 5 March

# Curing Piles with Petrol

A music student died after being caught in a flash fire while apparently trying to relieve his piles with petrol, an inquest at St Pancras heard yesterday.

Norik Hakpioan, 24, of Sloane Terrace, Kensington, was found naked with 90 per cent burns. The fumes from an open bottle of petrol had been ignited by a cooker hotplate.

His brother Hiak said relieving piles with paraffin was an old family remedy, but it was possible Norik had used petrol. The coroner, Dr Anthony Missen, said it must be 'the most lethal treatment recorded for many years.' Verdict: Misadventure.

REFERENCE

*Daily Telegraph* October 6, 1982

# Early Medical X-Rays

X-rays were discovered in November 1895 in Würzburg, Germany, by Wilhelm Röntgen. His publication *On a new kind of ray* was generally received with acclaim throughout the world and press headlines appeared stating *Electrical photography through solid bodies, Illuminated tissue, Searchlight of photography*, etc. There were only a few dissenting voices such as that of the *London Pall Mall Gazette* which was of the opinion that 'We are sick of the röntgen rays . . . you can see other people's bones with the naked eye, and also see through eight inches of solid wood. On the revolting indecency of this there is no need to dwell'.

One of the first X-ray photographs (skiagrams) taken by Röntgen was of his wife's hand, and the effect on the public as well as on the medical and scientific

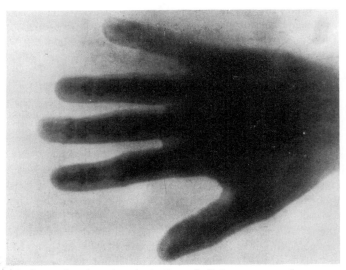

*Radiograph* taken by A A Campbell-Swinton on January 13th, 1896 and exhibited at the Camera Club, London, on January 16th. (Courtesy: The Science Museum, London.)

professions was so marked that skiagrams were soon being taken world wide. In England, the first skiagram of any part of the human anatomy was produced by Campbell-Swinton on January 13th, 1896 and in America, the first to appear in the literature was of a hand taken on January 17th, and published in the *New York Medical Record*. Sidney Rowland, who was commissioned by the *British Medical Journal* to write a series of articles reviewing the uses of X-rays, published an excellent summary beginning with the February 8th, 1896 issue and dealt with fractures, dislocations, detection of foreign bodies and examination of pathological conditions. Much later in 1937, Thurston Holland described his X-ray work book entries of 1896 for the centenary celebrations of the Liverpool Medical Institution. These included a matchbox containing various items, such as keys, deformities of the hand and other bony structures, the oesophagus of a boy who had swallowed a coin and an attempt to find if a set of false teeth had been swallowed. In total, 261 X-ray plates were taken from June to December 1896, perhaps the only documented work load of an 1896 X-ray Department.

Suggestions for therapeutic uses of X-rays were also forthcoming and the earliest appears to have been by an X-ray manufacturer, E H Grubbé in January 1896. J E Gillman, in Chicago, sent a breast cancer patient to him for treatment on January 29th and, on the following day, he also treated a patient with lupus vulgaris. Two other early treatments were of a cancer of the nasopharynx patient, February 1896 in Hamburg and a cancer of the stomach patient, July 1896 in Lyon. However, it is generally accepted that Freund in Vienna was the first to use X-rays logically and scientifically within the limits of the age. His first

*X-ray apparatus, 1897,* used to skiagraph a hand. The photograph was taken during a demonstration before the Royal Photographic Society.

patient, treated in December 1896, was a five-year old girl with hirsuites (naevus pigmentosus pilosus). Schiff, who was the doctor responsible for this patient, describes the case in which an artificial alopecia was produced over the whole of the back which had been exposed to X-rays two hours per day for some 16 days. The hair started to fall out after 12 days and after total epilation a violent dermatitis set in which healed only very slowly. Schiff remarked that 'this accident was full of instruction' and in future, he reduced the exposures to 10 minutes duration. This case history was the beginning of radiotherapy.

The epilatory effect of X-rays established them as a direct agent for producing biological change and evidence for the harmful effects of the radiation was also known within a few months of their discovery. For example, Edison reported in March 1896 that his eyes were sore after experimenting with X-rays, radiation burns to hands were reported by Stevens on April 18th, 1896, Gilchrist reviewed 23 cases of X-ray injury which had been reported in the literature before January 1897, in May 1897 Scott reviewed 69 reports of X-ray injury. In April 1898 the Röntgen Society appointed a committee to collect data on the harmful effect of X-rays and review papers discussing the hazards of over-exposure to radiation continued to appear in the radiological literature. However, in spite of these warnings of the dangers of superficial injuries through indiscriminate exposure to X-rays, the radiation protection facilities for many workers were rudimentary or non-existent for several years. This was no doubt partly due to the latent period between exposures and the appearance of permanent ulcers or malignancies. It is also noted that some of the patients were doubtful about the effect

41

of the radiation. 'An occasional slight erythema (a reddening of the skin) is the sole visible result of the treatment, so that patients were apt to become very sceptical of its success'.

The first successful treatment of proven carcinoma by X-rays is attributed to Sjogren in 1899 in a man with advanced epithelioma of the cheek.

In X-ray teletherapy, the X-ray machine is placed at a distance from the patient and the X-ray tube is then energised so that the patient is *exposed* to the radiation until the tumour has received the required *dose*.

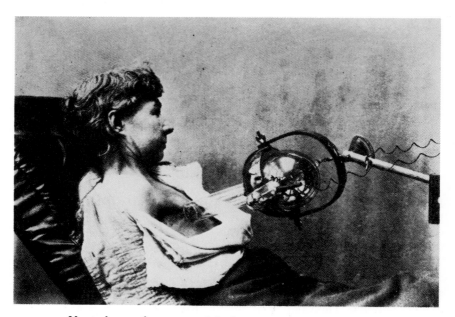

*X-ray therapy* for cancer of the breast in 1903. Note the lack of radiation protection features in the apparatus here and on the previous illustration.

In principle this is essentially what still happens today, although ideas on the definition and measurement of *exposure* and *dose* have altered over the years and the design of an X-ray machine has, of course, also altered. It is no longer true (as stated in 1912) that 'the X-ray tube is a remarkably fickle appliance, and it is quite impossible to estimate the magnitude of the dose by simply noting the amount of current passed through the tube and the length of exposure to the rays'. Differentiation was soon made between *hard* and *soft* X-rays by Keinböck and the influence of X-ray focus-to-skin-distance on the penetration of X-rays through tissue was also recognised.

The radiation dose at a depth in tissue increases with the energy of the incident X-rays and by 1915, Knox was stating 'up to quite recently it had been held that

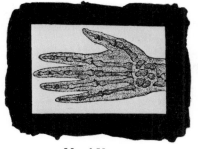

Hard X-rays

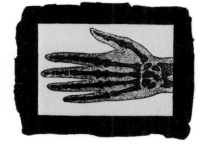

Medium-soft X-rays

Soft X-rays

*When X-rays* were termed hard or soft, the adjective referred to their quality, which described the penetrating power of the X-ray beam. This is demonstrated in the diagram given by Kienböck in 1905 in which the hard X-rays have a greater penetration through a hand than soft X-rays.

X-rays penetrated successfully to a depth of 1 cm or less, and anything deeper had been left alone' and 'with hard tubes, effects can be produced up to 10 cm'. Ten years later the 'maximum capacity of apparatus presently available is approximately 250 kV', a variety of machines of energies greater than 1 MV were in use in the United States from the early 1930s; and in the 1970s cobalt-60 teletherapy units and 5, 10 and 15 MV X-ray linear accelerators were commonplace.

Radiation protection during the early X-ray treatments for cancer was more convenient once the *self-protected* X-ray tube was developed, in which lead protective features were incorporated into the design of the X-ray tube housing. Prior to this when protective measures were considered, the X-ray tube was enclosed in a cumbersome lead-lined box or a huge protective cylinder which gave rise to the term *X-ray cannon*. Both staff and patient protection were considered by Knox in 1918, who stated that the tube must be properly enclosed in X-ray proof glass or lead rubber, and that it was 'not sufficient to enclose the X-ray tube in a shield closed only on half its diameter as X-rays escape from

43

*X-ray cannon of the 1920s*, Erlangen, Germany.

behind the tube'. He also listed the four 'dangers attendant on the use of X-rays and radium' as acute dermatitis, chronic dermatitis, late manifestations and sterility.

REFERENCE

Mould R F 1980 *A History of X-rays and Radium* (Sutton: IPC Business Press)

# Tavern of the Dead

The Cabaret du Neant was a music-hall illusion performed in Paris in 1896 and prompted by the discovery of X-rays in the previous year.

The principal interest centred on a stage conversion of a living man into a skeleton, which was achieved using a series of mirrors and Argand burners, a coffin and some blocks of wood. Two notices were displayed, 'Rest in Peace' and

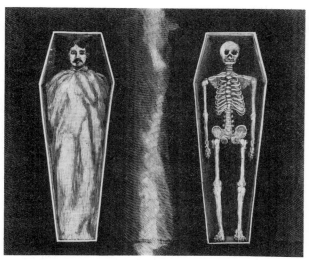

*The subject and his skeleton.*

'No Smoking' and the audience was informed that the latter was necessary because a perfectly clear atmosphere was required for the illusion. The method was based on an optical illusion known as Pepper's Ghost.

The Spirit World was also linked to X-rays in other than a music-hall manner as can be seen from a reference in the *New York Herald* on the use of X-rays to

*An X-ray illusion* upon the stage: conversion of a living man into a skeleton.

45

restore life. This was in contrast to another American suggestion, this time in the 1896 *Scientific American*, that X-rays could assist in determining whether death had occurred, since 'dead flesh offers more resistance to the penetration of the rays than living, and a glance at the radiograph of the person would determine whether it was that of a corpse or not'.

REFERENCE

1896 *Sci. Am.* March

# The Sufferer

French music halls have already been mentioned in 'Tavern of the Dead'. The British music hall was famous for its monologues with titles such as *The Green Eye of the Little Yellow God, The Lion and Albert* and *The Shooting of Dan McGrew*. Most of the monologues were intended to keep the listener in suspense until the last stanza and the one given below, called *The Sufferer* (written by L E Baggaley and Edwin John in 1934) is no exception. Others, such as *The Madman's Will*, which appears later, have a more melancholy aspect to them, reflecting the average working man's view of life in the nineteenth and early twentieth centuries. *The Sufferer* and *The Madman's Will* are two of only a few which relate to the medical profession.

> Now hospitals aren't cheerful places,
> Not even the best, you'll agree:
> But the one that fair gives you the horrors
> Is that at St. Earaches-on-Sea.
>
> It says o'er the gate 'English Martyrs,'
> A name that inspires a doubt,
> It's appropriate, though, because many go in,
> But very few live to come out.
>
> A cousin of mine, name of Arnold,
> Who went there for treatment last year,
> Says the bloomin' place very nigh killed him,
> And it all came about like this 'ere.
>
> He lay there one day in a dark dreary ward,
> Its name was the 'Angel of Death.'
> They'd all got such names, nice and sociable like,
> When you saw 'em you fair caught yer breath.

46

Now Arnold was not feeling robust,
His face was a nice shade of green.
He'd a mouth like a dustbin, and felt a bit raw
Around where his appendix had been.

A sour-faced Sister then entered
And said 'You're to come right away,
Operation it's got to be done o'er again.'
The patient looked up and said 'Eh?'

She retorted 'It's no good being stupid,
Them forceps has got to be found,
They must be inside you, so Doctor Scull says,
And that's where they are, I'll be bound.'

Our Arnold was always obliging,
'Twas a way that he'd had all his life.
So he pulled on his slippers and followed her out,
And was shortly put under the knife.

The next day a towel and bandage were missed,
And Arnold was opened once more,
By then he began to get slightly annoyed,
Life for him seemed to be one big bore.

The staff of 'The Martyrs' all laughed and agreed
That the series of mishaps was rummy
The Matron remarked 'When we slit him again,
We'd best fit a "zip" on his tummy.'

The climax arrived on a Tuesday
When they all gathered round him to gloat
And Matron said 'Come on, you're wanted,
For Doctor Shroud's lost his white coat.'

At that Arnold's patience fair failed him,
'Oh has he,' said he, 'where's my hat?
I don't mind a bandage or forceps and such,
But a jacket—I'm not wearing that.

An instrument-shelf and a cupboard I've been
Since I came in but three days ago.
But at being a cloakroom gratuitous like
I'm drawing the line—cheerio!'

REFERENCE

Marshall M (ed) 1981 *The Book of Comic and Dramatic Monologues* (London: Elm Tree Books/EMI Publishing)

*I am dying with the help of too many physicians.*

ALEXANDER THE GREAT

# Medical Diploma

In 1897 a patent was registered in London by G Izambard, which related to a method of reproducing a large number of copies of writing, or printing, by printing on photographic paper using Röntgen rays. Izambard intended the process to take the place of letterpress printing for newspapers, using an ink opaque to X-rays. This did not work! However, radiography of documents was attempted and a radiograph of a will within a sealed envelope is shown below. This was intended to demonstrate the possibility of reading sealed documents.

The *British Journal of Photography* of October 30, 1896 also reported on the subject of parchment radiography, but this time in relation to a medical diploma.

At an enquiry last week, as to the cause of the death of a person who had been attended for cancer by an alleged specialist, the question was again raised as to the genuineness of the diploma produced. While questioning one of the witnesses as to some clear, and some blurred names upon it, the Coroner remarked to him according to the reports in the Daily Papers: 'I may tell you that we have had the Röntgen rays through it, and so we know all about it'. The Coroner then produced a photograph of the certificate which showed that there was, or had been, some old writing under the names.

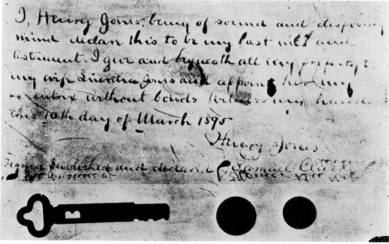

*Radiograph by William J Morton, 1896.* Copies of this radiograph were on sale for 50 cents from the American Technical Book Company of New York.

# Marriage and X-Rays

The *Archives d'Electricité Médicale*, August 25, 1906, reports that a case has occurred in New York of breach of promise of marriage, with a claim for 25 000 dollars. The young lady absolutely refused to be examined by an X-ray specialist at the request, or rather command, of her fiancé. There was nothing the matter with her, but the young man wished to be on the safe side before marriage. On her refusal, he broke off the engagement. This breach of promise case is the result, and it is suggested that the judge is in favour of the plaintiff on the grounds that by law no one can reasonably claim an X-ray examination before marriage and on its being refused break the engagement.

# Real and Fake Diamonds

A Radiograph of a diamond star and a paste brooch, by Charles Thurston Holland in 1896 is shown here. He later described how after a lantern-slide lecture in 1896, the audience were invited on payment of a fee for charity, to view their hands on a fluorescent screen. One 'very over-dressed lady' who found that stones in a large ring were densely opaque, and therefore fake, made quite unprintable and vitriolic remarks, and Thurston Holland observed that 'he had done someone a very bad turn'.

# The Madman's Will

In a work-house ward that was cold and bare,
The doctor sat on a creaking chair,
By the side of a dying madman's bed.
'He can't last much longer,' the doctor said.
But nobody cares if a pauper lives,
And nobody cares when a pauper's dead.
The old man sighed, the doctor rose.
And bent his head o'er the ricketty bed,

To catch the weak words one by one—
To smile—as the dying madman said:—
'Beneath my pillow when I am gone—
Search—hidden there you will find it still!'
'Find what, old madman?' the doctor asked,
And the old man said, as he died, 'My Will.'
How they all laughed at the splendid jest—
A pauper madman to leave a will.

And they straightened him out for his final rest,
In the lonely graveyard over the hill,
And the doctor searched for the paper and found
The red taped parchment—untied it with zest,
Whilst the others laughingly gathered round
To hear the cream of the madman's jest.
Then the doctor with mocking solemnity said,
'Silence, my friends,' and the Will he read.

'I leave to the children the green fields,
The fresh country lanes for their play,
The stories of fairies and dragons,
The sweet smell of heather and hay.
I leave to young maidens romantic
The dreaming which all maidens do.
And the wish that some day in the future
Their happiest dreams will come true.

To youth I leave all youth's ambition,
Desire, love, impetuous hate.
And to youth with years I leave wisdom,
And the hope that it comes not too late.
I leave to the lovers the gloaming,
The time when all troubles are old,
When true love, hand in hand, goes-a-roaming
To the heart of the sunset of gold.

To the mother I leave children's voices
And curly heads close on her breast,
The soft whispered prayer that rejoices
Her heart as she puts them to rest.
I leave to old people sweet memories,
And smiles that endure to the last,
With never a fear for the future,
And not a regret for the past.

I die without earthly possessions,
Without the last word of a friend,
To you all I leave good cheer and friendship
That lasts through all time to the end.
I leave to the wide world my blessing
In the hope that the long years will find
That my wishes shall grow like a flower,
And bring God's good peace to mankind.'

The ward doctor laid down the parchment,
His smile had gone—turned into pain.
The faces around laughed no longer,
But grew grave with regret that was vain.
'No wonder that he looks so happy,
Whilst we who derided are sad,
For the things he has left are the best things in life
I wonder if he *was* mad?'

REFERENCE

Marshall M (ed) 1981 *The Book of Comic and Dramatic Monologues* (London: Elm Tree
  Books/EMI Publishing)

*To the memory of Richard Richards*
*who by gangrene lost first a toe afterwards a leg*
*and lastly his life on the 7th day of April 1656*
  *A cruel death to make 3*
    *meals of one*
  *to taste and taste till*
    *all was gone*
  *But know thou tyrant*
    *when the trump shall call*
  *He'll find his feet*
    *and stand when thou shalt fall.*

Gravestone inscription, Banbury, Oxfordshire

# Gas Bulb Surrounded by Profanity

Cuthbert Andrews was a well known supplier of X-ray apparatus to the medical profession for many years and his firm is still in existence. The title is his description of an X-ray tube and the cartoon is one of many amusing advertisements which he issued.

## Phoque Rouge Model.

(Suggested by Prof. Jean Lemarcheur).

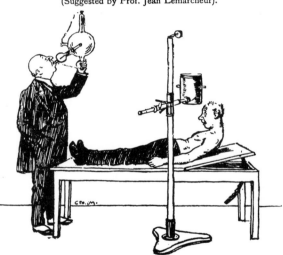

The chief feature of this superb machine is the ease by which it may be removed from the stand. and used in any position (as illustrated). The only precaution to be observed is that the operator should be always at a slightly lower level than the tube.

Price, with supply of Water .... .... £14 17 3
Do.     do.     Fine Old Hungarian Sherry £17 13 4
Do.     do.     Oxo .... .... £3 14 7
**Special.** The patent Whisky & Water-cooled
        pattern .... .... .... .... £314 7 0
Do.     with Patient-proof Safe .... £3147 14 3

The Department of Half-Seas Over Trade writes: "It is an hic-straordinary tube."

52

# Who Made Hospital Kettle Radioactive?

Scotland Yard continued over the weekend to investigate the mysterious case of the radioactive tea kettle at Charing Cross Hospital in London.

More than a week ago, it has now been established, a small amount of a radioactive chemical, iodine-125, was stolen from the refrigerated store of such substances in the biochemical laboratory in the department of oncology (cancer). This iodine is used to 'label' protein in samples of patients' blood for laboratory tests.

Radiation checks traced the iodine to a kettle in the laboratory, where it was found stuck around the inside.

While spokesmen for the hospital insist that the amount of radioactivity is extremely low, and definitely not a cause for great alarm, there are important security implications.

How could someone walk off with potentially dangerous material? Suppose he or she had taken enough to build up into a dangerous dose? The hospital's safety committee and administrators are urgently reviewing their security procedures.

The case has curious echoes of the recent television drama 'Hotter than the Sun,' in which a disaffected nuclear power worker smuggled out a radioactive source and finally swallowed it.

The hospital theft was discovered last Monday when laboratory workers were undergoing their usual weekly radiation checks. Whereas normal measurements are so small as to be negligible, this time they were higher than usual, and worrying enough to warrant an investigation.

The dose received by most workers was between 5 and 10 $\mu$Ci, which is lower than the 10 to 15 $\mu$Ci given to patients for various diagnostic tests.

With the two workers, however, the readings were 25 and 100 $\mu$Ci which, although higher than doses used for tracing within patients, is substantially lower than a treatment dose for someone with, say, a hyperactive thyroid or cancer.

Hospital officials are 'as sure as they can be that there will be no long- or short-term effect on the health of the workers.'

Investigations showed that the iodine had probably been removed from its lead case in the store on the previous Thursday, the glass phial—containing 3 mCi—broken, and the contents put in the kettle.

The turnover of such samples, which are used in a wide variety of tests, is high, and more powerful sources of radiation, which may require sophisticated handling procedures, are kept elsewhere.

There is a handling procedure for removing samples, but a hospital spokesman admitted that it would be easy to remove them unseen. However, the thief would have to know which samples to take. The glass phial has not been found.

The tea kettle, now with the police, was kept in the laboratory kitchen, a short distance from the store, and the implication is that those with higher radioactivity readings drank most tea or coffee.

The total number affected is not known, and people who might have entered the laboratory are being checked.

The Charing Cross Hospital incident scarcely seems an accident. The motive remains a mystery. Who wanted to give the workers radioactive tea?

REFERENCE

*The Observer* November 11, 1977

# Early Medical Radium

The phenomenon of radioactivity (the spontaneous emission of alpha, beta and gamma rays from certain natural heavy elements) was discovered in March 1896 by Becquerel. Following this, in December 1898, the discovery of radium was announced by Marie and Pierre Curie. However, although the existence of the new element had been established, it was an immense problem to refine it in any quantity and tons of the radioactive mineral ore, chemicals and water were required for only a few grams of radium.

Radium therapy, as distinct from X-ray therapy, began with the accidental radium burn received by Becquerel in April 1901, ten days after he had carried a tube of radium, loaned to him by the Curies, in his shirt pocket for some six hours. The effect was recognised as being identical with X-ray dermatitis, and prompted the Curies to loan some radium to St Louis Hospital, Paris. This was then used for the treatment of dermatological conditions, and, in Becquerel's Nobel Lecture of 1903, he remarked that the use of radium was being explored for the treatment of cancer and indeed, the first successful treatment was in 1903, of basal cell carcinoma of the face.

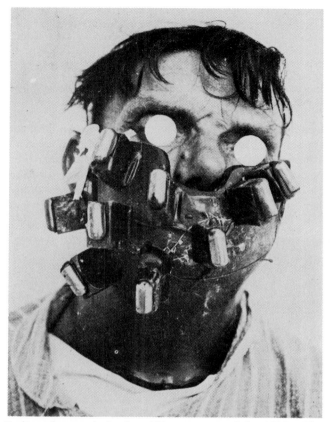

*Radium face pack (or 'mould') treatment of the 1920s.* The radium sources were mounted on a mask to treat this patient with cancer of the head and neck. Masks would be constructed for each individual patient.

A possibly unique early description of what it must have been like to accompany a patient scheduled to receive radium treatment, was related to me in 1979 by Mr Andie Clerk, who is now well over 80.

I recall an interesting experience many years ago when I was a young soldier before the First World War in 1914. Although I can't remember the name of the doctor, he was a Birmingham specialist struggling, shall I say, with radium, which he knew was powerful and was something he'd got to handle carefully as he touched his patients endeavouring to treat them with it. It was very early days and he offered to try what he could do for any patients who would accept his help. There was a chap with me in our Company whose young sister suffered with lupus. All kinds of things had been tried without success and our Medical Officer, hearing of it, arranged for her to let the Birmingham doctor try, and I went with the chap, his mother and the girl to the doctor in New Hall Street, Birmingham. We went several times and he had what seemed like a small stone at the end of a bamboo cane. The cane had been split to hold it and it had been squeezed in the two ends. Primitive indeed, wasn't it? He touched the skin with it very carefully, but it wasn't effective, rather defective. It made her slightly out of her mind though I believe she got that back in later years. But as I've said, he was handling something which he knew was dangerous but didn't know to what extent. A few years later he lost both his hands as a result of contact with it. Well, there are better means of doing things now. No more split bamboos. But there's a long way to go, isn't there?

The diseases first treated by radium were those of a superficial nature such as epitheliomata of nose, hand, cheek; keloids, lupus vulgaris, naevi and syphilitic ulcers; and those within a natural body cavity such as the uterus or vagina. In England, the development of radium teletherapy machines (for treatment using radium 'at a distance from the patient' as distinct from in contact with the patient) began in 1919 at the Middlesex Hospital using a 2.5 g radium source

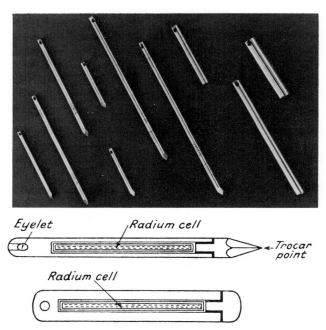

In the early years the design of the radium sources used in medicine varied widely, and some were small flat applicator sources whereas others were encapsulated in a glass or platinum tube. Later designs were more standard and consisted of radium *cells* containing a mixture of radium salt (e.g. radium sulphate) and an inactive filler. These *cells* were sheathed by platinum to either a tube design (for insertion into natural body cavities) or a needle design (for direct insertion into the cancer).

provided by the Ministry of Munitions. At about the same time similar developments were also made in New York, Paris and Stockholm. The early radium teletherapy units, often called *radium bombs*, provided only one radiation beam of large dimensions and multiple-beam treatments were impossible. Later designs incorporated higher activity radium sources (up to 10 g) and applicators which could be attached to the radium unit to provide a variety of radiation beam sizes. With the higher activity radium sources, the protection problems increased and these were in part overcome by housing the radium source when not in use in a separate lead safe some 3 m from the radium teletherapy unit. The source travelled from safe to unit by pneumatic transfer when the patient was in the treatment position, and was returned by the same method on completion of treatment. This type of design was for several years a standard piece of apparatus for treating head and neck cancers. However, problems did sometimes arise with these pneumatic transfer units, notably when the radium source refused to return to the safe. It was not unknown for a long broom handle to come to the assistance of the person who had to free the source from the treatment unit head! These were not everyday problems, but the sound of the radiation source bobbin thumping hard against the interior of the unit after transit instinctively made one wonder whether sometime there might not be the disaster of a shattered source. Fortunately pneumatic transfer systems of this design are now obsolete and the replacement of radium by cobalt-60 has reduced the potential radiation hazards.

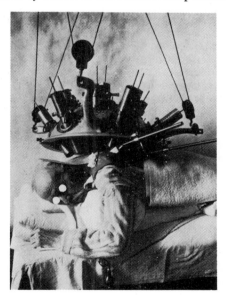

*Radium 'bomb' unit of the 1920s* known as the Sluys-Keppler apparatus. It was of a complicated design with 13 radium sources enclosed in identical applicators and fixed on a hemispherical cupola.

*Left, X-ray diagnostic picture of a frog, 1898; right, radiumgraph diagnostic picture of a mouse, 1904.* The poor quality of the radium picture compared with the X-ray picture is clearly demonstrated.

Radium was also experimented with for diagnostic purposes but the gamma-ray energy from radium is so high compared with that of diagnostic X-ray tubes that contrast was poor and another disadvantage recorded in 1904 was that 'it takes hours to represent an image'. There was no mention of radiation protection during such a long exposure time!

When radium decays it forms a gas called radon or emanation. Various treatments were on offer up to about 1915 which involved the use of emanation-activated water (see the advertisement opposite), injections of vaseline or water impregnated with emanation, radioactive mud packs, radium face powder, radioactive baths and other hazardous techniques for conditions which were unrelated to cancer. Fortunately, these are now only a relic of history, but the medical literature contains several case histories of radiation induced cancers diagnosed some 20 years after radiation treatment for cosmetic purposes only, in the early years of this century.

REFERENCE

Mould R F 1980 *A History of X-rays and Radium* (Sutton: IPC Business Press)

# RADIUM THERAPY

The only scientific apparatus for the preparation of radio-active water in the hospital or in the patient's own home.

This apparatus gives a <u>high</u> and <u>measured</u> dosage of radio-active drinking water for the treatment of gout, rheumatism, arthritis, neuralgia, sciatica, tabes dorsalis, catarrh of the antrum and frontal sinus, arterio-sclerosis, diabetes and glycosuria, and nephritis, as described in Dr. Saubermann's lecture before the Roentgen Society, printed in this number of the " Archives."

### DESCRIPTION.

The perforated earthenware " activator " in the glass jar contains an insoluble preparation impregnated with radium. It continuously emits radium emanation at a fixed rate, and keeps the water in the jar always charged to a fixed and measureable strength, from 5,000 to 10,000 Maché units per litre per diem.

SUPPLIED BY

**RADIUM LIMITED,**

93, MORTIMER STREET, LONDON, W.

Telephone: 4794 MAYFAIR.

# Sir William Gull

When this famous physician was once called in consultation about a young lady, it chanced that he passed her open bedroom door on his way to meet her mother. He walked straight up to this lady saying 'Madam, I congratulate you, your daughter will get quite well'. The mother was taken aback and said 'but you have not seen her'. He replied 'Madam, I saw her through the open door of her bedroom, that was quite enough'. Gull had been told that the patient had typhoid; he saw her sitting up in bed; any sufferer from typhoid who sits up in bed must be doing well!

# Tutankhamun's Head

This photograph of the mummified head of one of the most famous of all the Egyptian Pharaohs, King Tutankhamun was taken by the dentist Mr F Filce Leek. He accompanied members of a team, including some from the Isotope Production Unit at Harwell, to study Tutankhamun for a BBC program. The illustration has only previously appeared in a 1969 issue of the Ilford Ltd journal *X-Ray Focus*, on the topic of technique of dental radiography of mummified heads. This involved the use of the radioactive isotope iodine-125 and a 2 hour exposure. However, although Leek has examined over 3000 mummified heads and radiographed several, King Tutankhamun never yielded up his dental secrets. The Egyptian authorities were concerned that the layer of resin beneath the King's chin would be damaged if pierced for the isotope insertion and Leek was forced to use a different angle, one which did not give an informative dental view.

# The Yin and the Yang

The tradition of medicine in China has probably changed less than that of any other country over the centuries[1]. The Yellow Emperor Hwang-Ti, who lived around 2700 BC, originated Chinese medicine with his work *Nei Ching*. This was devoted to internal medicine yet scarcely mentioned anatomy. The Chinese approach to health and ill health was founded on two opposing principles: the yin and the yang. The yin was a negative, female quality. The yang was active and masculine. As with the Hippocratic humours, a proper balance between the two was considered essential to health. Illness developed when the breath, the life spirit, which governed the interplay between the yin and yang, became blocked. The purpose of acupuncture, the insertion of gold or silver needles into the skin in a particular pattern, was to prevent or ameliorate such blockages.

Chinese medicine boasts few great names, because there have been few specialists. The practice of medicine has been open to all, and diet, hygiene, massage and preventive measures have long been the dominant motifs. Yet Chinese physicians have discovered genuine, effective drug treatments for disease, including mercury for syphilis and arsenic for skin conditions. In the twelfth century BC, they also pioneered a crude form of variolation, inserting material·from smallpox pustules into the nostrils of healthy children in an attempt to combat epidemics.

Belief in the yin and the yang was not limited to China[2]. Gold and jade, sharers of the cosmological principle yang, preserve the body from corruption and according to the Chinese alchemist Ko Hung 'if gold and jade are placed in the nine apertures of the body it will be saved from corruption'. Later, in the fifth century it was also stated 'if on opening an old tomb the cadaver seems to be still living, one can be sure that inside and round the body, there are large quantities of gold and jade'. It is presumed[2] that this same idea is behind the fact that central American prehistoric tombs are also rich in gold and jade and that the indigenous peoples of Colombia buried their dead with golden objects so that the metal would ensure immortality. These objects were often nosepieces, small containers known as *poporo* which contained lime to be used in coca chewing to obtain the narcotic effect of this plant, pendants, necklaces and anthropomorphic figurines. Today, the Museum of Gold in Bogotá, Colombia, contains almost 25 000 pieces of this type of goldware and is the largest collection of its kind in the world. Included, are gold artefacts from near Lake Guatavita from where the legend of El Dorado, the Gilded Man, arises. A ceremony used to take place at this lake, some six miles from Bogotá, '. . . they take two cords long enough to span the

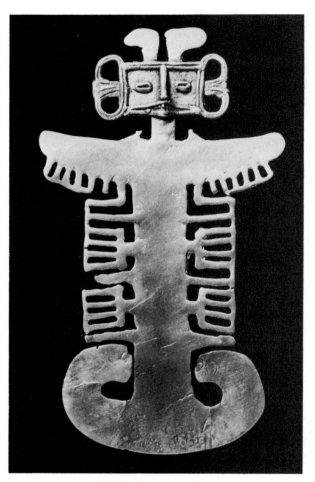

*Gold birdman pendant.*

lake (which is 2½ miles in circumference and 120 feet deep) and pass them over from one side to the other. They cross at the centre of the lake where the military officers and the person who was to make the offering went on rafts made of reeds bound together or tree trunks like a sort of boat in which three or four persons could ride. At the middle of the lake they threw in the offerings of gold and emeralds. Chief Guatavita, gilding his body, bathed in the lake. . . .'

REFERENCES

1  Dixon B 1978 *Beyond the Magic Bullet* (London: George Allen and Unwin)
2  Panesso A (ed) 1979 *'El Dorado', Museum of Gold* (Bogotá: Banco de la República)

# Circumcision and Birth of Twins, 2340 BC

These paintings (and also that of Surgical Instruments) were discovered on a wall of the tomb of Ptah-Hotep, a physician who died in 2340 BC and was buried in Saqqara, Egypt, near the famous step pyramid of Zoser—the first Egyptian pyramid to be built (around 2780 BC) which is older than the three famous pyramids and sphinx at Giza.

Reference is made by Gairdner to the circumcision of Egyptian mummies when he wrote his paper entitled 'The Fate of the Foreskin'.

Male circumcision, often associated with analogous sexual mutilations of the female such as clitoric circumcision and infibulation, is practised over a wide area of the world by some one-sixth of its population. Over the Near East, patchily throughout tribal Africa, amongst the Moslem peoples of India and of South-East Asia, and amongst the Australasian aborigines circumcision has been regularly

*Circumcision, 2340 BC.*

*Birth of twins, 2340 BC*

practised for as long as we can tell. Many of the natives that Columbus found inhabiting the American continent were circumcised. The earliest Egyptian mummies (2300 BC) were circumcised, and wall paintings to be seen in Egypt show that it was customary several thousand years earlier still.

Similar illustrations relating to the birth of children appear on ancient buildings other than the pyramid at Saqqara and there is, for example, a bas relief on the Temple of Esneh of the birth of Cleopatra's child. The *key of life* symbol appears there and also in the illustration shown here as the hand on the far right of the picture. The amazingly large size in which the child is represented is indicative of its royal birth, as is the royal falcon. The position taken by the mother is still used during childbirth by some primitive peoples.

REFERENCE

Gairdner D 1949 *Br. Med. J.* **2** 1433

# Medical Handwriting

As my doctor was penning my prescription I couldn't help commenting on his impeccably elegant handwriting. 'Yes, I know', he said, 'I'm a disgrace to the profession'.

*Reader's Digest*, October 1982. Contributed by Mrs J L Monson.

# Lead Poisoning in the Ancient World

Although it is clear that lead was used almost universally throughout the ancient cultures it remained very much a background metal and assumed a pre-eminent position only when the Romans devised their elaborate projects for providing their towns and houses with water. The Romans' lead technology was impressive. They manufactured sheet lead by casting onto flat sand beds and had ingenious methods of rolling and jointing pipes which were the basis of their water-carrying systems. The amount of lead consumed by the Romans was extraordinary. In building the great aqueduct at Lyons it had been estimated that 12 000 tons of lead were used on just one of the siphon units, and the description of the construction of the leaden water systems given by Vitruvius shows what major undertakings these were. The Romans were avid in their demand for lead, and after the conquest of Britain the native mines were extensively worked. They were a plentiful source, for Pliny described lead being found 'in the surface stratum of the earth in such abundance that there is a law prohibiting the production of more than a certain amount'.

The use of lead water systems represented a hazard to health, but both the Romans and the Greeks exposed themselves to a far greater risk. Pliny defines the problem when he writes: 'when copper vessels are coated with stagnum the contents have a more agreeable taste and the formation of destructive verdigris is prevented . . .'.

The Romans and Greeks found that by coating their bronze or copper cooking pots with lead, or lead alloys, not only was the leaching of copper from the pot prevented, thus avoiding spoiling the taste of the food, but also these were of great value in preparing wine and grape syrup which was used almost exclusively as a sweetening agent.

At source its danger would be minimal if modern experience is a guide, for galena, the principal lead ore of the ancient world, is relatively non-toxic. McCord, in his history of lead mining in America says that there were few cases of lead poisoning from the early workings when galena was mined. When lead carbonate was worked in Utah after 1870, however, it became an all too common hazard. Similarly, a more recent investigation of men exposed to galena dust revealed few indications of lead intoxication amongst them. The galena mines gave off sulphur dioxide, however, and on this account Pliny warned that the 'exhalations from silver mines [i.e. galena mines] are dangerous to all animals'. Once the ore was smelted, however, its dangers were apparent.

McCord quotes Durant as saying of the Athenian silver mine, 'Laurium pays the price of the wealth it produces, as mining always pays the price for metal industry; plants and men wither and die from the furnace fumes, and the vicinity of the works becomes a scene of desolation'.

The fumes given off from heated lead were well known to be poisonous. Both Vitruvius and Pliny gave warning of the danger. Describing the production of white lead, Vitruvius says, 'At Rhodes they place a layer of chips in a large vessel, and pouring vinegar, they then put lumps of lead on top. The vessel is covered with a lid lest the vapour which is enclosed should escape.'

Pliny is more dramatic: 'For medicinal purposes lead is melted in earthen vessels, a layer of finely powdered sulphur being put underneath it; on this thin plates are laid and covered with sulphur and stirred with an iron rod. Whilst it is being melted, the breathing passages should be protected . . . otherwise the noxious and deadly vapour of the lead furnace is inhaled; it is hurtful to dogs with special rapidity.'

Vitruvius was also at pains to point out that water found near mines was not free from hurtful effects. 'When gold, silver, iron, copper and lead and the like are mined, abundant springs are found, but mostly impure . . . . When the water is taken into the body, and . . . reaches the muscles and joints, it hardens them by expansion. Therefore the muscles swelling with expansion are contracted in length. In this way men suffer from cramps or gout, because they have the pores of the vessels saturated with hard, thick and cold particles.'

He was also critical of the custom of using lead for water systems. Not only were earthenware pipes considerably cheaper and easier to repair than those made of lead, but

water is much more wholesome from earthenware than from lead pipes. For it seems to be made injurious by lead because cerusse is produced by it; and this is said to be harmful to the human body. Thus if what is produced by anything is injurious, it is not doubtful but that the thing is unwholesome in itself.

We may take example by the workers in lead who have complexions affected by pallor. For when, in casting, the lead receives the current of air, the fumes from it occupy the members of the body and rob the limbs of the virtues of the blood. Therefore it seems that water should not be brought in lead pipes if we desire to have it wholesome.

The first unquestioned clinical account of lead poisoning must be accredited to Nicander who wrote in the second century BC. In his Alexipharmaca he describes in verse the symptoms arising from the ingestion of litharge and cerusse, including colic, constipation, palsy and a pallor which he fancifully likened to the dull colour of lead. Dioscorides, in the first century AD, described in his materia medica the ill-effects of litharge in graphic terms. 'The drinking of litharge causes oppression of the stomach, belly and intestines, with intense griping pains; . . . it suppresses the urine, while the body swells and acquires an unsightly leaden hue.'

The ancients were thus unquestionably aware of the dangerous character of lead and knew that it was poisonous when taken internally. Thus we find in Pliny, 'red-lead is a deadly poison and should not be used medicinally'; and, 'lead acetate is a deadly poison'. Celsus also knew of its toxic effects for he prescribed as an antidote to poisoning by white lead 'mallow or walnut juice rubbed up in wine'.

And yet the Romans and Greeks continued to expose themselves to the effects of a metal they knew to be harmful through their food and drink. The contamination of both was considerable. In an experiment, Hofmann found that a litre of must took up 237 mg of lead when boiled down with a plate of lead in it. A further experiment was conducted in which he treated Gebirgswein, Talwein and must according to the directions of Columella. He extracted 390 mg of lead from the Gebirgswein, 582 mg from the Talwein and 781 mg from the must.

This continual ingestion of lead from the diet resulted from time to time in epidemics of lead poisoning throughout the Roman world. Paul of Aegina described one such epidemic in the seventh century. He writes of a colic 'having taken its rise in the country of Italy, but raging also in many other regions of the Roman empire, like a pestilential contagion, which in many cases terminates in epilepsy, but in others in paralysis of the extremities, while the sensibility of them is preserved, and sometimes both these affections attacking together. And of those who fell into epilepsy the greater number died; but of the paralytics the most recovered, as their complaint proved a critical metastasis of the cause of the disorder.'

This is the first account there is of the great outbreaks of lead colic which occurred sporadically throughout history and which were known variously as the colic of Poitou, the entrapado of Spain, the Huttenkatze of Germany, the bellain of Derbyshire, the dry bellyache of the Americas and the colic of Devonshire.

The circumstantial evidence is strong, therefore, to support the hypothesis that lead poisoning was pandemic in Rome as Kobert and Hofmann in particular have shown. The position of these authors has been this: there is frequent reference to symptoms—particularly colic and constipation—which are concomitants of poisoning with lead, amongst the Roman and Greek authors; lead was much used; therefore the symptoms were caused by lead.

REFERENCE

Waldron H A 1973 *Med. Hist.* **17** 391

---

Here lies the body of William Jay,
Who died maintaining the right of way,
He was right, dead right, as he walked along,
But he's just as dead as if he'd been wrong.

Contributed to *The Lancet* **1** 164 by Rhode Hogg from America, 1923

# Inheritance Pattern of Death

We wish to communicate a striking observation which we believe will have broad implications for our understanding of life and mankind's tenuous mortality. This deceptively simple concept may well affect our very social structure and radically alter patterns of marriage.

Genetics has historically been a pragmatic and arbitrary branch of medicine, as would any field whose origins stem from a lonely and celibate monk staring vacuously at pea plants in a monastery courtyard, arriving at no more profound an observation than that some were tall and some were short. As the insufferable boredom of this behaviour became evident, even in a monastery, and the monk took to lasciviously and incestuously breeding all permutations of short and tall, the prurient interest of physicians was inevitably kindled, resulting in the incorporation of genetics into the realm of medicine.

In subsequent years, genetics has struggled to elevate itself from its ignominious origins. Like Maxwell's Demon, genes themselves are still impossible to observe directly, and all conclusions concerning their behaviour must be derived from indirect (some say irrelevant) methods. Therefore, much of a modern geneticist's time is consumed in the lengthy and often uncomprehending scrutiny of long pedigree charts, with occasional forays into the lab to gawk enviously at fruitflies and funny-looking mice copulating to produce even funnier-looking offspring.

We recently had occasion to engage in a bit of this dubious pastime, somewhat against our better judgement, when several pedigrees (of which the figure is a representative example) became available to us while awaiting the completion of a seemingly endless departmental meeting. As with most momentous discoveries, our original goal was totally unrelated to our final conclusion. The original goal, of course, was simply to ignore and thus survive the meeting, whose agenda ranged from confiscation of the residents' beer fund to the mysterious disappearance of the departmental chairman's pet hamster. Thus, we were serendipitously and seductively drawn to contemplation of the suddenly fascinating line-and-circle patterns of the pedigrees.

After many random, dead-end lines of reasoning, substantial free association and several short naps, it became apparent to us that a definite and dimly familiar pattern of genetic inheritance was present. We were unable to obtain a genetics consult at the time (it being past noon), but managed with the aid of incense and a small sacrificial lamb (later modified into a superb *noisettes d'agneau*) to discern a

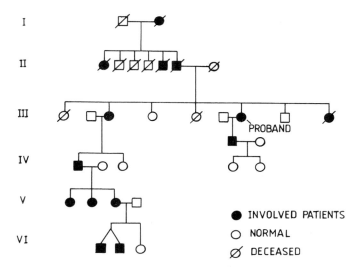

I

II

III
PROBAND

IV

V

VI

● INVOLVED PATIENTS
○ NORMAL
⊘ DECEASED

classic Mendelian autosomal recessive pattern for the phenotype of death.

We are confident that the discovery of a recessive transmission pattern (apparently with very high penetrance, often greatly delayed expressivity, and no racial or sexual predilection) for the trait of death will be recognised as highly significant from a purely scientific point of view. Of course, as with most elucidations of genetically controlled diseases, no practical significance is immediately evident, and no prospect for a therapeutic solution to death can be foreseen. However, in the grand tradition of science, we refuse to be discouraged by the basic irrelevance of our discovery.

There are, nevertheless, certain steps which individuals may consider to minimise the risk of death developing or perpetuating in their families. Now that an autosomal recessive pattern has been demonstrated, the families and ancestors of a potential spouse should be carefully screened for any incidence of death, as a positive history greatly increases the probability of the proband carrying the trait. Should the investigator's family also have a positive history of death, there is, of course, a 1:4 chance of communicating the trait to potential offspring—an inexcusable action in light of the new knowledge communicated herein. One should especially make certain that a potential spouse has not already expressed the gene, i.e. is not dead, since this elevates the probability of perpetuating the trait to proportions approaching 100%, as well as substantially reducing the prospects for a fulfilling relationship.

We hope that this discovery will aid knowledgeable people in minimising the risk of death in their descendants. Hopefully, spurred on by inspirations at future tedious departmental meetings, we may succeed in developing an enzymatic test to be done on all newborns for the presence of the death trait, though this research has already been vigorously opposed by the Amalgamated Morticians' Association (AMA) and the national Kounty Koroner's Konvention (KKK). We intend to remain steadfast against such blatant special-interest pressure, however, and with God's help and a National Institute of Health grant, we are determined to prevail.

Eastern J, Drucker C and Wold J E 1982 *J. Irreproducible Results* **28** 22

# Life and Death of Mok the Gorilla

Mok, the late lamented male gorilla at the London Zoo, came with his female companion, Moina, from the French Congo in 1932. At that time they had both been in captivity over two years and had been well cared for by a local district commissioner, M Capagorry, and his wife.

At this time Mok was about 3½ years old and Moina two years older. Both were tame up to a point, especially with Madame Capagorry, with whom they played for an hour or more every day. They were not found together but in different localities in the French Congo and at different times. Mok was a typical lowland gorilla (*Gorilla gorilla gorilla*), whereas Moina probably came from farther inland and is intermediate in form between the lowland type and the Kivu gorilla of the Belgian Congo (*Gorilla gorilla beringei*). The male of the latter species grows to a much greater size and weight than the lowland form. Mok, at his best in 1937, weighed 320 lb and Moina 252 lb, although when first brought to the Zoo in 1932 he weighed only 70 lb against Moina's 112 lb. An adult male Kivu gorilla died at the Berlin Zoo in 1937 weighing 568 lb. Had Mok lived he would have probably never weighed more than 400 lb.

Although he had the appearance of being immensely strong and healthy, he was never so resistant to disease as Moina and he had two attacks of pneumonia, the first in the winter of 1932 and again a year later. During the first of these attacks it was possible to examine Mok and to percuss and auscultate his chest in the usual manner, as he was still in the approachable stage, but by the following winter, when he had his second attack, this was not so easy. By that time both gorillas had been transferred to the new house which was specially built for them and had become so rough in their play that even their keeper could not go in with them. It is interesting to note that the first attack of pneumonia which Mok had was definitely lobar in type and unlike the usual broncho-pneumonia of young chimpanzees, which resembles the broncho-pneumonia of children.

The following list gives an idea of the varied diet which is necessary to keep a gorilla in health in captivity.

|  | Per week |  | Per week |
|---|---|---|---|
| Grapes | 14 lb | Porridge | ½ lb |
| Bananas | 42 | Tea | 2 oz |
| Apples | 14 lb | Sugar | 2½ lb |
| Oranges | 21 | Bread | 3½ hf-qtns |
| Prunes | 2 lb | Butter | 6 oz |
| Rhubarb (when in | | Eggs | 10 |
| season) | 3½ bdls | Milk (condensed) | 3 tins |
| Tomatoes | 2 lb | Milk (tubercle- | |
| Lettuces | 6 | free, irradiated) | 7 quarts |
| Greens | 5 lb | Cod-liver oil | |
| Onions | 2½ lb | cream | About ¾ |
| Carrots | 1½ lb | | pint |

In addition to the above, Mok and Moina used to get a boiled chicken once a week, and sugar-cane, corn-cob, melons, cucumbers, and other delicacies when in season. They also had products rich in various vitamins. The cost to the Zoological Society of keeping this pair in food alone was at least £4 per week.

The special house which was designed by Messrs Tecton is considered to be one of the most ingenious and modern of all animal-houses. It was arranged in such a way that during the winter half of the circular house is reserved for the animals and half for the public, direct contact with the public being prevented by a movable screen of glass in addition to the bars of the cage. The glass, of course, is to prevent direct infection from visitors and improper feeding. During the summer, in addition to this interior cage, the animals are given the other half of the house which the public use in the winter time, and the visitors then look on from outside, being separated from the outside cage by a barrier 8 feet away from the bars of the cage. At this distance any infection is absent, or at any rate so attenuated as to be negligible.

The first signs of illness in Mok were noticed at the beginning of November 1937, when he lost his appetite and became listless. These symptoms increased until Christmas time when a slight improvement was noticed. This improvement, however, did not last and by the beginning of the New Year he was refusing all solid food and taking very little in liquid form. Due to his strength and truculence, it was impossible to examine him in any way or even take his temperature, and apart from a certain amount of loss of weight and occasional diarrhœa he had no other symptoms of disease except that when he could be induced to open his mouth to take a drink, it appeared to be somewhat inflamed and infected. During the ensuing fortnight there were days when he seemed to rally and take a little more nourishment, while on other days he lay listless and refused even water. The question of giving him an anæsthetic for purposes of examination was considered, but was held not to be expedient owing to certain obvious difficulties in catching up the animal, with resulting exhaustion, and also because of slight lung symptoms which towards the end made themselves apparent. Mok died at 7.30 PM on January 14th. The body was transferred immediately to the prosectorium and opened up in order to prevent decomposition. During that night, the brain was removed and injected and certain other organs, such as the testicles, were preserved for research purposes.

A very full post-mortem examination was made by Colonel A E Hamerton, pathologist to the Zoological Society. The findings of the post mortem were surprisingly few, and macroscopically it could not be said that anything could be found sufficient to cause death, the only obvious changes being some fatty degeneration of the liver and some changes in the cortex of the kidneys. A full microscopical examination of the tissues is being undertaken and also cultures have been taken from the blood. It is possible that the results of these may throw more light on the cause of Mok's illness and death.

REFERENCE

1938 *The Lancet* 1 237

# Internal Combustion

The inflammable character of marsh gas or methane—the hydrocarbon $CH_4$—is well known, although less so than the similar quality inherent in colonic gas, giving rise to a torch singeing with which every schoolboy is familiar. The explosive quality of bowel gas is, in fact, due partly to its methane content, which, with hydrogen, predominates in patients on leguminous and milky diets. Since both hydrogen and methane are odourless it is obvious that other constituents are included in the orthodox composition of the winds that inhabit the nether regions. Nitrogen is ordinarily the chief of these fabrics of flatulence; carbon dioxide is alleged to constitute 10% and oxygen less than 1% of the whole.

This explosive quality of rectal gas, mentioned by Hayden in his progress report and discussed editorially in the *British Medical Journal*, constitutes a definite, if unusual, surgical risk. Less common in the operating room than the verbal explosions to which the surgical overlords of former times were addicted, such incidents, in these days of electric contrivances, entail a greater hazard to the patient, if not to the operating team. A number of cases have in fact been reported in which diathermy or the use of the cautery has resulted in intraluminal explosions with rupture or massive ecchymosis of the colon above the site of treatment.

It is assumed that insufficient carbon dioxide is normally present in the colon and rectum to blanket the fire, and that enough oxygen must be present to fan into life the smoldering flame. To supply oxygen air may be introduced through the anus via the endoscopic tube, or it may be cribbed during the induction of anesthesia. Swallowed under these circumstances it may reach the anus in half an hour. The particular lesson to be learned is that an awareness of the explosive possibilities should persist and all electric apparatus should be properly grounded.

Not only is the rectum a site of fire-damp explosions; gases eructated by sufferers from pyloric stenosis may sometimes be detonated in like manner and with disastrous results. This risk has not yet been used as ammunition in the cigarette battle; it might, in fact, apply more aptly to a drive against pipe smoking in which tobacco and matches are consumed in approximately equal quantities; nor has the possibility of such a conflagration yet been used as an argument against hot and intemperate words or other incendiary language.

REFERENCE

1954 *New Engl. J. Med.* **251** 872

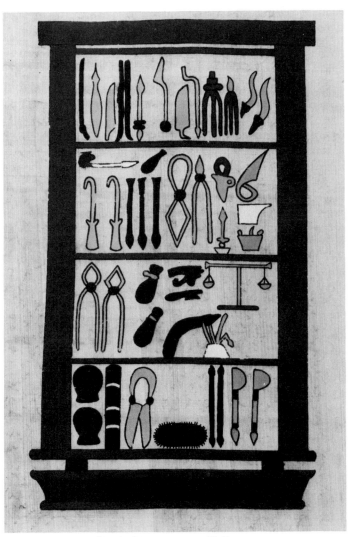

*Surgical instruments, 2340 BC.*

Many men have gained success at the expense of others. Fortunately, in my profession, medicine, success does not readily come by harming other people. What leads to success is pure gain to the community.

SIR ALEXANDER FLEMING

*Surgical instruments, 1739.* (Courtesy: Wellcome Institute library, London.)

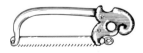

*Brigham Young*
*born on this spot 1801*
*a man of much courage and superb equipment*

Memorial inscription to the famous Mormon
leader who practised polygamy

*This photograph of a Siberian shaman* was taken during the 1914–1915 Czaplicka expedition to Siberia. (Courtesy: Wellcome Institute library, London.)

# Why the Pill's Success Rate is Low in Asia

Men who wore condoms on a finger or took the Pill meant for their wives were two reasons for contraceptives often being ineffective in Asia, a United Nations report said yesterday.

The report to the Third Asian and Pacific Population Conference related that remote Asian villagers had been shown how to wear condoms in demonstrations with a bamboo pole. When field workers returned several months later they were confronted by groups of irate pregnant women. Inquiries disclosed that men had been wearing condoms on a finger or keeping them on a bamboo pole.

Investigators also discovered that in many instances condoms had been boiled or swallowed. The most common pitfall with the Pill, the report said, was that men were taking it instead of women.

REFERENCE

Reuter *The Times* September 28, 1982

*An American Indian medicine man* attending a sick child. Oil painting by Valentine W Bromley, 1876. (Courtesy: Wellcome Institute library, London.)

*This photograph of Mapuche medicine women* was taken in the 1920s by the South American Missionary Society in Southern Chile. (Courtesy: Wellcome Institute library, London.)

# The 'Deceased' Left Funeral, Swearing

When grave-diggers shovelled the first spadefuls of earth into a grave in the village cemetery at Pecaya, Venezuela, the 'dead' man, Roberto Rodriguez, who had collapsed after a heart attack, burst open the lid of the coffin, scrambled out of the grave, and ran home shouting and swearing.

His mother-in-law, who was standing at the graveside, dropped dead from shock. She will be buried in the grave prepared for her son-in-law, after doctors have made absolutely certain there is no mistake this time.

REFERENCE

Reuter *Daily Telegraph* October 2, 1971

# Theory of Evolution

The theory of evolution, which underpins so much of modern medicine, has been the subject of fiercely held views and is, as a consequence, it seems, a happy hunting ground for fakers, forgers and frauds.

Darwin's original idea of how variation could be inherited was by a sort of blending of parental characteristics. When, in 1867, this was shown to be mathematical nonsense, he resurrected the inheritance of acquired features and inserted a Lamarckian chapter into the sixth edition of *Origin of Species*. Ironically, a more acceptable answer had already been published by Mendel two years earlier, but his work was ignored until 1900.

In the early part of this century a battle royal was waged between the Lamarckians and Mendelians. Champion of the former was Paul Kammerer, the Austrian zoologist, whose great work was to breed the midwife toad in captivity. Most frogs and toads mate in water. To get a firm grip on the female's slippery body the male toad develops in the mating season black horny spines on the hands, known as nuptial pads. The midwife toad mates on land and neither needs nor possesses nuptial pads. Kammerer claimed that in 1909 he induced the midwife toad to copulate in water for several generations and that they eventually developed nuptial pads as an acquired hereditary feature. The Mendelians would have none of it and harried Kammerer for 15 years. In 1926 the scientific world was shaken by the revelation that Kammerer's nuptial pads had been faked with Indian ink injections. Within two months Kammerer killed himself. Interestingly, ten years later Mendel's own results came under scrutiny. Sir Ronald Fisher, the famous statistician, proved conclusively that Mendel's published figures must have been doctored. They were so close to the expected ratio of 3:1 that it would have taken an 'absolute miracle of chance' to produce them. Mendel had the luck to be right, and history has treated him kindly. His gardeners were possibly to blame or perhaps he stopped counting accurately when he established his point.

For a long time the Darwinian viewpoint, although intellectually attractive to scientific humanists, suffered from a lack of supporting evidence. The great pathologist Rudolf Virchow had despatched Neanderthal man, the only candidate for the missing link, as Homo sapiens with rickets, a view that modern science has shown to be essentially correct. Virchow's former student, Ernst Haeckel, professor of zoology at Jena, set out to prove the case. He proposed the (now discredited) theory that an embryo retraces its evolutionary history *in utero*.

For example, the so-called gill slits in the human embryo were evidence of our supposed fishy origins. Haeckel not only altered his illustrations of embryos to support his case but actually printed the same plate of an embryo three times and labelled one human, the second a dog, and the third a rabbit 'to show their similarity'.

He was charged with fraud and convicted by a university court at Jena. He admitted that he had faked his results but defended himself in the following manner. 'I should feel utterly condemned and annihilated by the admission were it not that hundreds of the best observers and biologists lie under the same charge.' Haeckel retained his chair and, surprisingly, his reputation. Virchow, however, considered him a fool.

Haeckel's next endeavour was to invent Java man and send Eugene Dubois, his former student and at that time a doctor in Holland, to look for him. In 1891 Dubois found him on the Solo River at Trinil—the skull cap from an ape-like creature and, nearly 15 m away, a human femur and two teeth. Thus was born the *Pithecanthropus erectus*. Haeckel telegraphed his congratulations from the 'inventor of Pithecanthropus to his happy discoverer'.

In Virchow's opinion the skull cap came from a giant gibbon and the human femur had no connection with it whatsoever. Before his death in 1940 Dubois came round to the same point of view.


REFERENCE

Hamblin T J 1981 *Br. Med. J.* **2** 1671


# Sex, Age and a Filippino General


These two comments are attributed to General Carlos Romulo, the 84 year old Foreign Minister of the Philippines who is well known for introducing jokes into his speeches.

Old soldiers and Ministers of Foreign Affairs never die and nor do they fade away. They just carry on with their foreign affairs.

General Romulo visited his doctor with a request that the doctor lower his libido. He was told that this would be difficult since libido was all in the head. The General replied, 'Exactly, I want it lowered'.

# The Skoptzy Sect

In 1757 a member of a sect of flagellants, André Ivanov, founded the sect of the Skoptzy, or castrates, or origenists. He commenced by castrating himself, then he mutilated his 13 disciples, the first apostles of the new religion. Arrested, he was tortured with the knout, and deported to Siberia where he died.

The news of this punishment only inflamed the ardor of the proselytes, and under the impulsion of Kondrati Sselivanoff, 'the new Christ descended among men', the sect took on considerable vigour. In 1770, the year of the plague in Moscow, Sselivanoff preached his doctrine in Petersburg. He became the protégé of the mistress of the Tzar, the Baroness of Krudener, who considered him a saint. He died in 1832 at the age of 100 years.

The Skoptzy reject the majority of dogmas of the Greek Orthodox Church, in particular the fundamental dogma of Christianity, that of redemption of Christ. In their view Sselivanoff is the true Christ who taught the principle of mutilation for the redemption of mankind.

It would appear that this sect would destroy itself and could not persist. But the Skoptzy did not abolish marriage; which ceased only after the birth of a second child. Then the spouses had to submit to castration: removal of the testicles from the man and the clitoris from his wife. From the social point of view the Russian Skoptzy included numerous bankers, cashiers and money changers, handling large amounts of capital. 'If I were a banker,' said a Russian, 'I would have only a Skopetz for a cashier. For the cash register, as for the harem, a eunuch is the most certain guardian. In all embezzlements, all stealing of funds, in all irregularities of the accountant there is ordinarily a woman; with the Skoptzy one can sleep in peace.'

The chaste and sober life which the Skoptzy led, abstaining from games and feasts, never getting drunk, and their eagerness for monetary gain explains how many became millionaires. They used the money for incessant progaganda in favour of their sect.

In general the sexual mutilations take place with extraordinary ceremonies (privod). They are of two types. The little seal which consists in ablation of the testicles alone confers on the initiated 'the right to mount the spotted horse'—he no longer has the keys to hell, the testicles, but he retains the key to the abyss. The baptism completed, the second purification leads to the imperial seal which gives the victim the right to mount the white horse of the apocalypse. The two operations constitute the two degrees of the initiation and in the majority of cases they are made one after the other. The severity of the trauma is sensibly diminished. Almost all Skoptzy of imperial seal show two scars separated by an island of healthy skin.

Death has rarely followed castration performed in this manner. This is somewhat surprising considering the brutality of the operation and the carelessness of the operators. The choice of instrument is indifferent to them: a razor, a pocketknife, scissors, a pruning knife, a chisel, a piece of sheet metal, or a bone from a bull, sufficiently worked to suffice for the surgical demands of these fanatics. After the penectomy, a plug of tin or lead is inserted into the urethra.

According to Stein in 1875 there were official police records of 5444 Skoptzy: 1465 women and 3979 men, of whom 588 had had total emasculation. The

(Courtesy: Wellcome Institute library, London.)

majority were Greek Orthodox to the number of 5024; there were only 8 Roman Catholics; the remaining 12 were Lutherans, Mahometans and Jews.

All classes of Russians furnished members to the Skoptzy: nobles, employees, priests, soldiers, the bourgeoisie, workers. But the greatest number were found among the peasants—victims of ignorance and religious fanaticism.

The progressive invasions of the sect, its growing importance, and finally the complaints of innocent victims provoked many measures toward suppression, but the repugnant heresy persisted at least until the time of the Russian Revolution of 1917.

REFERENCE

Derbes V J 1970 *J. Am. Med. Assoc.* **212** 97

(Courtesy: Wellcome Institute library, London.)

*In memory of Mrs Phoebe Crewe*
*who died May 28 1817 aged 77 years*
*who, during 40 years as a Midwife in this city*
*brought into the world 9730 children*

Gravestone inscription, Norwich

**EXTIRPATION OF ALL KINDS OF VERMIN.**

Patronized by Her Majesty, H.R.H. the
Duchess of Kent, and the Corporation of the
City of London, and by Special Appointment
to the principal Courts of Europe

**Office: 69, King William Street, City, London.**

## Mr. J. A. MEYER,
### Practical Chemist,

Respectfully informs the Nobility, Gentry, Merchants, and the Public generally, that he is the only person who undertakes to cleanse, by Contract, for a term of years, or otherwise, (without the use of Poison,) Mansions, Hotels, Dwelling-houses, Estates, Ships, Warehouses, Granaries, Breweries, Hospitals, and all descriptions of Public or Private Property, from **Rats, Mice, Bugs, Fleas, Moths, Moles, Centipides, Scorpions,** and all kinds of **Vermin,** guaranteeing to keep the same free from one to five years, according to contracts.

#### CERTIFICATES.

I hereby certify that Mr. Joseph Meyer, Practical Chemist to the Courts of Russia, Prussia, &c. has cleansed the whole of the Prison of Newgate, from Rats, Mice, Blackbeetles, Bugs, and other Vermin, without the use of Drugs of any poisonous nature whatsoever, for which I deliver him this certificate.

*London, August 23, 1843.*     (Signed)  WILLIAM WADHAM COPE, Governor.

A similar certificate for cleansing the Mansion House, signed by order of the LORD MAYOR; and numerous certificates from Courts, Noblemen, Gentlemen, Merchants, Farmers, &c.

#### LIST OF PRICES;

| | |
|---|---|
| For Rats and Mice *(prepared in Pills)* | in Pots, at 10s., 20s., 50s. and 100s. |
| For Rats and Mice *(Herbs to drive away)* used in rooms | in Bottles, 10s. and 20s. |
| For Bugs *(Liquid)* | in Bottles, 5s., 10s. and 20s. |
| For Bugs *(Powder)* | in Bottles, 5s., 10s. and 20s. |
| For Fleas *(Powder)* | in Bottles, 5s., 10s. and 20s. |
| For Fleas *(Liquid for Animals, &c.)* | in Bottles, 5s., 10s. and 20s. |
| For Blackbeetles *(in Powder)* dry places | in Pots, 5s., 10s. and 20s. |
| For Blackbeetles *(Liquid)* damp places | in Jars, 10s., 20s. and 50s. |
| For Moths *(Powder)* | in Bottles, 10s. and 20s. |
| For Moles *(Powder)* | in Pots, 10s. and 20s. |

Dwelling Houses cleansed from £1 to £5.     Mansions and Estates from £10 and upwards.

Ships, Docks, Warehouses, Granaries, Breweries, &c., in proportion.

Contracts from One to Five Years, guaranteeing for the said term.

N.B.—The preparation with full directions for use, may be had at the Office, which can be effectually used without difficulty or any injury to property.

Cronland & Co. Printers, 2, Fenchurch-street, City.

(Courtesy: Wellcome Institute library, London.)

# The Wife's Relief or the Husband's Cure

SATURDAY, OCTOBER 16, 1736.

LINCOLN's-INN-FIELDS.
By Desire of Mrs. MAPP,
The Famous BONE-SETTER of EPSOM.
Acted but Twice these Twelve Years.
By the Company of Comedians,

AT the Theatre-Royal in Lincoln's-Inn Fields, this Day, October 16, will be presented a Comedy, call'd

The WIFE's RELIEF;
OR,
The HUSBAND'S CURE.

The Part of Cynthia, by a GENTLEWOMAN, (Being the Third Time of her appearing on any Stage;) Riot, Mr. Giffard; Volatil, Mr. Wright; Sir Triftrum Cash, Mr. Penkethman; Young Cash, Mr. Yates; Horatio, Mr. Havard; Valentine, Mr. Richardson; Spitfire, Mr. Norris; Slur, Mr. W. Giffard; Hazard, Mr. Hewitt; Howd'ye, Mr. Lyon. Aurelia, Mrs. Hamilton; Teraminta, Miss Hughs; and the Part of Arabella, by Mrs. Giffard.

To which will be added, a Pantomime Entertainment, call'd
The WORM-DOCTOR.
In which will be Introduc'd a new Scene of Action call'd,
HARLEQUIN
FEMALE BONE-SETTER.

Dr. Pestle, Mr. Dove; Harlequin, Mr. Lun, jun. Pierrot, Mr. Norris; Mortar, Mr. Penkethman; Dropsical Man, Mr. Ware; His Servant, Mr. Hamilton; Constable, Mr. Wetherilt; 1st Drawer, Mr. Harrington; 2d Drawer, Mr. Yates; 1st Chimney-sweeper, Master W. Hamilton; 2d Chimney-sweeper, Master J. Hamilton; Colombine, Mrs. Hamilton; Maid, Mrs. Dove; Bawd, Mr. Nichols; Nurse, Mrs. Wetherilt; Mrs. Lovepuppy, Miss Thornowets.
The FEMALE BONE-SETTER, by HARLEQUIN.
Bandage, Mr. Rosco; Callous, Mr. Ray; Vertebra, Mr. Lyon; Dislocation, Mr. Davis; South, an Inn-keeper, by Mr. Hewitt.
Concluding with an Entertainment of Dancing by Mr. HAUGHTON, Madem. ROLAND, Monf. Vallois, Mr. Lefac, Mr. Delagarde, Mr. H. Fayting, Mrs. Bullock, Mrs. Woodward, Miss Gerrard, and Miss Oates.

Boxes 5 s. Pit 3 s. First Gallery 2 s. Upper Gallery 1 s.
Places for the Boxes to be taken at the Stage-Door of the said Theatre.

To begin exactly at Six o'Clock.
N.B. Never Acted there before, on Tuesday next will be presented a Tragedy, call'd MITHRIDATES King of PONTUS. To which will be added, a Pantomime Entertainment (never perform'd there before) call'd HARLEQUIN SHIPWRECK'D: Concluding with the Loves of PARIS and OENONE.

(Courtesy: Wellcome Institute library, London.)

85

# TOXO-ABSORBENT
## THE GREAT DRUGLESS TREATMENT

The most important medical discovery in the world's history.

Diseases can be cured more promptly and with greater certainty without taking medicine in any form.

By the new treatment lingering sickness and premature death can be avoided and mankind can live to a good old age.

THE TOXO-ABSORBENT CURE can be relied on for the cure of any of the following diseases. If suffering from any one of them, write us at once. *See directions for treatment in this book*

| | |
|---|---|
| Asthma | Typhoid Fever |
| Bronchitis | Gastritis |
| Diphtheria | Ulceration of Stomach |
| Swelled Glands | Chronic Diarrhoea |
| Hay Fever | Catarrh of Stomach |
| Catarrh of Throat | Neuralgia of Stomach |
| Consumption | Kidney Diseases |
| Inflammation of Lungs | Bright's Disease |
| Congestions | Abscesses |
| Pleurisy | Fever Sores |
| Pneumonia | Varicose Ulcers |
| Malaria | Blood Poison |
| Congestion of Liver | Rheumatism |
| Biliousness | Cancers |
| Jaundice | Fibroid Tumors |
| Gall-Stones | Scrofula |
| Appendicitis | Erysipelas |
| Peritonitis | Chilblains |
| Ivy Poison | Syphilis |

*This treatment* referred to 'improved oxytoner oxygen', an apparatus for 'gas pipe therapy' marketed in Chicago in the early years of the twentieth century. From *Nostrums and Quackery* 1912 (Chicago: American Medical Association). (Courtesy: Wellcome Institute library, London.)

Here lie I and my three daughters
killed through drinking Cheltenham waters
had we but stuck to Epsom Salts
we ne'er would have been in these 'ere vaults.

Gravestone inscription

# Coles's Patent Truss

RUPTURES EXTRAORDINARY.—COLES'S PATENT TRUSS for single rupture has three springs, and for a double one six. The common, or German Truss, for double rupture, has but one spring generally. A mail coach or a gentleman's carriage when compared with a common waggon, is not more conspicuous to the eye, nor when used more sensibly felt. The following lines, taken from the Lancet, No. 195, but too plainly shew how difficult a task it is to suppress real merit, even by the most powerful agency.

> In London's great city, a place so renown'd
> For learning and arts, none so clever,
> Three lawyers of eminence, two had a son,
> One a nephew, only son of his mother.
>
> The last for advice to Sir Astley was bent,
> His mother and uncle to witness,
> He had then on his person Coles's Patent,
> When Sir Astley disputed its fitness.
>
> 'Perhaps,' says the mother, 'Sir Astley can tell
> Where we can procure something better,
> Though doubtless the other has done very well,
> Which appears by Hunt and Son's letter.'
>
> The second also to Sir Astley did go,
> His father most likely went with him,
> But finding a common Truss very so so,
> To Mr Coles straitway he took him.
>
> The latter a bandage preferr'd in his case,
> Which to him gave great satisfaction,
> When a second interview shortly took place,
> He related the former transaction.
>
> 'What could Sir Astley,' quoth he, 'mean by saying,
> 'Don't get a Patent Truss for your son,'
> Surely my money for nought I've been paying,
> The truss has been alter'd four times since 'twas done.'
>
> Dr. Birkbeck's advice in the third case was given,
> The remedy was found quite complete,
> Tho' Travers' advice was previously taken,
> Neither patient had worn them a week.

Manufactory, No. 1, London Bridge.

(Courtesy: Wellcome Institute library, London.)

# Tobacco Therapy

Writing in the Argentine journal *Accion Médica* for February 10, N Hughes mentions that Nicot used the leaves to cure his cook's cut finger, and that later the prior of Lorena, who received a plant from Nicot, gave some to Catherine of Medici, whose hobby was medicinal herbs. It was not long before snuff was made in France for curing headaches and removing 'humours'. Smoking was used to cure asthma, and even infusions of tobacco were taken. Tobacco was regarded as an antidote to the poisonous arrows of the Indians, although tobacco was an ingredient in that very poison. Ferdinand II of Aragon, anxious to preserve the lives of his explorers in America, instituted pharmacological tests. A dog was wounded with a poisoned arrow, and tobacco juice poured on the wound saved the dog's life. Cataplasms were applied to relieve indigestion and torticollis. Powdered finely, tobacco was used as a disinfectant, and during the plague of 1665 the scholars of Eton were thrashed if they did not smoke before coming into the classrooms. Tobacco was used as both a stimulant and a sedative, until it was found to have a toxic action. Enemas of tobacco smoke were administered through special bellows, and tobacco was used as an adjuvant to mercury in the sweating baths for syphilis.

REFERENCE

1939 *The Lancet* **1** 968

# Transmigrations of a Cow

Some three months ago an 'ancient' cow suffered from a wasting disease at Glastonbury and was then killed, we suppose, to prevent its dying. The carcase was sold for 7 s to a gentleman named Chivers and was despatched by him to a meat salesman named Harrington, of 99, Charterhouse-street, in a hamper labelled 'cat's meat' per the London and South-Western Railway, being delivered at his (Harrington's) premises with the label attached. From cat's meat the carcase entered upon a higher incarnation in the house of one Robinson, an East-end sausage-maker; but at the moment of its being cut up for sausages on December 5th, 1895, a sanitary inspector entered, with the result that Robinson was fined at Worship-street Police-court in the sum of £50. The meat having been traced to Harrington, he was charged at Clerkenwell Sessions and sentenced to three months' hard labour, about as much as a vagrant gets for that mysterious 'offence' known as 'sleeping out'. The scoundrels who sell bad meat for human food ought to be treated as the poisoners they really are. We would suggest the revival of the pillory. Messrs Robinson and Harrington standing in this machine in St Paul's Churchyard labelled in large letters with their names, addresses, and offences, with fragments of the cow or the sausages round their necks, would be a most wholesome sight.

REFERENCE

1896 *The Lancet* **1** 570

# Medical Spooks

Dr Cecil Henry Coggins is one of a small but distinguished group of medically qualified secret agents whose double-life has recently been described by Eugene G Laforet (*Journal of the American Medical Association*, 1980 **243** 1653). Although posted to the obstetrical department of the United States Naval Dispensary, Long Beach, California, in 1933 at the age of 31 years, he was soon picking up clandestine radio transmissions from Japanese fishing boats and by 1941, had been transferred to the staff of C-in-C Pacific Fleet as Chief of Counterespionage. Later in the war, he was in charge of a special unit responsible for psychological warfare and research. Coggins was a splendid choice for, as Laforet points out, 'the intellectual processes by which a physician collects, evaluates and synthesises information to arrive at a diagnosis are identical, he [Coggins] believes, to those employed in intelligence work.' This is borne out by the career of Dr Charles Dent, later professor of human metabolism at University College Hospital Medical School and a fellow of the Royal Society who, before qualification in 1944, worked in the censorship department where his chemical knowledge was used in the detection of secret enemy messages.

Apparent or actual medical practice can also be used to cover the activities of a 'mole' or 'sleeper'. In 1933 Dr Grigory Burtan, a Russian-born New York physician, was arrested for passing forged banknotes. He was the contact man for the Soviet resident director in the United States, and his activities were part of a plan to flood the United States with counterfeit money.

Also practising in New York in the 1930s was Dr Ignatz Theodor Griebl who, after qualifying in Munich in 1922, came to America in 1925. He was at the centre of a Nazi organisation and a search of his safe revealed microfilms of state documents and a list of prominent American Jews who were targets for assassination. Another German doctor, Bernard Julius Otto Kuehn, went to Hawaii in 1941 with a cover story that he was writing a book about the East. He transmitted information about ship movements to Japanese trawlers and was arrested during the attack on Pearl Harbor just before he could be evacuated by a Japanese submarine.

America was a target for friend and foe alike and in 1940 Dr Robert Soblen and his brother were allowed to leave Lithuania on condition that they formed an espionage network. Robert Soblen practised as a psychiatrist between 1940 and 1945 but also obtained secret documents from the Office of Strategic Services, information about the atomic bomb project and photographs of the nuclear development centre at Albuquerque. Another Russian doctor, however, was helpful to the West. Dr Michael Bialoguski had settled in Australia and was working for his adopted country's security services. Vladimir Petrov, then in the Soviet Embassy, Canberra, was given permission by his superiors to recruit him, doubtless with the hope that they could turn him. Bialoguski was actually planning to turn Petrov whose defection caused a world-wide sensation in 1954.

REFERENCE

1980 *The Practitioner* **224** 775

# Spark Coils and Spy Scares

The Great War had been raging for a few months and the reservist Royal Marine responsible for work in the plating shop in the small factory near the present BBC building in Portland Place, in London, was already on his way to France earning his Mons Star. X-ray units would be wanted for the Forces and ten and twelve inch spark coils crackled away on their test benches, whilst your writer was engaged in the delectable task of cleaning mercury 'breaks', or interrupters as they were more technically known, surely the dirtiest and most smelly job in Christendom. The spy scare was on, neighbours' alert ears 'knew' that spies were rampant in the factory in the heart of the residential West End. Could they not hear the Morse Code going out over the air?

A burly policeman entered the factory, suspiciously he looked round. These crackling induction coils under test could well be used to signal Zeppelins, dropping their load of bombs in the still night air. He was not convinced when matters were explained. More visits from policemen, each time the status and number lifted. Only the War Office could placate them, and placate them they did. At last we were left to our noisy spark coils and open gas tubes suspended in the middle of a small workshop un-X-ray-protected!! Far more potent than Zeppelins did we innocents but know it. Well, yes, sometimes they were laid in a lead glass shield (for radiation protection) on test, but who cared, there was a war on?

REFERENCE

1965 *J. X-ray Technol.* 24

# *Orders for Infirmary Inpatients, 1720*

1. No patient to be out above 2 hours without leave.
2. No Inpatients of this Infirmary to be abroad after nine of ye clock.
3. All Visitors to ye Inpatients to leave ye House by nine of ye clock and none to be admitted after that hour unless on Extraordinary Occasions.
4. Ordered that: No person be admitted to visit any of ye Patients in this Infirmary on Sunday till Divine Service is over in ye afternoon and that the Matron take strict care that no strong Liquor be brought into this Infirmary from any Publick House. And that all be obliged to quit ye House by Eight of the Clock.
5. Every One Discharged Cured from this Infirmary be enjoined by the Chairman to give Publick Thanks in their Parish Churches.

REFERENCE

Humble J G and Hansell P 1974 *Westminster Hospital 1716–1974* (London: Pitman Medical) 2nd edn

# Shearer's Delineator

In 1916 Sergeant James Shearer of the Royal Army Medical Corps described in *British Medical Journal* his results for the delineation of internal organs by an electrical method. The most impressive illustration was a brain picture from a gunshot wound case.

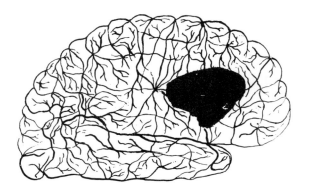

The BMJ noted that the new work appeared to succeed where X-ray photography fails, in producing pictures of structures hidden far below the surface of the body. An eye witness account of the technique stated that the exposure lasted 60 seconds, a little clicking was heard in a cupboard nearby, no darkened rooms, flashing lights or crackling spark gaps were required and the record of the organ was conveyed to a wax sheet before being printed on photographic paper. Over three pages of print were devoted to Shearer's Delineator, and he also had the approval of the Director General of the Medical Services in France. His downfall came later, when he made the further claim for his apparatus that it was able to detect and identify enemy airplanes flying over the trenches at night. The Army Intelligence Department succeeded where the BMJ editors had failed, and exposed the work as fraudulent. Shearer's reward was a court martial.

REFERENCE

1916 *Br. Med. J.* **2** 459, 468

# Morbidity in Assistants at Surgical Operations

Various horror stories (not always true) circulate about the unfortunate fate of certain patients undergoing surgery. More rarely are the dangers to assistant surgeons and to nurses documented, and the following, culled from the *Canadian Medical Association Journal* and written by two doctors from the University of Oslo, may therefore be of interest to readers.

The first case history is a nineteenth century account of a famous surgeon who was once going to perform an amputation on a patient with only alcoholic beverages as anesthesia and premedication. By a single cut with his sharp knife the surgeon performed amputation of the leg, as well as three fingers of the assistant and his own coat-tail. Cases have also been reported where the operator's knife cuts were so strong that in addition to amputating structures from the patient, he managed to perform self-circumcision!

There then followed case histories in three categories: mental trauma, minor physical trauma and major physical trauma.

A young, ambitious doctor was working in a large hospital under a great surgeon, who was so above his assistants that they rarely communicated with him. After one year's hard work, the young doctor was going to leave his job. He found it correct to say good-bye to the great surgeon and, therefore, asked the head nurse if she could arrange an audience with him. The young doctor, whose request was complied with, approached the great surgeon and said very politely 'I should like to say good-bye and thank you for the time at your department'. The surgeon peered over his spectacles and replied: 'Thank you. I hope you have fully recovered and are satisfied with the treatment you received'.

An intern, male, 27 years old, in a surgical department complained of regular frontal headache after prolonged operations. The same kind of headache was also experienced by the consultant, who was a feared and battle-worn surgeon. By anamnestic investigation it was found that this headache occurred after operations where both these persons participated.

Further inquiries revealed that when the intern bent forward during the operations to obtain a better view of the operation field, he was immediately butted back by the surgeon. This usually happened repeatedly during each operation. Both of them had become so accustomed to this procedure that they did not realise that prolonged butting may induce headache. The intern was advised by his family doctor (GP) to wear a motorcycle helmet during operations.

A surgeon had the habit of throwing scalpels when he became angry in the operating theatre. One day he exploded during an operation and threw a scalpel without looking where it went. The knife penetrated the assisting nurse's big toe, with resultant heavy bleeding. She was laid on the operating table in the neighbouring room and sutured under local anesthesia. Afterwards she always wore steel laminated protective shoes at work.

The final case history defied any categorisation.

Two very ambitious young interns, prior to a major and very important surgical intervention, could not agree who should be the first assistant and who the second. Both thought they were the better qualified and would not give way for the other one. As the operation began, they were standing beside each other, each trying to push the other away. One suddenly shoved his elbow into the other's stomach, and the other hit back in the same way. This continued until one of them had to go out and vomit. The other reigned victorious for the rest of the operation.

REFERENCE

Laerum O D and Skullerud K 1974 *Can. Med. Assoc. J.* **110** 632

# An Artist, a Gun and a Cigarette

A case of gunshot injury to the femoral vein, remarkable not only for its surgical interest, but also for the peculiar way in which it was caused, is related in the *Lyon Medical* for 1895. An artist aged thirty six used to keep a shell in his studio as a relic of the war of 1870, believing it to be uncharged. On August 3rd, 1894, having lighted a cigarette, he threw down the unextinguished match, when by chance it entered the narrow mouth of the projectile, which very soon afterwards exploded, one of the fragments impinging on the heedless smoker's right inguino-crural region and producing violent haemorrhage.

REFERENCE

1896 *The Lancet* **1** 246

*Saint Apollonia* is the patron saint of toothache, and is usually shown holding the pincers with which she was martyred. The painting above is seventeenth century Spanish, from the school of Zurbaran; the saint is seen to hold the pincers and a tooth in her right hand and a martyr's palm in her left hand. (Courtesy: Wellcome Institute library, London.)

# A Bucket of Pills—and it's Help Yourself

A doctor piled all the drugs in his surgery into a bucket in the waiting room and attached a label telling his patients to help themselves and not bother him.

Even when the General Medical Council did act it got the oddest answers—the doctor with the pills in the plastic bucket simply retorted that his treatment was no more random than that of other doctors.

REFERENCE

*London Evening Standard* April 16, 1975

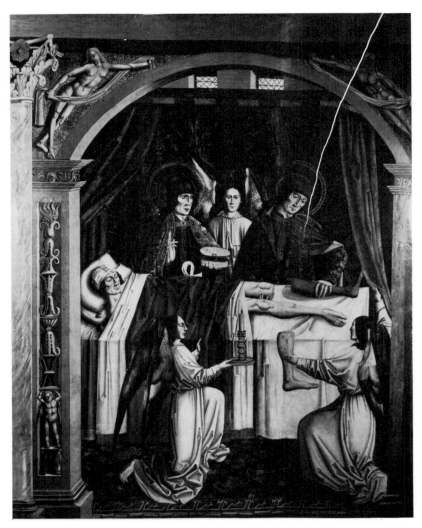

*Saint Cosmos and Saint Damian* are patron saints of physicians and surgeons. The fifteenth century Spanish oil painting by Alonso de Sedano shows their most famous miracle—that of replacing a cancerous leg with a sound leg from the dead body of a Moor. In a different painting of the same miracle, the dead Moor is also shown, and it is clearly indicated that the black corpse on the floor has one white transplanted leg whereas the live patient on the operating couch has one transplanted black leg. (Courtesy: Wellcome Institute library, London.)

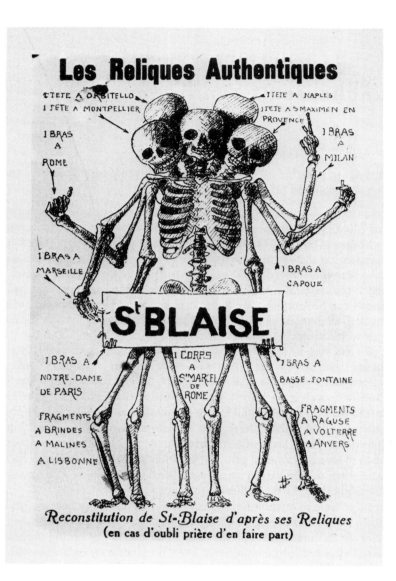

*A caricature of Saint Blaise* reconstructed from his relics at Montpellier, Milan, Rome, Marseille, Paris, Lisbon and elsewhere. St Blaise was a martyred bishop who is invoked against diseases of the throat. (Courtesy: Wellcome Institute library, London.)

# Munchausen's Syndrome

Richard Asher[1,2] has described several cases of Munchausen's Syndrome and one of these is reproduced below.

A woman of 41, giving the name Elsie De Coverley, was admitted to the Central Middlesex Hospital on February 7, 1950, having collapsed in a bus. She gave a history of one day's severe abdominal pain with vomiting of dark blood. On examination she was apparently in severe pain. Her abdomen was a mass of scars and her veins showed marks of many 'cutting down' operations. Her heart showed a murmur. She told us that in the last five years she had had two operations for perforation, one for gastro-enterostomy and one for intestinal obstruction, all done at the Royal Devon and Exeter Hospital, where she said she was due to return for a partial gastrectomy.

At first she was diagnosed as a case of a bleeding ulcer with probably a small perforation, but despite her severe pain there was no abdominal rigidity and radiography showed no gas under the diaphragm. Her haemoglobin was 96 per cent and her stools gave only a weakly positive benzidine test. A telephone message to the Royal Devon and Exeter Hospital revealed that she had only once been there in 1944 and had never had an abdominal operation there, but that many other hospitals had made inquiries about her. The patient continued to complain of severe abdominal pain, and a barium meal and gastroscopy had been arranged when she discharged herself against advice on February 15, saying that nobody really thought she had pain.

Since that time an attempt has been made to find out about her past, and here the Royal Devon and Exeter Hospital have been most helpful, because they have been able to trace much of her progress round the country by noting the hospitals that asked after her. More inquiries at these hospitals have each disclosed more admissions at other hospitals, and the complexity of the patient's wanderings grew in the manner of a snowball. I had not the time or the patience to pursue the meanderings completely and the fact that she was found to use nine different names made the project more formidable. The following list of some of her activities is probably incomplete.

In February, 1944, she was admitted to the Royal Devon and Exeter Hospital as Elsie De Coverley from Exeter Prison, with epistaxis.

On April 9, 1947, she was admitted to the Croydon General Hospital as Miss Joan Morris, a shorthand typist, giving a long history of dyspepsia and haematemesis. While being treated with rest and diet she discharged herself two days later.

From October 31 till November 16, 1947, she was in the Royal Sussex Hospital, giving the name Joan Summer, and suspected of having a bleeding leaking ulcer; but at operation the findings were subacute obstruction from adhesions and scarred duodenal ulcer. Gastrojejunostomy was performed. She discharged herself on November 16 against advice, still complaining of severe abdominal pain.

On December 10, 1947, she was admitted to the Croydon General Hospital with acute abdominal pain, 'doubled up in agony', as Mrs Elsie Layton, a district nurse. She said she was awaiting a partial gastrectomy in Birmingham Hospital. Efforts made to contact her relatives showed that the names and addresses were all false and the patient discharged herself on December 12. She was admitted to Redhill County Hospital a few days after leaving Croydon. She again gave a similar story and discharged herself within a few days.

From January 29 to March 29, 1948, she was in Paddington Hospital, again as Joan Summer, and a laparotomy for bleeding duodenal ulcer was performed; later she complained of severe anginal pain and discharged herself after giving much trouble in the ward.

In March, 1948, she was in the West London Hospital as Joan Lark with alleged pain and bleeding, and discharged herself.

From April 6 to 9, 1948, she was in Fulham Hospital as Joan Summer giving a history of having all her previous operations in York (inquiries at York failed to trace her), and again presenting with alleged abdominal pain and haematemesis. On April 9 she was recognised by a surgeon as Joan Lark of the West London and discharged herself that day.

On July 10, 1948, she was admitted to Guy's Hospital as Joan Malkin, but later said this was an assumed name and gave the name Joan De Coverley, and said all her previous operations had been done at Edinburgh (Edinburgh denied knowing her). At Guy's she was suspected of having a perforated ulcer, but a laparotomy on July 10 showed nothing abnormal except a mass of adhesions, and she discharged herself on July 13, only three days after the operation, and never returned to have her stitches out.

On January 6, 1950, she was admitted to the Royal Free Hospital as Elsie De Coverley, with the usual story. She complained of continuous pain and discharged herself on January 11, saying her sister had just died. (At many hospitals she has given a story of a dying sister.)

On January 18 she was admitted to St Mary Abbots Hospital still as Elsie De Coverley, having collapsed in Kensington High Street with alleged pain and vomiting of blood. On January 20 when told she was not going to have an operation, she pulled out her stomach tube and demanded her immediate discharge.

Later the same day, she was admitted to University College Hospital, and she had a laparotomy on January 25 for suspected perforation. Nothing was found except adhesions. The usual self-discharge followed on January 30.

The same day she was admitted to St Bartholomew's Hospital and discharged herself on February 2.

From February 7 till 15 she was in the Central Middlesex Hospital—see the description at the beginning of this story.

Since leaving us she has remained active, because on March 7 she arrived at the Elizabeth Garrett Anderson Hospital as Jean Hops and was transferred via the Emergency Bed Service to the Royal Free Hospital. The transfer was unfortunate for her because she was recognised by the ward sister and registrar at the Royal Free, where she had been under another name in January that year, and she discharged herself immediately.

On March 27 she was admitted to Middlesex Hospital as Elsie De Coverley, with the usual story of pain and haematemesis, where she remained for three days. She discharged herself when she learnt that inquiries were being made about her at the Royal Devon and Exeter Hospital.

On April 12 she was admitted to the Croydon General, saying she was awaiting a partial gastrectomy at the Royal Devon and Exeter Hospital, and giving a history of severe abdominal pain and bleeding. The Croydon General rang the Royal Devon and Exeter Hospital and while the telephone conversation was being conducted she got out of bed, dressed, and made off.

On April 17 she turned up in Hackney Hospital as Elsie Shackleton, and that is the last that has been heard of her. Probably she is now careful to avoid mentioning the Royal Devon and Exeter Hospital, who have repeatedly exposed her in the past, and, with new names and a slightly different story, is continuing to deceive the few remaining hospitals she has not yet visited.

REFERENCES

1   Asher R 1951 *The Lancet* 1 339
2   Asher R 1972 *Richard Asher Talking Sense* (London: Pitman Medical)

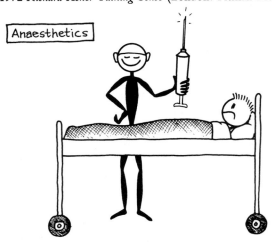

# Is Hypnotic History a Hoax?

The use of hypnotism to attempt to regress a person back through time to a previous life experience has recently excited professional psychiatrists and the public alike. But a simple search through bookshops in Liverpool has revealed cheap, readily available sources of the facts claimed to be revealed only under hypnosis regression.

Most of the media and professional commentary to date has involved a discussion on the hypnotic aspects of the experiences, with mention of reincarnation, clairvoyance, telepathy and genetic memory. The psychologist Hans Eysenck questioned the authenticity of one character under regression—a 17th century Jesuit priest—who could not say any Latin prayers and didn't know his Pater Noster. But not much else has been queried about the availability in the 1970s of other facts from earlier centuries.

There are seven case histories in a book, *Encounters with the Past: How Man can Relive History*, by Joe Keeton, one of the best known hypnotists in this field, and Peter Moss. But it is the case of Ann Dowling, a Liverpool housewife, who regresses to a 19th century Everton waif, Sarah Williams, which has attracted the most media attention. Some of the items of local news in the 1840s and 1850s which are produced under hypnosis are apparently so obscure that to quote Moss and Keeton at the end of their chapter on Ann/Sarah (entitled Real Reincarnation?) '. . . there seems no physical way in which Ann (in the late 1970s) could have collected the factual material unless memory has in one way or another been transmitted'.

Three facts in the case history of Ann/Sarah who lived for some time in the 1840s in Shaw Street, Everton—a suburb of Liverpool—appear to be almost impossible for most people to know readily. These are that in 1846 Prince d'Musigna, son of Lucien Bonaparte visited Liverpool; that during the 1846 visit of Prince Albert in the royal yacht, he made his harbour tours of inspection in the royal tender, the Fairy; and that in 1850 Jenny Lind, the famous opera singer, sang in Liverpool. In particular, the reference to the Fairy appears so obscure that the British journal *World Medicine* and the *Liverpool Weekly News* imply that this lends great weight to Keeton's investigations.

However, all three facts are reasonably readily available in a 50 pence booklet published by the Scouse Press in Liverpool, entitled *An Everyday History of Liverpool*, which is a reproduction of entries in the 1895 *Gore's Directory* for that city. The information concerning Prince d'Musigna and the Fairy appear on the

101

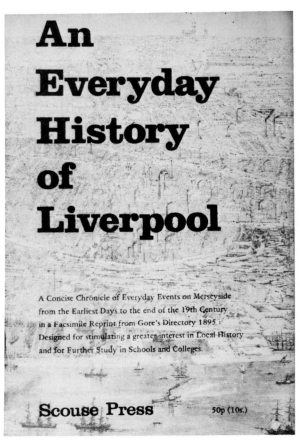

A cheap booklet, available to all, gives away the secrets of regression back to the nineteenth century.

same page of this book and that for Jenny Lind a few pages later. Moreover, it was produced for 'further study in schools and colleges' and not for the attention of a select group of knowledgeable scholars. It would appear that the necessary factual material is not so secure as might be first thought.

The more general information given by Ann/Sarah relating the 19th century life would be applicable to most big Victorian cities, and throughout Liverpool bookshops a paperback reprinted book entitled *Her Benny: A Tale of Victorian Liverpool* written by Silas K Hocking in 1879 is also available (for 90p). This provides much general information as does the later local history book by Andie Clerk called *Arab: A Liverpool Street Kid Remembers*.

REFERENCE

Mould R F 1979 *New Sci.* **84** 504

# Caterpillar Erythema

Many of the common hairy caterpillars in this country when handled 'sting' with their hairs just as the nettle and primula do. The hair penetrates some layers of the skin and breaks off, leaving a little piece projecting. If the hand is passed over the face the stumps of hair projecting from the skin sting the face in a similar manner. Soon an intolerable itching begins, more extensive than the parts actually irritated; vigorous rubbing goes on until in a short time the arms, face, neck, and often chest become covered with a red eruption in which minute nodules can be detected. The hairs and the poison may even be transferred (as Mr Lawford has shown in the case of Bombyx Rubi) to the conjunctiva, the hair ultimately becoming encapsuled in a little nodule of fibrous tissue. The rash is often attended with considerable swelling and slight rise of temperature, and, as Dr Dukes of Rugby pointed out, may lead to an erroneous diagnosis of rötheln. It would be well if medical officers of schools warned the boys of the caterpillars which should not be handled and of the risk they run in rubbing themselves when the skin is irritated. The chief British offenders are the common 'woolly bear', the caterpillar of the tiger moth (Arctia caia), and its near relations; almost all the Bombyx group, including the oak eggar, fox moth, drinker and lappet moths; and the Liparis group, including the common 'gold tail' and 'brown tail'. The hair of some foreign caterpillars has much more severe effects. Gangrene has been reported from contact with the hairy Indian Shoa Pokâ, and the Australian Lasiocampa vulnerans is credited with producing fatal results. On the Continent there is a very common caterpillar, the Bombyx processionea, which fortunately does not appear to have reached this country. The erythema resulting from contact with it is followed by severe general symptoms, and in the case of one boy in the south of France, whose case was detailed in *The Lancet*, after a number of caterpillars had fallen on his chest while climbing a tree, there ensued violent irritation, general swelling, fever, somnolence, delirium, and death in a few hours in spite of energetic medical treatment. Although fortunately we do not meet with such severe cases in this country, the effects are so unpleasant to the victim and the diagnosis is often so difficult for the practitioner that a word of warning to those most likely to suffer will not be superfluous.

REFERENCE

1896 *The Lancet* **1** 1239

# Who is Least Likely to Develop Atherosclerosis?

At a symposium in Athens in 1966 Dr Howard, in describing the person *least* likely to develop atherosclerosis, is reported to have described her as:

> a hypertensive, bicycling, unemployed, hypo-β-lipoproteinic, hypolipaemic, underweight, premenopausal female dwarf living in a crowded room on the island of Crete before 1925, and subsisting on a diet of uncoated cereals, safflower oil, and water.

He could also, had he wished, have reasonably described her equally non-atheromatous consort—an ectomorphic Bantu, who works as a London bus conductor, spent the war in a Norwegian prison camp, never eats refined sugar, never drinks coffee and always eats five or more small meals a day. He is taking vast doses of oestrogens to check the growth of his cancer of the prostate.

All these phrases mark correlations established in the last few years in a field of medical research which, in volume at least, is unsurpassed. The conflict of evidence is unequalled as well.

REFERENCES

Norton A 1973 *Drugs, Science and Society* (London: Collins/Fontana)
Howard A N 1966 *International Medical Tribune of Great Britain* June 30

# Alligator Bite

The following has been adapted from a letter in the *Journal of the American Medical Association*.

> A 49 year old farmer complained of having been bitten either by the devil or an alligator. From numerous versions of the event, the following facts are given

104

credence. One hour before appearing in the emergency room, the farmer was returning home on foot from a prayer meeting in a rural island area, on a path which led through a swamp. Suddenly, to the accompaniment of hissing and scuffling on the muddy bank of the path, the patient felt his right thigh locked in the jaw of some large beast. The patient was able to extricate his leg from the apparent alligator by beating it about the head with the family Bible.

Cleansing of the wound, and open drainage resulted in a successful outcome. Psychological trauma was offset by the patient's recent attendance at the prayer meeting and a shot or two of *spiritus frumenti* en route to the emergency room.

*The alligator* (Courtesy: Charleston Evening Post.)

The authors commented that the ultimate lethal effect was upon the biter and not upon the bitten!

REFERENCE

Doering E J, Fitts C T, Rambo W M and Bradham G B 1971 *J. Am. Med. Assoc.* **218** 255

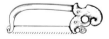

*I dressed his wounds, and God healed him.*

AMBROISE PARÉ

# Heart Disease as an Accident

Reddish *v*. London, Midland and Scottish Railway, decided at the Manchester county court last week, was one of those interesting cases where death from heart disease is found to be an accident within the Workmen's Compensation Act. An engine-driver collapsed and died on the express from Manchester to London. The court had to consider whether his employment had caused a strain which in turn had caused death. The widow contended that the heart attack was due to her husband's effort in helping to turn the turn-table at Manchester station before beginning the journey: it was aggravated, she said, by his action in closing the regulator of the engine when the train was running down from Haddon Tunnel to Rowsley. The judge went in person to the railway station, entered the cabin of an express engine, and tested for himself the requisite degree of exertion. He came to the conclusion that to pull the regulator over when the train was at rest meant no serious effort to an engine-driver in good health, but that, if the train was travelling down an incline with considerable oscillation, the effort would be a strain to a man in Reddish's state of health. The work contributed to, and materially accelerated, death; the employers were therefore liable to pay compensation.

REFERENCE

1936 *The Lancet* **1** 799

# The Genius of Claude Bernard

One hundred years ago Claude Bernard, the greatest physiologist of modern France, stepped aside from the path of scientific research to survey the principles upon which research should be conducted. His *Introduction à l'étude de la médecine expérimentale* was published in 1865 after he had accomplished his pioneer studies on digestion by the action of the pancreatic juice, the 'internal secretions'—a term he invented—and the mechanisms behind the nervous control of the vascular system. All this and much more had been achieved before he was fifty. Ill-health turned him away from the laboratory to his library, where he wrote his study on the methods of experimentation, an essay regarded by many as his most noteworthy contribution to the advance of medicine.

First, he set aside all notions of 'doctrines' or 'systems', which, he believed, did 'not exist in Nature but only in men's minds'. Good experimental standards were only possible if all previously accepted formulations were put out of mind. The true scientist, moreover, should have no fixed starting point for his research; only on observed facts should he frame a hypothesis. Then the tests for accuracy or fallacy would follow as a logical sequence. Finally, imagination, although essential before and after the experiment, had no place in the experiment itself. His often quoted attitude towards scientific investigation is best summed up in his own words:

> Put off your imagination, as you take off your overcoat, when you enter the laboratory; but put it on again, as you do your overcoat, when you leave the laboratory. Before the experiment and between whiles, let your imagination wrap you around; put it right away from you during the experiment itself lest it hinder your observing power.

REFERENCE

1965 *New Engl. J. Med.* **272** 1292

*Dr Comicus selling his pills*, from an engraving circa 1800. (Courtesy: Wellcome Institute library, London.)

# Wat Po Traditional School of Medicine

In Bangkok, at Wat Po, one of the city's most popular temples among tourists because of the enormous golden Reclining Buddha, there is also a school of traditional medicine. There treatments were effected through the use of mixtures of dried herbs, powdered horn and other such medicines. This medical school,

which looks rather like a pharmacy from the outside, with its dried or powdered substances and coiled snakes in glass jars, can be seen by visitors to the temple grounds. The illustrations show views inside and outside this remarkable medical school in Bangkok.

# Punk Rocker's Lung

This is a disease reported in *British Medical Journal* and refers to a 21 year old builder's labourer who until three months before being seen at the hospital had been a full-time drummer in a punk rock band, leading an irregular existence which ended when he ignited his flat with a cigarette. He smoked 60 cigarettes a day and had taken drugs since he was 16, often swallowing them with beer. As part of his act he used to fill his mouth with turpentine or paraffin which he would then blow out and ignite. He also occasionally inhaled powdered drugs such as cocaine through a rolled-up pound note.

Two months later he was well with no treatment and his lung function (normal) and chest radiograph were unchanged—the latter showing pulmonary fibrosis in this drug-snorting fire-eater.

REFERENCE

Buchanan D R, Lamb D and Seaton A 1981 *Br. Med. J.* **283** 1661

# The Wit of Oliver Gogarty

One of the most pleasant things about Dublin is its good conversation. This national characteristic has seen better days; it suffered greatly from television but is now recovering. Before television, one of my pleasures was to nestle in a corner of a quayside or Liberties pub and there to let the sweet music of conversation soothe a restless mind.

The best conversations that I have overheard or participated in have had a raconteur—a rare bird—with the ability to reminisce accurately on something worth recalling, the sensitivity to capture the ambience of the distant occasion, a facility with language to appreciate nuances of expression, and the wit to know when to shut up. Happily, I know one who fulfils these criteria and more—Niall

Sheridan (Brinsley, in Flann O'Brien's *At Swim-Two-Birds*), and with characteristic generosity he has allowed me to recount the following tale.

Dublin's skyline in the thirties was marred by only two large electrical signs: one for BOVRIL in College Green cast its iridescent message towards Westmoreland and O'Connell Street; and the other, for OXO, commanded the attention of the denizens of Nassau Street. The latter celestial sign was situated on a tall building beside Fanning's Pub—now the Lincoln Inn, but then owned by Senator Fanning. One day the first O in OXO failed to light, and an electrician named Joe, known not only for his professional prowess but also well liked for his wit and geniality, was summoned to rectify the fault. But having, perhaps, dallied a little too long in the convivial bar of Fanning's, he slipped from the rooftop and was killed. That evening there was a general air of gloom in Fanning's, where Brinsley MacNamara, Austin Clarke, Fred Higgins, and Seamus O'Sullivan were among the gathering. Oliver St John Gogarty joined the company and the bartending senator—to mark the sad occasion—stood a free drink on the house. Raising his glass to the proprietor, Gogarty bowed and, adroitly misquoting Milton, intoned: 'They also stand who only serve and wait.' Sheridan and the others then suggested that he should write an epitaph for the late electrician. The senator provided a pencil and a brown paper bag—of the type designed to carry half-a-dozen Guinness—and after some thought Gogarty wrote:

> Here is my tribute to engineer Joe,
> Who fell to his death through the O in OXO
> He's gone to a land which is far far better,
> And he went, as he came, through a hole in a letter.

REFERENCE

O'Brien E T 1976 *Br. Med. J.* **1** 828

*Glys-terpipe Fillpacket, Peregrino Mountebanko and Timothy Mouth*, from an engraving circa 1720. (Courtesy: Wellcome Institute library, London.)

# Was Sherlock Holmes a Drug Addict?

The personality of Sherlock Holmes has excited so much interest that for years scholars in England and America have engaged in acute controversy on various points on which the text of the Holmes chronicles is obscure, deficient, or, apparently, contradictory. Was Holmes an Oxford or a Cambridge man? Was he dependent on his professional earnings? Did Watson marry one, two, or three wives? On these and similar questions the most learned authorities are at variance. No attempt has, however, yet been made, so far as I am aware, to examine a more important question: Can we accept Watson's statements that Holmes was once a victim to the cocaine habit?

It seems that the two men first met late in 1880. We know that Watson joined the army soon after taking his MD degree in 1878; that he was wounded at the battle of Maiwand, which was fought on July 27th, 1880; that he was some months in hospital before going home on sick leave; and that shortly after returning to London he and Holmes began their joint occupation of the rooms at 221B, Baker Street. At an early stage of their acquaintance he drew up a list of Holmes's chief characteristics, which, he reminded Holmes some time later, contained the item 'self-poisoner by cocaine and tobacco'. No wonder that Holmes 'grinned' on hearing this remark; for the list, which is set out in 'A Study in Scarlet' contains no such item. This discrepancy, which seems to have passed unnoticed by Holmes scholars, is but one instance of Watson's inaccuracy in recording events.

The alleged addiction is first mentioned in 'The Sign of Four' which described events that took place in July, 1887. Watson states that he saw Holmes give himself a hypodermic injection of what Holmes told him was a 7 per cent solution of cocaine; and he adds that this performance had been seen by him 'three times a day for many months', and that Holmes said he took cocaine to escape from boredom when not occupied with his cases. Watson records no other administration of the drug; but he refers to Holmes's cocaine habit in 'The Scandal in Bohemia' and 'The Yellow Face', and, for the last time, in 'The Missing Three-quarter', a case, probably investigated in 1897, which, Watson states, followed one of those periods of inaction he had learnt to dread: 'For years I had gradually weaned him from that drug mania which had threatened once to check his remarkable career . . . but I was well aware that the fiend was not dead but sleeping.' He describes his horror when he saw Holmes holding a hypodermic syringe, and his relief on finding that it was to be used to squirt aniseed solution on the hind wheel of a doctor's car to enable a draghound to trace it to its destination.

Though Watson was on intimate terms with Holmes he watched the daily dosing for a long time before making any attempt to check it. Not until 'many months' had elapsed did he venture, after a luncheon that included an unspecified quantity of Beaune, upon a mild remonstrance, which was taken in excellent part by the supposed addict. The history of Holmes's alleged drug addiction is, in short, that it began 'many months' before July, 1887, and was gradually discontinued under Watson's treatment, extending over 'years', some time before 1897.

Now Holmes was not one of those men who are unable to occupy themselves unless some definite task is presented to them. He was a man of immense mental resource and initiative. He was an expert chemist, an accomplished linguist— with an intimate knowledge of Goethe and Petrarch—an assiduous student of Black Letter texts, a capable performer on the most exacting of all instruments— the violin—a composer, and an authority on the music of the Middle Ages. That a man with such resources should be driven by ennui to seek distraction in cocaine is so improbable that nothing but the strongest evidence could make it credible.

Nor was the great detective in other respects of the stuff of which cocaine addicts are made. The victim to cocaine is not, like the opium addict, a solitary self-poisoner; he is sociable and prefers to take his dope in company. The first effect of the drug upon him is to make him lively and voluble in disconnected talk. He tries to make jokes and shine as a brilliant person. The opium addict may be a man of exceptional powers: Coleridge and De Quincey are examples; but the victim to cocaine is lacking in mental capacity—incurious, vacuous, needing the gross stimulus of a drug to rouse his interest in life. The continued use of cocaine leads to degeneration—physical, mental, and moral. All this is the antithesis of what we find in Holmes, who was the most unsociable of men: reserved, self-controlled, self-sufficient. There was no falling off in his mental powers, his physical activity, or his character. The effects of his alleged dosings, as described by Watson, are not the manifestations of cocaine poisoning.

Moreover, the gradual 'weaning' treatment that Watson adopted is unusual in such a case. The sudden discontinuance of cocaine does not give rise to the distressing withdrawal symptoms that follow the sudden discontinuance of morphia. This must have been known to Holmes, who was far more deeply versed in the effects of poisons than this medical friend. Watson was an inexperienced practitioner. There is no evidence that he was in civil practice before he joined the army, and he was invalided out after some two years' service. For about seven years afterwards he apparently made no attempt to improve his professional proficiency either by post-graduate study or otherwise, but led an idle life, except for the assistance he gave Holmes in the cases in which they collaborated and for his literary work in recording a number of cases. It is most unlikely that a man of Holmes's overpowering personality would be influenced by Watson in a matter so personal as the discontinuance of a drug habit. On the other hand, we have Watson's statement that he actually saw Holmes inject himself with what he said was a solution of cocaine, and that his forearm and wrist were 'all dotted and scarred with innumerable puncture points'. How can this statement be explained?

The explanation is, I suggest, that Holmes was pulling the good Watson's leg. He had a sense of ironic humour, and when he disguised himself for professional purposes he was delighted at Watson's failure to penetrate the disguise. He was a consummate actor, and in 'The Adventure of the Dying Detective' he shammed a severe illness with such skill that Watson, completely deceived, was prevented only by extreme measures from rushing out for assistance from Harley Street. The puncture marks could easily be counterfeited by a man so proficient in make-up as Holmes; and we have no evidence beyond Holmes's own statement that the bottle Watson called the 'cocaine bottle' ever contained cocaine. It was like Holmes to enjoy mystifying Watson and watching his attempts to screw up his courage to protest against his friend's self-poisoning. What happened when Watson, emboldened by Beaune, did protest is consistent with the hypothesis of a leg-pull: 'He did not seem offended. On the contrary, he put his finger-tips together, and leaned his elbows on the arms of his chair, like one who has a relish for conversation.'

The hypothesis explains also Watson's account of his 'weaning' treatment. Holmes, who had a genuine affection for his Boswell, was doubtless intensely amused at the 'weaning' efforts, which probably began when Watson after his marriage purchased a small practice in Paddington. Holmes naturally wished to encourage his friend in his new rôle of general medical practitioner by allowing him to think he was successfully treating a difficult case of drug addiction.

All we know of Holmes's alleged addiction can be explained if we assume that he did not actually take the drug, but mystified Watson into believing that he did. The facts can be explained on no other hypothesis. This conclusion reflects unfavourably on the professional competence of Dr Watson; but the interests of truth are paramount, and this contribution to the literature that has gathered round Holmes is offered in justice to the most famous character that has appeared in English fiction since the great days of Dickens.

REFERENCE

1936 *The Lancet* **2** 1555

*Who shall decide when doctors disagree?*
ALEXANDER POPE in *Moral Essays*, Epistle 3

# Human Rocket

One of man's primitive urges, since he started having them, has been to get his feet off the ground and go places. The successful passage of Daedalus over the Aegean Sea seems to lend substance to the hypothesis, despite the fate of Icarus as well as that of Phaeton when he also tried to emulate his father in the sun god's own chariot. The number of Indian leaps around the country suggest that the wily red-skin too had made abortive attempts at soaring.

Four-footed animals have been more successful than man in blasting off under their own power. Kangaroos and wallabys can leap to almost terrifying heights and cover extraordinary distances without even letting their forelegs down, and nearly everyone is familiar with Mark Twain's champion jumping frog of Calaveras County, capable of attaining an enviable trajectory. It jumped less well after an unprincipled gambler, just before a contest in which the stakes were high, filled its belly full of birdshot. Such zoonic feats of leaping have no reference to the innocent victims of man's scientific curiosity who have been put into orbit without their advice or consent. The cow that jumped over the moon is totally unrealistic and probably purely fictional.

Man's first authentically successful excursions into the atmosphere were attained in balloons, which, so light as to be involuntarily air borne, had to be pegged down to prevent them from taking off on their own responsibility. The heavier-than-air flying machines of which Darius Green's contraption was an unsuccessful prototype, came later; like most new inventions, they had to fail a few times before being considered practical. Before World War I a local college flying club owned a machine that came to be known as the Harvard groundhog. True, it never went beneath the surface of the earth but was never known to rise above it, probably a fortunate thing for its owners.

Eventually, it came to be realised that one of the essential factors in successful flying is sufficient power to take off and remain aloft; one bold designer claimed that he could fly a kitchen table, given enough power, and that is practically what is being done.

Unrelated to air-borne vehicles, there has remained a peculiar fascination with the idea of taking off practically on one's own for a trip through the medium that surrounds us. Mr Thomas, of Boston University, and a very few others have leaped a head higher than a tall man's head, but at considerable personal effort and for limited flights only.

And now, as described by John Lear in *Saturday Review's* science section, man has been able to blend into the age of rocketry by making himself a sort of cross between a jet-propelled missile and an animated pogo stick. The inventor of the steam-powered corset that makes such a metamorphosis possible is Wendell F Moore, of Bell Aerosystems Company's test facility; the test pilot has been Harold M Graham, of the same laboratory. The power plant, whether garment or other device, consists of a fiberglass corset with a fuel tank of hydrogen peroxide. When this substance is released into a gas generator it mixes with a catalytic agent that decomposes it into steam, which rushes out through two downward-directed rocket nozzles. So great is the force that the corset, with its human cargo, rises.

Distances covered have so far been an unpretentious 360 feet, traversed at 30 miles per hour. A fire truck has been cleared at one hop. Such achievements are modest, but the possibilities are limitless with Mercury's winged sandals as a model and the exploit of Puck, who girdled the earth in forty minutes as a goal.

An oversized umbrella might be carried for double duty as a parachute, in case of power failure during a rainstorm, and the chance of nose dives would need to be eliminated.

REFERENCE

1961 *New Engl. J. Med.* **265** 295

*Quacks at Tien-Sing, China*, from an engraving by P Lightfoot circa 1800. (Courtesy: Wellcome Institute library, London.)

# As Big as a Piece of Wood

Some of us, young or long-memoried enough, may recall receiving the above teasing reply from the grown-up pestered by our oft-repeated question, 'How big is it?' A moment's reflection showed us that the answer was unsatisfactory and our shrill questioning would soon begin afresh. Readers of the January issue of the *Journal of Pathology and Bacteriology* received a pleasant surprise in the form of an inset essay, On the Quantitative Study of Tumours, which shows that the clinical investigator, like the difficult adult, is still disinclined to give plain answers to plain questions. Rather than use a millimetre rule, he prefers a vital system; for 'has he not around him countless convenient measures . . . from his own head, which is sometimes swelled, to the head of a pin . . . ?' In the course of one paper the writer found tumours compared in size to 'a walnut, a lentil, a pea, a large pea, a golf ball, a small hen's egg and a pigeon's egg'. Many of these obviously fall within the same unsatisfactory category as the piece of wood; and with the golf ball there is the added danger of describing it as large or small which 'would inevitably lead, sooner or later, to friction between the Pathological Society of Great Britain and the Royal and Ancient Club, two bodies who, if never friendly, have up to now always lived in mutual forbearance'. For the oölogical method of measurement there is more to be said, though even this method is not free from objection. The range of size in eggs, from the ostrich's on the one hand to the humming bird's on the other is wide

enough for all ordinary tastes and, when in doubt, there is always the cuckoo's egg which apparently varies with the size of the egg of the foster-parent. Nevertheless there are large and small eggs in every species and it must be remembered that 'a small hen's egg' is not necessarily small. The writer amusingly suggests that a standard egg might be kept in a strong room in the Houses of Parliament. We should prefer the National Institute of Medical Research.

REFERENCE

1937 *The Lancet* **1** 491


# The Hardier Sex

In the imaginary war between men and women, sometimes referred to by the persistently facetious, the usually gentler sex has had the long-term advantage. Throughout the entire ambisextrous life cycle, as revealed by the studious statisticians of the Metropolitan Life Insurance Company, females have had a significantly better mortality record than their virile complements. It has long been recognised, in fact, that muscularity and a more general hirsutism may be the only evidences of physical superiority that can be found; women have frequently demonstrated the possession of greater stamina than their bulkier mates.

Masculine vulnerability, the statisticians have discovered, varies with age, the smallest disparity in the mortality rates showing itself in the early years of childhood. Thus, between ages one and four, the death rate is approximately ⅙ greater for boys, a disparity that increases with age. From five to fourteen the rate for males is 1½ times that of females, and at fifteen to twenty four the vulnerability of the dominant, if not the numerically preponderant, sex reaches its highest ratio of more than 2½ to 1. This is the golden age at which adventurous youth, trying out its spurs, is most susceptible to accidents.

These misadventures, in the group from twenty five to forty four years of age—the lusty prime of life—acquire heart disease as their lethal ally, as man, in his maturity, becomes more cautious; the ratio of male to female mortality, although somewhat lower, is still 1¾ to 1. From forty five to sixty four the male death rate is about twice that of the female, with heart disease now leading,

118

having long since replaced consumption as Bunyan's captain of the men of death.

After sixty five, the wage earner's mortality drops to an insignificant 40% over that of the homemaker's. A woman, Browning maintained, is 'always younger than a man at equal years'.

REFERENCE

1965 *New Engl. J. Med.* **272** 1078

# Fildes' 'The Doctor'

The following discourse on this famous painting is adapted from an article by Gifford.

One of the most popular paintings at the turn of the century was 'The Doctor'. More than a million engravings of it appeared in the United States in parlors and physicians' waiting rooms. Painted by Sir Luke Fildes, RA (Royal Academy), in 1891 for Sir Henry Tate, it still hangs in the museum Sir Henry founded, the Tate Gallery in London. Hailed as a Victorian masterpiece when it was painted, the reactions of the public and the profession, at the time and throughout the years, help us understand the patient's image of the physician and the physician's image of himself.

There is little hope in this sad picture. The child is seriously ill, quite near to death. The mother is sobbing in the background, comforted by her husband. The child rests on a roughly made bed of two chairs, and the doctor sits brooding before her. There is no more he can do for the child. The light from the kerosene lamp shines on the child and there is a faint glimmer of morning light in the single window—symbolically mustering some hope. The black and white reproduction does not show the sparse use of colour in the original: the blue and white cup, the green shade of the lamp, the brown bottle of medicine on the table, and the glint of light on the flowers on the window sill.

Gifford remembers the picture hanging in the office of his family doctor when he was a boy: A reverie of smells, colours and feelings so peculiar to that office return when I think of it: the smell of alcohol, the taste of Scott's emulsion, a bleeding wart turning yellow-brown when swabbed with nitric acid, the taste of ammoniated mercury for impetigo, recollections of a broken collar bone, the white agate trays, and the injections from the ever gentle nurse 'who gives

(Courtesy: The Tate Gallery, London.)

needles better'. I can picture the waiting room with the stacks of *Saturday Evening Post* and *National Geographic*; the great oak bookcase with glass doors packed with the thickest books I had ever seen; the diplomas on the walls; a World War I discharge from the Army Medical Corps; the polished brass nameplate and the door chimes; but above all, I remember the engraving of 'The Doctor'—to me a visual representation of the Hippocratic oath.

The picture was reproduced many times during the years 1949–1952 when the American Medical Association was campaigning against socialised medicine. A public relations firm distributed copies of the painting to thousands of doctors, medical societies, and hospitals throughout the country. 'The Doctor' was reproduced on a postage stamp in 1947 'issued to honour the physicians of America', on the 100th anniversary of the American Medical Association.

The picture has not lost its message, even in these days of changing social patterns, professional status, and patient attitudes. In a recent issue of *Time* was an advertisement by a pharmaceutical company showing the family group from 'The Doctor' accompanied by an appropriate quotation from William Osler, 'Amid an eternal heritage of sorrow and suffering, our work is laid.'

REFERENCE

Gifford C E 1973 *J. Am. Med. Assoc.* **224** 61

# Medical Intelligence

The world of medical literature is crowding even more rapidly than the world of people. There seems to be so much that has to be written and published that new journals must be created and older journals must grow thicker. And, to find his way to what he needs in the thicket of pages, the searching reader must be guided by abstracts and journals of abstracts. Soon there will be guidebooks for the journals of abstracts, to be called *indices indicum indecorum*. The medical writer who wishes to see his work imprinted must apply to boards of editors whose policies or interests may be as mysterious as the reasons for rejection by one or acceptance by another.

Pity the authors. Pity the editors. Pity most the reader. He must have his information. Yet one must agree that he must read too much to gain too little. In other words, too much is written, and too much is published. I am dismayed to see the same material rewritten by the same authors published in as many as four different journals. To be sure, there may be rearrangements of paragraphs, and there is usually a rewording of the title, to confuse the index, but not the discerning reader. The order of the authors' listings is always changed. I would not examine here the many reasons why republication is so common a practice, but I admit to my suspicions that venality is among the motives.

Another abuse of excessive publication is the expanded report of a single case, which adds pages of repetitious citations of material already worked over by armies of case reporters. Why must one fear to refer but once to the last collected review, and briefly add one's own recent experience? Must one always put to page proof that he has read what he ought to read?

Ah, but these are times when a career in medicine grows from seeds planted in journals. Before he is appointed to a hospital, to a medical society or to a faculty in a medical school the doctor lists his publications. Those who review his credentials *count* his list. Which of those judges actually reads or criticises the articles or books on that list? Only the rare and perhaps compulsive critic takes the time for it. I suppose some sort of balance scales may become necessary to compare two candidates for the same position whose numbers of publications are equal but whose weights are not. In that way it has become easier to pass judgment on the doctor with an extensive credit of by-lines than on his less literary colleague.

These paragraphs are meant less to discredit the worth of the writer—for, after all, he exposes his flanks to the critics—than to praise his silent partner. Because posterity has few memories that are not supported by bibliographic references, the immortality of memory is chancy. Still there is no evidence to support the belief that it profits the man alive what his successors think of his articles.

In every generation physicians have been adjudged great by their colleagues for skill, gentleness, honesty and exemplary performance in the eyes of their patients and their pupils, who have left but little in writing. Some have no hands for the pen, others no time. And some, like Montaigne, whose words opened this essay, consider the excess of writing an evil, and 'In a time when to doe evil is common, to doe nothing profitable, is in a manner commendable.'

Boards of trustees, committees for review of credentials and those who pick new members for professional societies should be reminded that there are great physicians and surgeons who do not write articles for medical journals. These men should be praised for what they are. They deserve membership in faculties, staffs of hospitals and societies. Although they do not write they may suffer no incapacity to communicate their lore or to teach their colleagues by speech and example.

The great historical precedent for the nonwriting surgeon was Hugh of Lucca. Recently, Campbell and Colton gave us a fine translation of Theodoric's 'Surgery'†. Theodoric, later Bishop of Cervia, was a surgeon of the Bolognese

†*The Surgery of Theodoric ca AD 1267*. Translated from Latin by E Campbell and J Colton. Vol 1 (Books I and II) 223 pp (New York: Appleton) 1955. (Issued under auspices of New York Academy of Medicine.)

school: a careful observer, an original-thinking surgeon, he was the teacher of Henry of Mondeville, and the respectful contemporary of William of Salicet. Mostly, however, he is notable for the record he made of Hugh of Lucca, his own teacher. Theodoric's book, written in 1267, reports the thoughts and practice of the wonderful Hugh, who worked in Bologna during the first half of the thirteenth century. Hugh wrote nothing that has come down through the centuries. But if Theodoric is accurate Hugh should be revered as one of the greatest surgeons. He disputed the authority of dogma whenever his own observations led him to contrary opinion. That alone—his Baconian courage to

*The blind Physician*; a sixteenth century French painting. (Courtesy: Wellcome Institute library, London.)

123

flaunt dogma in a time when defection from precedent was punishable—marks him above nearly all his forebears and contemporaries. He had confidence in the reproducible results of methods *he* had tested, when most of his generation of surgeons dared not test at all. He understood the biology of wounds, and if he knew not how to make them heal, he knew better than the others how not to prevent healing. Theodoric, the writer of a textbook, was constrained to cite the traditions of Graeco-Roman and Arabic surgery. On the other hand, the blessedly honest old Bishop remarked time and again that 'my master, Hugh' did it differently, and his results were better. The modern reader can understand why that was so in the light of his broader knowledge. Indeed, a few hours with Hugh and Theodoric are recommended for the good of the soul. There is nothing there to dilute the strength of these arguments, although I cannot deny that Theodoric demonstrates that it would not do for all doctors not to write.

REFERENCE

Rosenman L D 1957 *New Engl. J. Med.* **261** 1333

# New Diseases in 1982

1982 was another notable year for new diseases, syndromes and maladies. Some, such as *roller disco injuries*[1] and *electronic war space video game epilepsy*[2] were sadly predictable.

More interesting are the vomiting and other toxic effects of daffodil eating—a woman thinking they were shallots, used them for her *coq au vin*[3]; face flambé, a restaurant hazard first delineated clinically in *Journal of the American Medical Association*[4]; the pipe smoker's eczema, produced by phosphorous sesquisulphide in 'strike anywhere' matches[5]; serum sickness due to hair straightener[6]; and the pneumomediastinum which, fortunately rarely, can follow an aggressive bout of fast bowling on the cricket field[7].

Most intriguing of all was the diagnosis eventually made in the case of a 69 year old Canadian lady admitted to hospital complaining of chest pain and confusion[8]. Her only measurable abnormality was an arterial oxygen tension of 57 mm Hg. Puzzling at first, this made perfect sense when the clinician learned that she spent much of her time in a local bingo hall where almost all of the 300 customers smoked heavily. He devised the label 'bingo brain' accordingly—and the woman recovered completely after three days in a smoke-free environment.

REFERENCES

Dixon B 1983 *World Medicine* 42, January 8 from which this article has been adapted
1  Wilkinson A J 1982 *Br. Med. J.* **284** 1163
2  Daneshmend T K and Campbell M J 1982 *Br. Med. J.* **284** 1751
3  Litovitz T L and Fahey B A 1982 *New Engl. J. Med.* **306** 547
4  Achauer B M, Bartlett R H and Allyn P A 1982 *J. Am. Med. Assoc.* **247** 2271
5  Steele M C and Ive F A 1982 *Br. J. Dermatol.* **106** 477
6  Kotowski K E 1982 *Br. Med. J.* **284** 470
7  Clements M R and Hamilton D V 1982 *Postgrad. Med. J.* **58** 435
8  Watson W C 1982 *Can. Med. Assoc. J.* **126** 1266

# Quack Medicines

To most people, undoubtedly, the great collection of recorded law amassed at the Public Record Office would be the last place they would expect to find the materials wherewith history is made pleasant. But to the antiquary, and to one who will trouble himself to look for them, the Records afford glimpses at the ways of the world that is gone, as interesting and amusing as do even the old chroniclers, quaint old Stow, or gossipy Froissart. The Records of the State Paper Office, consisting as they do chiefly of letters, are extremely prolific in these curiosities, and a few of them augmented by some from other departments, have been noted down here. Their miscellaneous character is astonishing. In one bound-up volume of old State Papers the reader will find recipes for gout, theological disquisitions, plans of fortifications, a novel, extracts from a play, Latin proverbs, and a score of other things equally alien to the original function of the State Paper Office proper, but still interesting, dirty and difficult to read.

Some of the recipes for quack medicines are very amusing. Here are two which Lord Audley sent to Cecil, Elizabeth's secretary, hearing that he was ill, and which, he says, he and his wife have 'proved upon herselfe and me bothe'. The first, which he calls 'a good medycen for weknes or consumpcion', runs thus:

> Take a sowe pygge of ix dayes old and fley him, and quarter hym, and putt hym in a styllytorie, wythe a handfull of spere mynt, a handfull of red fenell, a handfull of lyverworte, half a handfull of red nepe, a handfull of ciarye , and 9 dates, clene pyked and pared, a handfull of great reasons, and pyke oute the stones, and a quater of a nounce of mace, and 2 stykes of good synamù, bressed in a morter and sett yt yn the sonne 9 dayes and drinke 9 sponfulles of yt at ones when yowe lyst.

Lord Audley does not say what course should be pursued when there was no sun, nor does he explain to Cecil the peculiar merits of a liquefied sow pig, but

125

the recipe is at least ingenious. The next, which is termed simply 'a composte', is even more extraordinary:

> Take a porpin, otherwyse called an Englyshe hedgehogge, and quarter hym in peces, and put the said beste in a styll, wythe thys ingredient. Itum, a quart of red wyne, a pynt of rose water, a quart of sugar, senamum, 2 grete reasons, 1 dete, and 2 nepe.

No directions are given as regards the taking of this, but the result will presumably be liquid after the distilling process has been gone through.

A very old recipe is to be found among the Exchequer Treasury of Receipt Miscellanea [No 53/13]. It is for an ointment, and is, to judge from the handwriting, of the date of Richard II, or thereabouts:

> Tak wormwode, and lemp, and sange, in even porcion. Take butter, a good quantite, and a litil fresh gresse, and oyle de bay, stampe the erbes smalle and myng all this togedur, and bray them in a mortar. When ye have done so, put them in a clene potte, and stope it that none ayre may enter, and sett it one the aymbres of coles, and stire it frome the potte bothum thrisse in an hour. And then take it of and let it kole, and sett it on a wysp of stree til it be colde, and then sett it on the fyre agayne, and stire it well to it bole and then take a string, embosse and draw it thorow in al a besyne, and let it stond a day, and kole. And then take and make it the syse of ointment holes a iij or iiij. And then set it don on the to syde and let the water rynne owte, and when the water is rune owte take the ontment and put it in a box.

This is the earliest recipe that has been discovered among the records, to the writer's knowledge, and no doubt in its day has been efficacious. A medicine, however, composed principally of butter, wormwood, and grease, would not, it is to be feared, recommend itself to our modern and perhaps fastidious tastes.

The next and last recipe to be given is as extraordinary as any, and is interesting on account of the great patient it was intended to benefit. It occurs in the series of *Irish State Papers*, vol xxxi, No 40, and is contained in a despatch from the Archbishop of Armagh to Lord Burleigh. Having informed his correspondent of the state of affairs in Ireland, his Grace goes on:

> I am sorofull for that your honour is greved w^ch the goute frome the w^th I besech Almighty God deliver you and send you health. And y^t shall please your honour to prove a medicen for the same w^ch I brought owt of Duchland, and have eased many with it. I trust in God it shall also do yow good. And this it is. **R** ij spaniell whelpes of ij dayes olde, scold them, and cause the entrells be taken owt, but washe them not. **R** 4 oz brymstone, 4 oz torpentyn, 1 oz parmaceti, a handfull nettells, and a quantyte of oyle of balme, and put all the aforesayd in them stamped, and sowe them up and rost them, and take the dropes and anoynt you where your grefe is. And by God's grace your honour shall fynd helpe.

Other curious recipes and compositions could be given, but enough have been described to show the fecundity of the Records in this respect.

REFERENCE

Hewlett M H 1881 *The Antiquary* 3 256

# Mr Port & Mrs Green

# MR. PORT,

### HAS REMOVED TO

---

## CANCERS, POLYPUS, & FISTULA,

### CURED WITHOUT CUTTING.

---

## 𝕻ort's 𝕮elebrated 𝕍egetable 𝕻ills,

### ALSO

**Specific Remedies for Diseases of the Spine, Chronic, and Acute Rheumatism, White Swellings and Scatica, Corns and Bunions, Scrofula, and all cutaneous disorders of the system.**

---

### TESTIMONIALS.

I hereby certify, that Dr. PORT has extracted two Cancers one from each of my breasts, within the last two years, and I am quite well and free from pain.

|  |  |  |
|---|---|---|
| WITNESSES | MARTHA EDWARDS, MARGARET JONES, ANDREW HALL, MARY HALL. | ELIZABETH DOYLE, Bridge End, Birkenhead. |
|  |  | *Aug. 23rd,* 1855. |

The above named person is now living in the same place hearty and well.  E. J. PORT.
Nov. 5th, 1867.

---

Miss HUBBERSTEY, of Clayton Green, near Chorley, was cured of a Cancer in the breast, in June, 1848, and remains free from pain or any other ill effect.

Miss RADLEY, of Bidston Mill, near Birkenhead, was cured of a Cancer in the upper part of the left arm, in February, 1849, she having been cut three years previously.

---

*This advertisement* is from an article on quacks and quackery, published in the *British Medical Journal* in 1911.

127

That 1911 issue of the BMJ also described Mrs Green's *Wonderful Ointment*, advertised like that of Mr Port as not requiring surgery 'Amputation avoided. No cutting or burning' and stating that it 'Also cures poor animals'. If her claims were to be believed, then the medical profession would have been put out of business since Mrs Green categorically stated that she could cure all kinds of diseases with her ointment and listed the following under the caption 'Miracles Will Never Cease'.

> Diseases of the skin, Old wounds of long standing and deemed incurable, (Ladies suffering from Bad Breasts), Old Ulcers, Abscesses, Cancers of all descriptions, Tumours, Polypus, Piles (blind or bleeding), Fistulas, Scrofula, Gun-shot wounds, Bad legs, Arms, Whitlows, Boils, Burns, Scalds, (King's Evil), Scurvy, all kinds of Poisoned Wounds, (will draw out Splinters, Needles, or Broken or Diseased Bones), Erysipelas, Stings, Venomous Bites, Scurf, Itch, Ringworms, Chilblains, Chapped Hands, Cracked Lips, Cuts, Gathering in the Ears, Deafness, Inflamed Eyes, Sore Throats, Eczema, Mumps, Bruises, Neuralgia, Rheumatism, Gout, Stiffness and Swelling of the Joints, Paralysis (Limbs drawn up, Crippled), Sprains, Lumbago, Pains in the Leg and Back, (Hip Diseases), Salt Rheum, Corns, and etc.

These claims were typical of many other quack medicines, and in addition, the advertisers often used to state who should never be without their wonderful ointment, which in the case of Mrs Green was 'Homes, Nurseries, Travellers, Mechanics, Horse Owners, Stables, Horse Keepers and Farmers'.

# John St John Long

John St John Long was an Irishman of lowly origin whose therapeutic equipment consisted chiefly of a liniment which he declared to have a selective faculty between sound and unsound tissue. This when applied to a healthy part produced no effect, but when applied to a surface under which there might be hidden disease, it caused a sore. He undertook to cure all sorts of diseases and acquired a large practice in Harley Street, London. Nine out of every ten of his patients were women. Long eventually ended up in the dock of the Old Bailey, was fined £250 on one occasion and given the benefit of the doubt by the judge on another. However, he continued to flourish, posing as a martyr suffering persecution at the hands of the professionals, until he died in 1834 at the age of 37, refusing to the last, to be treated by his own remedies.

His grateful patients erected a monument to him in Kensal Green Cemetery with the following inscription.

It is the fate of most men
To have many enemies and few friends
This monument pile
Is not intended to mark the career
But to show
How much its inhabitant was respected
By those who knew his worth
And the benefits
Derived from his remedial discovery.
He is now at rest
And far beyond the praises or censures
Of this world
Stranger, as you respect the receptacle of the dead
(As one of the many who will rest here)
Read the name of
John Saint John Long
Without comment.

*John St John Long.*

REFERENCE

1911 *Br. Med. J.* **1** 1273

# Unicorn's Horn

The following commentary on unicorn's horn, or alicorn as it is sometimes called, is adapted from an article by Ellerbroek on *The Unicorn*.

The unicorn was a fabulous beast who inhabited only the realm of mankind's imagination. But the student of human thinking will find it well worth his while to pursue such trails as the unicorn offers—not hoof marks in the forest, but black print on white or yellowed pages: the lair of *this* beast is in the library stacks, and not in the fields or on the wooded hills.

It was held that putrid or contaminated water was purified if stirred by the unicorn's horn; this was called water-conning. The size of the beast gradually decreased from the Talmudic enormity to that in certain Christian versions in which the animal, although uncapturable and ferocious, was only the size of a kid. It was said that when closely harassed, the unicorn would run to the edge of a cliff, leap, land impaled upon its horn—unharmed—and then run away, leaving behind its perspiring pursuers; the horn was stated by most to be rigidly set in the animal's forehead, but Rabelais and Bulfinch concur, with phallic overtones, that the horn might dangle limply, but that on occasion for use it became straight as an arrow.

The length of the horn was usually estimated in cubits, and since no one could say nay, the writers over the years, starting with a respectable one cubit length, added a bit here and there, until what passed for medieval common sense called a stop to the practice when the length reached about ten feet—simply on the practical basis that no one could handily figure how such a horn could be carried by less than Behemoth himself. A note of rationality might not be out of place at this point: in 1292 Marco Polo actually *saw* a rhinoceros, related it to the fabled unicorn, and wrote of their pig-like heads, their wallowing in the mud, and pointed out that—if this was indeed the unicorn—that it was quite different than had been thought, and postulated that neither unicorn nor virgin maid would find propinquity enjoyable.

Although periodically an author would question the existence of the unicorn, the *monokeros* came hale and hearty into the Middle Ages. Da Vinci, von Guericke, Mercati, and Leibnitz wrote of the unicorn, and expressed no doubts. None of the classical medical authors, however, mention the unicorn, until St Hildegarde of Bingen in the twelfth century. Until that date, the prime value was in the use of the horn (by this time pieces, fragments, and entire horns were accumulating) in testing food and drink at royal or wealthy tables; slabs of horn

130

were assembled to make goblets, and entire horns were treasured possessions of fortunate churches, abbeys, and principalities.

In earlier centuries the dread word 'poison' alone was enough to produce panic in any heart. Little was actually known of the nature of poisons or poisoning techniques, and thus any vagrant breeze might bear a noxious vapour; a saddle could be anointed, a glove could be permeated—and almost any sudden death could be blamed on poison. Further, since rather specific techniques *were* available for producing confessions, there was no dearth of poisoners to execute. These executions complemented many deaths due to bacteria, viruses, or perhaps perforated ulcers.

The unicorn's horn, or alicorn, as it came to be called, was for hundreds of years considered infallible in the detection of poison. It was alleged to sweat, or to cause the liquid to boil without heat, if poison were present. Fabulous prices were paid for this most potent of protections. The question arises, what explanation was offered when death occurred in spite of the magic. Certainly either the horn in question was false or, in God's opinion, the time had come for the man to die, and therefore nothing could prevent his demise.

The natural scarcity of this nonexistent creature led to argument about his possible locale. Some thought that with a creature so splendiferous God entrusted it to the earth only one at a time. Others believed that all the unicorns had perished in the flood. Their opponents insisted that this was blasphemy, that since God had created such a beast to exist in nature, and since everyone knows that nature abhors a vacuum, it was not possible that something once created should be totally eliminated. Some held that the subsidence of Atlantis had taken all remaining unicorns to the bottom of the sea. A few more prosaic souls supposed that relatively unexplored areas of the earth's surface were its habitat.

In spite of the lack of the actual animal, the horn was usually available—at a price. The alicorns were never extremely rare, nor were they common; the supply allowed about ten whole horns throughout Europe in the Middle Ages, but by 1671, the coronation of a Danish king on a throne made of unicorn's horns indicated that the supply had increased. The source of the horns was various: prehistoric ivory and fossils were the main source, but the tooth of the narwhal possessed the requisite spiral turnings, and became the accepted article. The value of the horns varied with the supply, and at the peak of the market, the powder and pieces brought up to ten times their weight in gold. Entire horns, more rare, might be worth double this amount. King Edward I, in 1303, did his part to discourage crime after a Westminster monk made off with a unicorn horn. The king literally made an example of him by tacking his entire skin to a wall near the spot where the theft had occurred.

As might be expected, there was no dearth of attempts to falsify the 'unicornum verum'; counterfeiting a valuable fake has always been worthwhile, and the ambitious were busy straightening walrus tusks and searching for other substitutes, such as the bones of domestic animals, stalactites, pieces of whale-bone, and even limestone. Soon tests in profusion were advised for protection of

131

the buyers; typical was the soaking of the 'horn' in water, then drawing a circle around a scorpion with the water. If the scorpion remained inside the ring, genuineness could be presumed.

By the time of Paré, a moderate amount of sophistication was exhibited. Paré himself urged his king, Charles IV, to test a bezoar on a cook who was due to be hanged; the cook took the corrosive sublimate gladly, and Paré did the autopsy confirming the inutility of the bezoar. He wrote at length against the unicorn's horn as a prophylactic against poison but, due to the religious sanction, did not question the actual existence of the unicorn. He concluded rather sadly, 'If it were not for the witness of Holy Scripture to which we are obliged to adjust all our beliefs, I should not think that such a creature as the unicorn had ever existed . . . thus it is necessary to believe that there are unicorns.' Fortunately, the Bible had said nothing about the medicinal value for unicorn's horn, and Paré was able to attack the idea that the unicorn horn was a panacea.

Through the sixteenth century, the horn was advised for epilepsy, impotence, barrenness, worms, and the plague, as well as for smallpox and other assorted ills, and the explorers of Africa and the Americas each took his piece of alicorn with him as carefully as travellers today get their inoculations prior to departure. Although by this time a good many physicians doubted the efficacy of the horn in medical practice, they were obliged to use it, or they could expect no respite from the relatives if a patient died. A pharmacopoeia of 1649 quoted the virtues of the horn as 'Sudorifick, Alexipharmacal, and Cordial, thus good against Poysons, infectious diseases, etc . . . and profitable in the Epelepsie of Infants'. Mad dog bites and the plague were almost mandatory indications. The horn was used both alone and in mixtures, in pieces, and as powder; the poor were not left out, for a horn placed in water transferred its power, somewhat weakened, perhaps, to the water, which the poor could either purchase or on occasion obtain from a chapel or cathedral that made such bounty available. The official drug lists, or items to be kept in stock in London by all registered pharmacists, included unicorn horn from 1651 to 1741, but it was deleted from the 1746 edition.

Until the alicorn went fully out of style, there were many who distrusted it. Gesner wrote in 1551 that the powdered horn had to be freshly scraped in order to be effective against poison, but elsewhere stated that 'he could not help but prescribe alicorn if his patients insisted . . . but that he neither forgot nor neglected' to order also other effective medications simultaneously.

Perhaps current medicaments are more to the point than frog's blood, toad eyes, and baby fat—by the by, has anyone checked these items lately for therapeutic activity?—But it is frightening to recall that for generations the left hind hoof of the elk was considered specific for vertigo and allied ills, even up to 1700. Whence came such an idea? It was thought that 'the elk was a chronic sufferer from vertigo . . . when pursued . . . he has to sit down and place his left hind foot in his left ear to cure himself of dizziness before he can run away'. How fortunate that the pharmacopoeia of today is free of elks hooves! Or is it?

The stage was gradually set for the increasing disbelief in the unicorn as an animal, and in the horn as a panacea. Paré attempted to allow the unicorn to remain in the universe, because of the Biblical entries, but Fundamentalism was jarred badly when faith in the unicorn began to fade, for if any part of the Bible can be questioned, then a shadow of doubt is thrown over its infallibility. The unicorn, by 1700 AD, had run most of its race, and was ready to retire to an inactive position. As late as 1775, the Japanese were still buying alicorn for the treatment of impotence, but elsewhere the market had collapsed, and horns became of interest only to museums and curiosa collectors.

REFERENCE

Ellerbroek W C 1968 *J. Am. Med. Assoc.* **204** 33

*No doctor lives long enough to write a reliable book on prognosis.*
SIR JAMES MACKENZIE

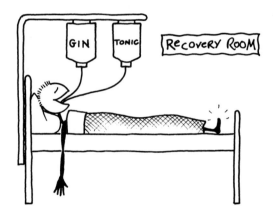

# Moscow Case History, 1983
## A true story by R F Mould

The circumstances which led to the events of October 1983 in a hotel room on Gorky Street, less than a mile from Red Square, were at least partly related to the death of Leonid Brezhnev the previous year and the succession of ex-KGB chief Yuri Andropov. It was under Andropov that the crackdown on corruption amongst officials in the USSR had begun to bite. A visiting UN lecturer would not have expected this to affect him, could not have anticipated interrogation and pressure to write a statement and even to make him spy for specific nuclear information.

### 1980 AND 1981

In Brezhnev's later years a blind eye had been turned to many practices which no doubt would have been considered to be good examples of capitalist decadence. Among these were various methods used to obtain Western currency. Foreign currency is a necessity if a Russian wants to buy goods at the special shops called *Beriozka* which cater for foreign tourists and supply goods not normally available to the ordinary Muscovite shopper.

Since 1980 a Danish colleague, Knud, and I have spent a week each year in Moscow under the auspices of United Nations organisations to participate in a nuclear medicine training course at a Moscow University Institute. All visitors are met on arrival at Sheremetievo Airport by a member of the International Department of the State Committee responsible for the course, taken to their hotel, and subsequently 'looked after'. In my case the 'looking after' varied between 1980 and 1983.

In 1980 and 1981 the 'system' was similar in that Valery K, the State Committee member, handed over a sealed envelope in the airport car on the way to the hotel. The envelope contained 100 roubles (approximately £80) in small denominations and I was told to repay this at a later date when the hotel bill would be settled. Knud was given the same instructions. We assumed, however odd, that this was standard procedure.

Lectures and experimental demonstrations were scheduled and most evenings were free as were the occasional morning and afternoon. This was fortunate since Moscow is justly proud of its theatres, circuses, art galleries, museums and cathedrals and there is no discouragement for foreigners wanting to make visits.

Indeed, hotel Service Bureaux are most helpful so long as advance notice is given. No State Committee help is required, only a passport, visa and a few roubles. A visitor can also obtain tickets direct from a theatre Administrator.

In 1981 Valery K organised a visit for us to the ancient Russian town of Suzdal, with its churches, museum, handicraft centre and monastery where the Russian Tsars used to lock up their unwanted wives before marrying a new one. He also organised a tour to the Suzdal prison where German Field Marshal von Paulus and his staff were confined following their capitulation after the battle of Stalingrad. Suzdal, which is 230 km from Moscow, is on an official Intourist tour, but for our trip Valery K arrived with a chauffeur-driven car.

When our lecturing was complete and it was time to settle the hotel bill, the 'system' defined by Valery K brooked no argument. Both Knud and I were informed that he had personally paid the bill in roubles and now required reimbursement by separately accompanying him to a number of *Beriozka* shops to make purchases to the tune of the hotel bill plus the 100 roubles given to each of us on arrival. His main requirement was for Scottish mohair wool, the second item on his shopping list being small gold jewellery.

During my first visit in 1980, Valery K asked for a specific present to be brought the following year. But this was only a token gift of an English book on mushrooms. Similar requests may have been made of other visitors, and from my Scandinavian colleague the small annual gift was a joint of Danish bacon.

## 1982

In 1982 Valery K was not present at Sheremetievo and I was met instead by his colleague Vladimir Z. An envelope was again handed to me on the journey to the hotel. The contents were 100 roubles and a letter. I passed the book on mushrooms to Vladimir Z with a short note in reply to Valery K. The week passed by uneventfully except for a visit with Vladimir Z on Tuesday to the Baku restaurant. This was notable because of a loud verbal outburst from him.

'I am not a member of the Communist Party.'
He then proceeded to try to justify this on the basis of party membership statistics. He vanished for the remainder of the week.

The next time I heard from Vladimir Z was on Saturday, the day before departure, when he telephoned to say that, as he was in hospital with a slight heart attack, someone else would take Knud and myself to the airport, and finally, in an emphatic manner, 'you will come again next year'. There was no mention of *Beriozka* shops and the hotel bill was settled in a more conventional manner.

## SUNDAY–WEDNESDAY, OCTOBER 1983
In October 1983 I arrived with my lecture notes and slides and yet another book on mushrooms. My colleague brought with him another joint of bacon. We were again met by Vladimir Z, but this time no letter was handed over. The

Mr Mould.

I apologize that I cannot meet you at the airport myself just now I participate at the International Conference in Vienna I'll be back on 20ᵗʰ September 1982.

This letter is given to you by my friend Vladimir Zaporozhko, We are working in the same department. He will give you 100 roubles — for you for meals. Hotel is payed by myself, you only kindly invited to pay me back ~~through~~ Beriozka Shop, i.e. £.40 per night. Vladimir will show you what you can by.

If you will bring a book about mushrooms — tell Vladimir how much he is needed to pay, to you for it. He will pass me this book later also.

Vladimir will accompany you through the Customer than you will come back to ~~t~~ London again I feel upset for not seeing you personally

Hope to see you next ~~year~~.

Many thanks

Valery Krivtsov.

136

hotel Service Bureau provided even more ballet and opera tickets than in previous years. On the Wednesday evening I went to Rossini's *The Barber of Seville* at the Kremlin Palace Theatre. On leaving the theatre, the opera-goers were ordered by police with hand-held megaphones to keep off the roads, from which all traffic was absent. An evening rehearsal of the November Day parade was taking place; all of a sudden the pavements shuddered to the roar of military scout cars, followed by troop carriers, tanks and rocket launchers. It carried on for hours and was a most impressive display of military might and precision. A group waiting for their Intourist bus thought the Third World War had started!

THURSDAY AFTERNOON, OCTOBER 1983
It was four o'clock in the afternoon. My day's lectures were over and I looked forward to the training course's closing banquet at seven o'clock. I was passing the time in my hotel room by reading. I was tired; for the first time ever in Moscow I had forgotten to lock the door to my room, and was thinking about going to bed for a siesta before the banquet. There was a knock at the door and a voice called 'Dr Richard Mould?' I thought it was someone from the Institute and answered 'Yes'. Before I could even reach the door two strangers entered and I found myself expertly and lightly pinned to the wall by my arms.

I was too shocked even to panic and just stood by the wall whilst one of the men rolled up the mattress on the bed and sat on the corner of the wooden base, the other sat in a chair. I returned to the chair in which I had been sitting, still numb with shock. The younger of the two men was introduced by his partner as Yuri. He wore a nondescript light brown suit, tinted glasses and carried a thin document case. Apparently, he did not understand or speak English. The older man said he would interpret for Yuri and introduced himself as Tommy.

Tommy's accent was impeccable with no trace of the Russian habit of rolling Rs. What was most noticeable to me, was his suit. It was identical to a blue pinstripe that I had bought from Marks & Spencer. With this suit Tommy wore a white shirt, what looked like a Horse Guards red and navy blue tie, a blue felt trilby and he sported a very British-looking moustache. He looked so out of place in Moscow that I had visions of Russians in London stripping the Oxford Street Marble Arch store to clothe him and his chums!

I had little time to decide how I would act but I determined to stick to a set of ground rules. I would volunteer no information unless I was positive that they already knew it or that it was useless to them; I would cover my true feelings of panic and present to them a hard exterior; I would avoid answering really awkward questions by changing the subject if at all possible and introduce material to waste time: to maintain my spirits I would try to think of amusing associations with anything bizarre or stupid they might say; I would give the impression of cooperation as far as this was possible without telling lies and try to memorise all events during the remaining 4½ days before my scheduled return to London.

I had the distinct impression from the moment of their arrival that anything other than cooperation would lead to an even more fraught situation and the probable removal to somewhere far worse than my hotel room. I also had the feeling that any request to make a telephone call to the British Embassy in Moscow would be both refused and considered an unfriendly act. I was about to speak and ask them what this visit was about when, before I could begin, Yuri rose from his perch on the end of the bed, fixed me with a look which said 'sit up and shut up', and walked over to shut the window in my fourth-floor room. This was unnerving since a case had been reported very recently of a British merchant banker who fell to his death from a block of flats in Moscow.

The whole tenor of these Thursday afternoon events was one of controlled menace on the part of Tommy and Yuri.

I was not given a chance to ask my question before Tommy spoke: 'We believe you know Valery K.'

'Yes.'

'He is under arrest and awaiting trial very soon. He has named you as his collaborator in the West for breaking currency laws and stated that you went with him to *Beriozka* shops in 1981 and bought gold jewellery and wool and that he took you to wild parties. It will be a big trial.'

It was a ridiculous claim that I was some sort of criminal financial mastermind participating in orgies when visiting Moscow. It seemed that K was trying to extricate himself from his current problems by laying unwarranted blame elsewhere. I explained about the *Beriozka* shops not only for 1981 but also for 1980, which surprised them. They chattered away in Russian which, apart from a few phrases, I do not understand. They did not appear to know that Valery K operated his 'system' in both years.

I ignored the accusation of being a wild party attender. I had certainly never had any invitations to such parties and the Russian equivalent of British fish and chips and a bottle of beer (sturgeon and sauté potatoes washed down with cognac at the Hotel Ukraina) hardly qualifies.

Tommy then mentioned the visit to Suzdal as an example of Valery K making unauthorised use of a State Committee car. I commented on the mushroom book requests, fortunately as it turned out, and showed them the intended 1983 gift of a paperback. I offered it to Yuri to keep as evidence but he quickly got out of the way to avoid touching it as if it had the plague. A strange reaction, I thought.

The tone of the questioning now mellowed, but only slightly, as the implication that I was fated to share the dock with Valery K changed to one of requiring me as a prosecution witness.

Tommy began, 'Your name will be mentioned in the papers for the trial. This will be bad for you. We might not be able to keep your name out of the newspapers.'

Not for one minute did I believe that Tommy couldn't stop anything he wanted, but I wondered, if I was safe in London, why did it matter if my name appeared in *Pravda* or *Izvestiya*. However, I had got it wrong, as I learnt when Tommy continued,

'The Western press already know of this and have copies of the papers,' and he reiterated 'we might not be able to keep your name out'.

I must have looked unworried to Tommy and Yuri at this stage since they again gabbled away in Russian. Then, 'Yuri says' (it was quite often to be *Yuri says*) 'would you be willing to return to the Soviet Union for the trial.'

I wanted to say something much stronger, but replied, 'No, my hospital schedule in London and overseas is far too heavy. I would never be able to make it at short notice.' I then told them of impending visits to Cairo and Munich.

'Would you be willing to make a statement at the embassy?' said Tommy. That was safer ground, 'Yes, in London.'

They weren't wearing that though, 'No, no, no, not the London Embassy, say nothing there.'

I followed that, tongue in cheek, with 'British Embassy in Moscow?'

I think that by this time Tommy felt he was getting nowhere fast so he said that wouldn't be necessary. After more conversation with Yuri, he continued, 'When K's flat was searched we found a letter from you.'

My memory was working overtime at this point and I remembered it.

'Oh yes, that would be my short note written in 1982 about the mushroom book.'

I described the contents.

Yuri then unzipped his document case and produced a photostat of my 1982 note to K. This was obviously meant to have a shattering effect, until I scanned the note I did feel the onset of paralysis. Fortunately, though, the note was almost exactly as I had described it. This must have disappointed them since their intention throughout was clearly to make me crack and agree to whatever it was they wanted. In the hiatus which followed my unexpected reaction to the photostat I took the opportunity to tell them that I felt I ought to explain why I was in Moscow in 1983. Then, intentionally in great detail, I went through my lecture notes for the training course and through a United Nations organisation draft report on nuclear medicine which I had brought with me. This told them nothing of value for their purposes and they must have felt they could not stop my monologue when I said that their customs officials were always interested in papers and books taken into the Soviet Union. I received a twisted smile from Tommy.

'We are bureaucrats, not investigators.'

'Of course,' I replied, since the two occupations are not mutually exclusive! More conversation between them.

'Yuri says, will you be prepared to help us?'

'What do you mean,' I replied; the choice was wide!

'Make a written statement.'

'I've no writing paper,' although I knew that would not get me far. Yuri handed me three sheets of paper and Tommy said aggressively,

'You will write that you will say no bad things about the Soviet Union.'

It was obvious that I had to write something, and could not refuse without some dire consequences, so I limited my statement to one page only, using large handwriting.

Hotel Minsk, Gorky Street, October 27, 1983
In 1980 and 1981 Mr Valery Krioukov of the State Committee for the Utilisation of Atomic Energy told me that I must pay my hotel bill to him in *Beriozka* shops because he had already paid the bill for me. Krioukov gave me no alternative. I had to go to *Beriozkas* with him to buy Scottish mohair wool and small gold jewellery items. I do not break laws of a foreign country.

When I got as far as this I asked, 'Will that do?'

'You have not written: I will say no bad things about the Soviet Union.'

This seemed to me to be standard phraseology. I fully intended to record events in print if I ever got back to London and this did not strike me as a convenient sentence to write! Tommy was a snappy dresser and had a high opinion of himself and his capabilities, probably justified, so I drew on his vanity about his command of English and said,

'That sentence is bad English. It is more correct to write: I will not criticise the Soviet Union in these matters.'

This was accepted by Tommy and I completed my statement on one sheet of paper, signed it and handed it over. My choice of this last sentence was intentional because I felt that recorded facts are not criticism and therefore I could write and publish the events.

This seemed to bring the proceedings to a close, except for Yuri visiting the bathroom and rattling the wastepaper basket, presumably looking for interesting papers. He returned to collect the two sheets of paper remaining on the table. This robbed me of my opportunity to obtain a copy of the statement by shading over the indentations of the pen with a pencil after Tommy and Yuri had left.

Tommy then issued a parting warning, delivered most effectively.

'We shall go now, but you will not tell your family; you will not tell your friends. This conversation never took place. You will not see us again.'

I hoped so but didn't exactly believe them.


THURSDAY EVENING, OCTOBER 1983
Two hours had passed since that knock on the door. I certainly needed to recover my wits in the hour left before the banquet.

Exactly how I managed to eat, drink and talk my way through the next few hours is still a mystery to me. But it must have been helped by the fact that the ordinary Russians present were no different from other nationalities; friendly, anxious to please foreign visitors and eager to know about famous London landmarks such as Westminster Abbey as well as about English food, clothes, customs and family life. The genial atmosphere helped me relax after the events of the afternoon.

The foreign students were also very friendly and told some amusing stories about themselves or about the Russians. One in particular which was enjoyed by all was about the student who arrived in Moscow and was asked his impression of Russian women. He replied, 'Cylindrical.'

However, there was one particular story told about the Lenin Mausoleum, fortunately not within Russian earshot, that after my recent experiences I felt bound to tell the speaker, a physician,

'For goodness' sake be quiet and talk about something else.'

He had got as far as 'I'm convinced that Lenin is a waxwork from Madame Tussauds. The light reflections from the skin are not those you would observe from embalmed skin but are more what you would associate with a wax surface. Although the body has been lying in the Mausoleum for many years it is not under a protective glass case. Even though the temperature is controlled in the room you would have expected to see at least some deterioration. There is none.'

I had heard enough and was thankful I managed to get him onto another topic of conversation.


FRIDAY EVENING, OCTOBER 1983

I was in my room at the hotel Minsk at six o'clock when the telephone rang.

'This is your friend, Tommy. We invite you to dinner at four o'cock, tomorrow.'

Not again, I thought and replied, 'I have a schedule for tomorrow evening. Can we meet earlier?'

'That will not be possible.'

I realised that I could not escape another session with him so tried to agree a venue which I knew and which would be crowded.

'All right, where are we eating, at the Aragvi restaurant?'

'No, we shall call for you at four o'clock.'

Tommy rang off.

I thought that a 22 hour waiting period before being taken to some unspecified place was probably an attempt to make me panic. I determined to do just the opposite—if I could!

I began to make preparations before four o'clock on Saturday to arrange a schedule for that evening. Luckily a request had gone to the hotel Service Bureau earlier, and Knud and his son had obtained three tickets for a concert at the Kremlin Palace Theatre celebrating the anniversary of the birth of Verdi, 1813–1901. The concert started at seven o'clock.

On the Friday evening I was also with Knud, watching two one-act ballets at the Bolshoi Theatre. I used this opportunity to prime him about recent events and told him that if I did not appear by seven o'clock on Saturday, or on Sunday for the return to London, then he should return to Copenhagen without making any enquiries in Moscow. He should then go promptly to the British Embassy and also notify the United Nations organisations. This seemed the best arrangement to make on the spur of the moment.

At four o'clock there was the expected knock on the door and I was escorted to a waiting car. Yuri sat in the front with the driver; I was asked to get in the back and did so. The door on the other side would not open and Tommy was left to struggle with the lock. Where we were going I did not know, or ask, but we turned down Gorky Street away from Red Square and drew up outside another hotel, the Hotel Pekin, where I had stayed in 1982. I got out of the car and walked straight into the main restaurant. I was pulled back and taken to a side room which was panelled from top to bottom. There, set to one side, was a table laid for three. A waiter stood nearby.

'What would you like to drink?' asked Tommy.

By this time I detected that a subtle change had come over my hosts; the menace of Thursday afternoon had been replaced by what, on the surface, appeared to be a friendly manner. First the stick, then the carrot; I wondered why. However, I was far from at ease and still of the opinion that something nasty was in store for me. Nevertheless, I thought that if I didn't start being positive now, I never would. Instead of replying vodka, as I thought was expected, I said Armenian cognac knowing this was the best and most expensive of the Soviet brands.

A look of consternation appeared on all their faces. The waiter returned full of apologies because the hotel did not have any Armenian cognac on the premises, but would Georgian do since when young it was very good.

'Certainly,' I replied, 'so long as it is not from Azerbaijan. That is very rough.'

With what I hoped was old-fashioned British courtesy, I thanked my hosts for this pleasant invitation, said that I hoped they did not mind but I had an appointment for seven o'clock at the Kremlin Palace and would they give me a lift to the Kremlin gate nearest the theatre. This would save me time. They seemed somewhat surprised but agreed.

To fill in some of the time with safe topics I spent the first ten minutes showing them my family photographs. The discussion moved on to football, and then it was time for toasts.

'To England,' proposed Tommy.

'To Russia,' I responded.

'To Liverpool.'

'To Moscow Dynamo.'

We proceeded with the meal and some Kiev cutlets were placed in front of us.

'Yuri says, who did you vote for in the last election?' asked Tommy.

'The Conservative Party and Margaret Thatcher.'

'Why?'

'I don't like a lot of the policies of the Labour Party.'

'Which ones?'

'Well, for example, the Labour Party destroyed the grammar school system in my country and I believe that this system favoured the able child from a poor or middle-class home.'

'Yuri says, what about Tory Party policy?'

Recollecting the recent newspaper reports of the Conservative Party Conference I tried to paraphrase Mrs Thatcher's comments on *dialogue* and *detente*.

'We would like Mrs Thatcher to come and talk to us, but she won't.'

After this I decided that I needed a break, and that a call to nature might relieve the pressure. However, Yuri also rose from the table and followed me like a bloodhound into the gents' toilet, standing on guard right behind me during the proceedings. Talk about unwanted company! On my return Tommy continued.

'Yuri says, what do you do in your job?'

This allowed me a long discourse on nuclear medicine, radiotherapy, computers, radiation protection, cancer statistics, radiation history. They must have been bored! In addition I told them of a letter I had written to the *London Evening Standard* about the 'gold grain gun' and the poisoned umbrella murder of the Bulgarian Georgi Markov on Lambeth Bridge in London. Since this was published on December 4, 1978 it would hardly be news to them.

THE GOLDEN GUN

In your November 24 issue of the *Evening Standard* you claimed that the murder weapon in the Georgi Markov case had been pinpointed by detectives as a so-called *cancer gun* and you also stated that the poisoned ball in Markov's leg was 1.7 mm in diameter. The *cancer gun* which you illustrated in your paper is used to implant radioactive gold grains into tumours as a treatment procedure. The implantation gun (a much better term than your emotive description) holds a magazine containing 14 small grains, each of length 2.5 mm and diameter 0.8 mm. The barrel of the gun is therefore too small to accommodate a 1.7 mm ball. Your reporter stated that the gun could be 'reduced to unrecognisable fragments within seconds'. This statement is untrue and it would take several minutes, using a screwdriver, to dismantle the gun.

'You did not publish that,' said Tommy.

'Yes I did, England is a free country and it is quite possible to publish, check if you don't believe me.'

It was time for more toasts.

'To our work,' proposed Tommy.

'To investigators,' I responded, since at last I was beginning to get a little courage back and with it a sense of annoyance at being sent to work on a course by United Nations organisations and being landed in this mess.

I started to talk about the Moscow museums and mentioned the exhibit of hundreds of German Iron Crosses in the Historical Museum in Red Square.

'These Iron Crosses are from the Battle of Stalingrad, I believe.'

'No, Volgograd,' said Tommy. I had forgotten that Stalin is still a 'non-person'.

I could no longer stand the suspense. There had not been any mention of Valery K and I still thought that they could be cooking something up in this area. I decided to raise the subject myself.

'It was extremely bad that a State Official like Valery K could be allowed to embarrass a United Nations lecturer.'

'We closed all that yesterday,' said Tommy.

They no longer wanted to talk about Valery K, but it soon became apparent what was the real reason for this dinner invitation and why they were being friendly.

'We would like you to help us,' said Tommy.

'I have already talked to you about Valery K.'

'We would like you to find out in England about the pill given to radiation workers in power stations which stops any effects of the radiation they receive,' continued Tommy.

'That is ridiculous, no such pill could exist. It sounds like a contraceptive pill for radiation! It does not make sense. In any event if it did exist, how do you know it is in England. Why not in the United States?'

'No, no, England. Yuri is a chemist and he is told from England.'

Then, when I was obviously not convinced, Tommy said in an almost inaudible tone.

'We will pay you money.'

This tone was sufficiently low for me to ignore the question completely.

The emphasis on an English source for Yuri was very clear and although I could not put any pertinent questions, it perhaps explained why my colleague Knud, who had also written a short note to Valery K in 1982, was left alone by Tommy and Yuri in 1983.

There was more discussion about this *radiation pill* and its relevance, if it existed, to the treatment of cancer. Then the waiter arrived with the dessert and there was a lull in the conversation.

'If it were to exist, then details should be published for all countries,' I stated.

'Your Government would never let you publish,' replied Tommy.

We had reached an impasse but fortunately half past six was approaching and it was time to leave for the Kremlin Palace Theatre. Tommy made one final remark as we were about to leave the hotel.

'Mrs Thatcher would never come to visit us at the moment because of the Korean airliner.'

I did not reply, just got into the waiting car.

SATURDAY EVENING, OCTOBER 1983

Even in the car, as we drove along Gorky Street Tommy still did not stop talking about this *radiation pill*.

'You can give the pill to Professor — in — in — weeks time,' said Tommy. He wouldn't give up and had the effrontery to assume that I would actually try to get a *pill* for him.

'Even if it existed it would be unrealistic to expect anyone to obtain it in a matter of weeks. It could take at least a year.'

Finally, as I was getting out of the car, Tommy said,

144

'You must not tell the British Government, you must only tell us.'

'It was a great relief, both to myself and to Knud and his son, that I made my seat with ten minutes to spare before the concert started. I still managed to enjoy Verdi's *Rigoletto* though, thankful for having survived the last 2½ hours of 'verbal chess'.

## LONDON POSTSCRIPT

I can hardly believe that I am the only person who has been approached in the manner described and questioned about a *radiation pill*. But for what particular reason was I singled out in October 1983? Because I was in Moscow? Because I lecture on radiation protection in hospitals? Because one of my medical publications has been misinterpreted?

In the treatment of cancer the use of chemotherapeutic drugs is now well established. These drugs act preferentially on the mitotic cycle of a malignant cell compared with the mitotic cycle of normal cells. A radiosensitising drug increases the radiation sensitivity of malignant cells which are treated using x-rays or gamma rays. Possible radiosensitising drugs or drug combinations are tested in clinical trials and one type of trial can involve comparing the effect of the drug with that of a non-active material (a placebo). A trial in which I was involved for cancer of the head and neck was reported in 1978 in the *International Journal of Radiation Oncology Biology and Physics* with unexpected results. It was found that the drug actually reduced the radiosensitivity of the tumours. What had been observed was a radiation protection effect. Of course, this result can only be taken within the framework of this single trial for head and neck cancer and only for the drug prescription used. There are no far-reaching implications or extrapolations—in spite of what might be thought by Yuri and friends.

To the best of my knowledge a *leukemia prevention pill*, as described to me does not exist and is certainly not given to power station workers in the United Kingdom. There are of course certain drugs in pill form which are used for the treatment of cancer and others which are used symptomatically to alleviate the side effects of radiation treatment experienced by some patients. I suppose Tommy, Yuri and the general public might call these *radiation pills* but the medical profession would not use this description.

If a *leukemia prevention pill* did exist then its applications are obvious. It would be relevant for any population far enough away from the hypocentre of a nuclear explosion not to be expected to die from the blast or the severe radiation effects including burns, but who nevertheless receive a radiation dose which could eventually lead to leukemia. So perhaps there are radiation pathologists specialising in blood disorders associated with nuclear fallout who are working on such a pill and it is not in the realms of science fiction. WHO knows!

# Andrew Lang on Statistics

'HE USES STATISTICS AS A DRUNKEN MAN USES LAMPOSTS – FOR SUPPORT RATHER THAN FOR ILLUMINATION'

Andrew Lang (1844–1912)

'He uses statistics as a drunken man uses lamposts—for support rather than for illumination.' *(Andrew Lang, 1844–1912)*

# Do Storks Bring Babies?

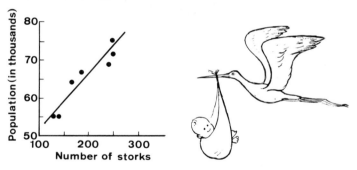

A plot of the population of Oldenburg in Germany at the end of each year against the number of storks observed in that year, 1930–1936.

Confusing correlation, just because it exists, with causation can lead to peculiar conclusions with either serious or funny implications:
'Anyone who draws the incorrect causal conclusion that storks bring babies and proceeds to shoot storks in the hopes of reducing the poulation will be disappointed!'.

146

# Nurse and Patient in the Fourteenth Century

Nurse bathing a patient. Fourteenth century manuscript illustration from a collection in the Royal Palace Library, Lisbon, Portugal.

*(Courtesy Wellcome Institute Library, London.)*

---

# The Red Cross, 1870

The *Graphic* was a nineteenth century popular magazine, somewhat similar to the *London Illustrated News*, and from time to time carried various illustrations of a medical or nursing nature, such as that of the Red Cross Nurse in 1870, which was by a Mr H Weigall, and was 'put on exhibition for the benefit of the Distressed French Peasantry'. The recent war referred to by the *Graphic* was the Franco–Prussian War which the French lost, and after which Emperor Napoleon III was exiled and died in England. The second illustration is from a later edition of the *Graphic* (September 1876) and is entitled *The Red Cross Quarters in the Field*.

The sketch represents an English lady, one of the numerous band of the 'Red Cross' sisterhood, who, during the recent campaigns, have devoted themselves to ministering to the sufferings of the sick and wounded victims of war. Many a refined and delicately nurtured woman has left the luxuries and comforts of her peaceful English home—a voluntary exile, following in the footsteps of a foreign army,

(*Courtesy Mansell Collection.*)

spending the long sultry days of summer, and the cold dreary nights of winter, in the wards of an army ambulance, often beyond the reach even of letters or news from her native home. Bright and graceful, and withal calm and simple, as the fair subject of our sketch, these true gentlewomen of England have soothed the dying agony, or cheered the weary hours of slow convalescence of many a humble hero. No scenes have been too trying, no work too hard, no services too menial for these devoted women. Wherever the Red Cross Banner has waved, they have been found, women of whom England may well be proud, women of whom doctors and patients speak with equal enthusiasm; gentle, obedient, tender to the sick, with neither theories nor crotchets

*(Courtesy Mansell Collection.)*

to work out, simply doing their duty under direction with loving patience and faithfulness. Our sketch depicts a room in one of the many stately old French chateaux, but a few short months ago the home of a prosperous family, the scene of festive gatherings, now the refuge of pain and suffering. The lofty 'salons' converted into hospital wards; camp bedsteads and the rough appliances of a field ambulance replacing the luxurious canopies and inlaid cabinets. The delicately furnished boudoir and 'Chambre de Madame' converted into store rooms, heaps of coarse woollen clothing and piles of bandages filling the wardrobes where gorgeous robes were wont to hang. On the shelves the varied appliances of a fashionable lady's toilet ruthlessly cleared away to make room for neatly-arranged stores of medical comforts and necessaries. The young Englishwoman has evidently with her own hands been opening one of the large well-packed bales, sent by her generous countrywomen. She is now carrying a bundle of bandages into the adjoining ward, where the surgeon is doubtless waiting for her gentle aid to bind up the shattered limb of a patient sufferer, who a few hours ago was marching into battle in all the pride of his youth and strength, now laid low on a bed of pain, grateful for the soothing sympathy and tender care of the English Infirmière. To him she seems a ministering angel, throwing a ray of light upon his sufferings, recalling to him in his hour of agony the fair and loving bride, whose soft-toned voice he longs, with all the longing of a dying exile, once more to hear.

REFERENCE

The *Graphic* 'English Infirmière in a French chateau' 14 January 1871 p. 34

## Murder by Digits

Swedish authorities have revealed that a computer has 'killed' thousands of citizens. It will take hundreds of hours and cost several million pounds to resurrect the slain Swedes, from Boraas county. The accident happened when a computer operator innocently keyed a request to the county's computer to update its records, and take account of recent deaths.

Unfortunately registers of people in Sweden are based on a six-digit number and a four-digit number, each linked by a hyphen. In defence of its mass murder, the computer only did what it was instructed to do. It interpreted the hyphen as a minus, subtracted the second number from the first, and was left with an unintelligible figure. The computer rejected these figures as obsolete and erased the records from its memory.

The records, and lives, of all these Swedes are now being reinstated manually.

REFERENCE

*New Scientist* 16 August 1984

---

## Qualities Needed by a Roman Surgeon, 30 AD

The following blueprint for a surgeon was given by Cornelius Celus whose eight volumes 'On Medicine' (*De Medicina*) were written about the year 30 AD in ancient Rome.

A surgeon must be youthful, or at any rate nearer youth than old age. He must have a strong, steady hand, which never trembles, and be no less ready to use the left hand than the right; his eyesight must be sharp and clear, his spirit undaunted. He must be compassionate, so that he wishes to cure the patient in his care, but not to the extent that he is moved by his cries either to go faster than is desirable or to cut less than is necessary; he must do everything just as if the cries of the patient cause him no emotion.

150

## Voltaire on Physicians

A physician is a man who pours drugs of which he knows a little into a body of which he knows less.

---

## The Original Siamese Twins, Chang–Eng Bunker, 1811–74

Eng and Chang were born in Siam about May 1811. Their father was of Chinese extraction and had gone to Siam and there married a woman whose father was also a Chinaman. Hence, for the most part, they were of Chinese blood, which probably accounted for their dark colour and Chinese features. Their mother was about 35 years old at the time of their birth and had borne four female children prior to Chang and Eng. She afterwards had twins several times, having eventually 14 children in all. She gave no history of special significance of the pregnancy, although she averred that the head of one and the feet of the other were born at the same time. The twins were both feeble at birth, and Eng continued to be delicate, while Chang thrived. It was only with difficulty that their lives were saved, as Chowpahyi, the reigning king, had a superstition that such freaks of nature always presaged evil to the country.

They were really discovered by Robert Hunter, a British merchant in Bangkok, who in 1824 saw them boating and stripped to the waist. He prevailed on the parents and King Chowpahyi to allow them to go away for exhibition, and they were first taken out of the country by a certain Captain Coffin. The first scientific description of them was given by a Professor Warren who examined them at Harvard University in 1829. At that time Eng was 5′ 2″ and Chang 5′ 1½″ in height. They presented all the characteristics of Chinamen and wore long black pigtails coiled three times about their heads.

After an eight-week tour of the eastern United States they went to London, but a tour of France was forbidden since the French considered that they would possibly cause the production of *monsters* by maternal impressions in pregnant women. After their tour of 1829 they returned to the United States and settled down as farmers in North Carolina, adopting the name of Bunker. When 44 years of age they married two sisters, English women, 26 and 28 years of age respectively. Domestic strife soon compelled them to keep their wives

151

in separate houses, and they visited each wife on alternate weeks. Chang had six children, Eng five, all healthy and strong.

*Oil painting by Edward Pingret, 1836. (Courtesy Wellcome Institute Library, London.)*

In 1869 they made another trip to Europe, ostensibly to consult the most celebrated surgeons of Great Britain and France on the advisability of being separated. It was stated that a feeling of antagonistic hatred after a quarrel prompted them to seek surgical separation but the real cause was most likely to replenish their depleted finances.

They were never in fact separated, and by 1869 both had diseased arteries, Chang had marked spinal curvature and they were both partially blind in their two anterior eyes. This was possibly due to looking outward and obliquely for so many years. The point of junction was about the sterno-xiphoid angle, and was a band of cartilage extending from sternum to sternum which in 1869 measured 4½″ but during early life was probably not greater than 3″.

The twins died in Januaray 1874, aged 62 years, and at autopsy it was found that there was a hepatic connection through the band to their two livers with slight intercommunication of the blood supply to the livers, but that they had independent peritoneal cavities and intestines.

The term *monster* was used for many years to describe the union of two foetuses and the x-ray picture shows what was termed a *skiagram of a double monster,* published in 1896 in the *Archives of Clinical Skiagraphy, London,* the forerunner of the *Archives of the Roentgen Ray* and the *Journal of the Röntgen Society.* It is therefore the world's first published radiograph of Siamese twins.

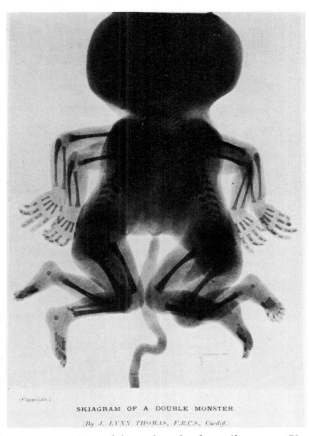

*(Copyright.)*

SKIAGRAM OF A DOUBLE MONSTER.

*(By J. LYNN THOMAS, F.R.C.S., Cardiff.)*

Not all foetuses are joined by a band of cartilage, as Chang–Eng,

153

and the possible variants include:

    (*a*)  union of two distinct foetuses by a bony junction of the heads;

    (*b*)  union of two distinct foetuses in which one or more parts are eliminated by the junction;

    (*c*)  fusion of two foetuses by a bony union at the pelvis;

    (*d*)  fusion of two foetuses below the umbilicus into a common lower extremity.

In addition, there are also instances of *bicephalic monsters* which are single foetuses with the exception of having two heads and two necks and *parasitic monsters* which consist of one perfect body complete in every respect, but from the neighbourhood of whose umbilicus depends some important portion of a second body.

These unfortunate specimens of humanity were invariably placed on show, as the advertisement of August 1838 illustrates.

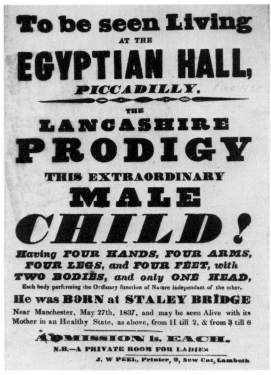

*(Courtesy Wellcome Institute Library, London.)*

REFERENCES

Gould G M and Pyle W L 1900 *Anomalies and Curiosities of Medicine* (Philadelphia: Saunders)

Rogers A 'Siamese twins' *Hospital Doctor* 12 March 1981

# Gorky Park and Gerasimov

In the well known Russian-set thriller *Gorky Park*,[1] which was also turned into a successful film of drama, intrigue and smuggling, three bodies are found in Moscow's Gorky Park with the soft tissues of their faces completely ripped away to avoid identification. Later, they were reconstructed by a Russian professor as an integral part of the plot.

Most readers and film viewers probably assume that this professor was a figment of the author's imagination, but in actual fact he really did exist and was Professor Mikhail Gerasimov, who until his death in 1970 was the Director of the Laboratory of Plastic Reconstruction at the Ethnographical Institute of the USSR in Moscow. His work was confined mostly to the faces of the paleolithic man and historical figures such as Ivan the Terrible,[3] but he did work on some one hundred forensic reconstructions and some of his achievements are described below from the book by Smyth.[2] The photographs were taken in Moscow and are published by courtesy of the Novosti Press Agency.

When he died in July 1970, Professor Mikhail Gerasimov was world famous not only in the field of palaeontology—his own discipline—but also in that of forensic medicine, for it was through his efforts that Russia came to the fore in the facial reconstruction branch of criminal science. As far back as 1950 Professor Gerasimov had been instrumental in the setting up of the Laboratory for Plastic Reconstruction, which forms part of the Ethnographical Institute of the USSR Academy of Sciences; since then the laboratory has assisted Russian investigators in the solving of hundreds of crimes.

Gerasimov was born in St Petersburg—later Leningrad—in 1907 and took a precocious interest in archaeology and palaentology. As early as 1920, when he was 13, he had become a part-time student at the University of Irkutsk, in Siberia, attending lectures in archaeology, ethnography, zoology and medicine. He was also fortunate in gaining the patronage of Professor A.D. Grigoriev, holder of the chair of forensic science, who encouraged the youthful Gerasimov in his interest in the morphology of the human skull.

The task which faced Gerasimov was a complex one; firstly, he had to establish some form of set measurement from those parts of the skull which remained more or less constant—where, that is, the flesh lay thinnest on the bone. Then, much more difficult, he had to discover a system of determining the muscular structure of an individual head. As he put it:

The essence of the programme was that not only definite information about the thickness of the soft parts must be found,

155

but also morphological features of the skull which could serve as clues for the reconstruction of the different parts of the face—nose, mouth, eyes and so forth.

*Mikhail Gerasimov selecting material for the second issue of his* Atlas *which includes portraits of ancient people (Paleolithic and neolithic) who lived in the Soviet Union, (Photography March 1961. Courtesy Novosti Press Agency.)*

Gerasimov spent two years measuring and dissecting the heads of corpses in the medical faculty of the university before attempting a reconstruction. He was disappointed with his first attempt and showed it to no-one. In 1925, however, he produced a portrait based on a contemporary skull which encouraged both him and Professor Grigoriev. That year he was appointed scientific technical assistant to the Irkutsk Museum, and was later put in charge of its archaeological section; there he reconstructed fossil heads, while at the same time persevering with his experiments on 'fleshed' heads.

It was not until 1935, 15 years after his first tentative steps had been taken, that an opportunity presented itself to prove to his colleagues that he was not driven by, as he put it, 'mania and fanaticism'. In the medical school he found the skeleton of a Dr Kolesnikov, who had left his body for research. Gerasimov had not known the doctor during his lifetime and had never seen a picture of him; however he was told that Kolesnikov's family had snapshots of him taken just before his death.

Carefully, Gerasimov set to work and produced on the skull the features of a man with a high, narrow forehead, a long nose, deep-set little eyes, a large though receding chin and protruding ears; despite this unprepossessing appearance Gerasimov judged that the muscular structure must have given the man a slightly smiling appearance, which 'lent the whole visage a surprising charm'. When Dr Kolesnikov's mother saw the reconstructed head she was at first shocked then deeply moved; her collection of photographs showed the scientist how accurate he had been, and he presented the mother with the portrait.

The same year, Gerasimov's work received what at first he thought to be a setback. He was handed the skull of a Frenchman, one Loustalot, a famous boxer and fencing master who had taught athletics at Leningrad. Gerasimov constructed a likeness in his usual fashion, and then went to compare it with Loustalot's death mask, which was preserved at the Anatomical Institute in Leningrad. To his profound disappointment, only the forehead and twice-broken nose tallied with the death mask.

> The whole of the lower part of the face was different and I would say even alien—strange. The face was noticably broader, as though bloated. The blurred eyes projecting from their orbits and the swollen mouth with thick, shapeless lips were quite unlike those of the face modelled by me.

To Gerasimov's relief there was an explanation. Loustalot had dropped dead in the street, and some time passed before the body was identified. By the time his identity was established, decomposition was fairly advanced, though a death mask had been taken from the decaying features nonetheless. When the portrait was compared with pictures of Loustalot in life, it was found to be as accurate as Dr Kolesnikov's had been.

Professor Grigoriev was delighted with his pupil's progress, and in 1939 gave him his first criminological task. In a wood outside Leningrad numerous bones had been found, scattered over a fairly wide area. They bore the marks of animal teeth, and it was assumed that they were the remains of someone who had been attacked and killed by wolves. However the skull and lower jaw told a different tale; when Gerasimov examined it he found the marks of what he considered to be a small hunting hatchet, along with a fracture which had been made by a blunt instrument. Adhering to the crown were a number of reddish-blonde hairs, which had been cut short some days before death.

The incomplete bone formation, the open sutures on the skull vault, the absence of wisdom teeth and the light wear on the adult teeth suggested to Gerasimov that the individual had been between twelve and thirteen years old. Sexing the head was difficult in view of the age, but the strongly developed supraorbital region and the large mastoid processes among other features seemed to point to the victim having been a boy.

After carefully refitting the lower jaw into place, Gerasimov modelled the most important masticatory muscles—these determine the whole shape of the face. Little by little he remodelled the head taking

157

into account such factors as the individual pecularities in the relief of the skull, the nasal bones and the chin eminence. The result was a snub-nosed, chubby-cheeked boy with a high forehead, a thick upper lip and slightly projecting ears. Short, red-blonde hair was finally added.

Meanwhile, the police had been searching their missing persons files for likely identities. Among them was a folio on a boy who, his parents believed, had run off to Leningrad six months previously. So as not to alarm them unduly, photographs of the reconstructed head were mixed in with 30 photographs of boys of similar age. The father immediately recognized the reconstruction as his son, and when the photograph was circulated the boy's movements were traced. As a result his attacker was caught and brought to trial: 'the first attempt to use a skull reconstruction in a criminal case was thus brilliantly successful' wrote Gerasimov, with justifiable pride.

Throughout the 1940s Gerasimov's fame grew in legal circles, and triumph followed triumph as his techniques became less empirical and more exact. Soon he was training students and by 1950 he had sufficient skilled pupils to establish his Laboratory for Plastic Reconstruction. With each case, he learned a little more about the morphology of the human face, but perhaps one final example suffices best to show how far he had come since his first tentative efforts in 1925.

Shortly after the setting up of the Laboratory, Gerasimov was called upon to reconstruct the face of an elderly woman whose skeleton had been found in a hut in remote woodlands. About a year earlier, the wife of a forester had disappeared in odd circumstances; her husband had said that she had set off to visit their son in a neighbouring town, but neither arrived at her destination nor returned home.

When Gerasimov examined the skull he found it severely damaged; the lower jaw was missing and only three molars in the upper jaw remained intact, the broken roots of the others remaining embedded in the skull. The roof of the mouth was peppered with small round holes, and in the base of the skull Gerasimov found several lead pellets. From this evidence he easily came to the conclusion that the woman had been killed by a shotgun blast from very close quarters.

Careful measuring of the remaining skull and comparison with other measurements in his files gave Gerasimov an idea of the shape of the missing lower jaw, and he was able to build this up in plaster, along with reconstructed teeth. His finished head resembled the missing wife so accurately that the forester confessed.

He had set out to drive his wife to their son's home in his horse-drawn wagon, but on the road there had been an argument. The forester had jumped from the wagon in a fury, telling his wife to drive herself; because the woods were haunted by wolves he had dragged down his double-barrelled shotgun from the back of the wagon, but in his irritation he mishandled the gun, and both barrels went off, one tearing away the bottom of his wife's face, the other hitting her in the chest. In a panic he had hidden the body. After hearing his story and reconstructing the crime, police were convinced that he was telling the truth and he thus avoided a capital charge.

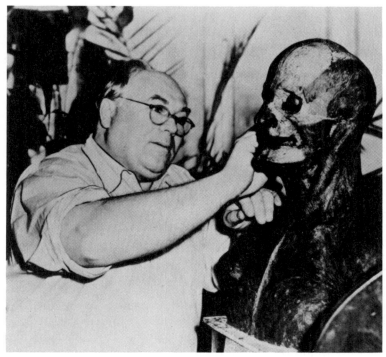

*Gerasimov reconstructing the head of the Tadjik poet Rudaki, who died in the tenth century. Basing his work on a detailed study of the way in which the muscles and cartilage of the face and neck were developed, he was able to build fleshed heads that corresponded closely to the living figure. He applied this technique both to heads of which portraits survived and to the heads of prehistoric and historic men and women. (Courtesy Novosti Press Agency.)*

Despite Professor Gerasimov's success, Western criminologists have so far seldom followed the lead he has given—a fact which puzzles the more progressive among them. In the course of the past twenty years or so, the widespread use of such techniques as 'identikit' and 'photo-fit' to reconstruct the faces of wanted or missing persons from their description have made the processes familiar to the general public. They have also been used in the identification of corpses and there is little question that their effectiveness would be increased enormously were they to be taken in conjunction with the Gerasimov system.

What is perhaps more extraordinary is the fact that Western forensic odontology has lagged behind the work done by the Russians; Gerasimov attached great importance to teeth as a factor in criminal

159

identification, and in fact the history of dental identification is ancient. Agrippina had an enemy killed, and satisfied herself that the right man had died by examining his teeth. In more recent years the victims of two disastrous fires—at the Vienna Opera House in 1878 and the Paris Charité in 1897—were identified by their teeth. And yet one of Europe's few experts, Professor Gosta Gustafson, writing in the *Criminologist* magazine in 1969, had to report that only three major forensic odontology departments existed in any of the world's dental schools—one in Norway, one in Denmark and one in Cuba.

REFERENCES
[1] Smith M C 1981 *Gorky Park* (London: Collins)
[2] Smyth F 1980 *Cause of Death* (London: Orbis)
[3] Mould R F 1984 'Ivan the Terrible's Skeleton' *Mould's Medical Anecdotes* (Bristol: Adam Hilger) p. 8

---

# King Philip of Macedon and The Manchester Mummies

Gerasimov is not the only person to attempt facial reconstructions, and the group at Manchester University under the direction of Dr Rosalie David and including the medical artist Richard Neave is famous for their work on ancient Egyptian mummified remains. In addition, Neave and his colleagues have provided a fascinating reconstruction of a skull from a royal grave in Vergina, Macedonia, which they have proved beyond reasonable doubt to be that of King Philip II of Macedon, the father of Alexander the Great. This is their story.

## THE TWO BROTHERS

The size and shape of a human head is determined largely by the underlying bone structure, as are the facial features, the skull being the matrix upon which the head and face are built. Detailed examination of the skull can generally reveal the age and sex of the individual, and characteristics peculiar to race or ethnic group may also be apparent. The condition of the teeth provides valuable information. The configuration of muscle attachments provides evidence of size and strength, although in the face these are confined mainly to the base of the skull and the lower jaw. It is therefore possible, in principle, to reconstruct the major features of the human head with

160

some degree of accuracy. However, a reconstruction can reveal only the type of face that *may* have existed, the position and general shape of the main features being accurate, but reconstructions of subtle details such as wrinkles and folds being inevitably speculative as there is no factual evidence as to their form or even their existence. Nevertheless, the achievement of the Manchester group is most impressive.

The interest of the Manchester Museum in Egyptology first became apparent to the general public via the press of the day in 1907 when Dr Margaret Murray, Egyptologist to the museum, undertook one of the earliest scientific investigations of Egyptian mummies when she unwrapped and dissected the mummies of the *Two Brothers*, before an invited audience in a large lecture theatre at Manchester University. Her experiment was unique in that she headed an interdisciplinary team whose members were specialists in the fields of anatomy, chemical analysis and the study of textiles. They all contributed knowledge to the detailed examination of these mummies, and the results of their investigations were published in 1910 the book entitled *The Tomb of the Two Brothers*. In the 1970s for the first attempts at facial reconstruction, the skulls of these two brothers were chosen.

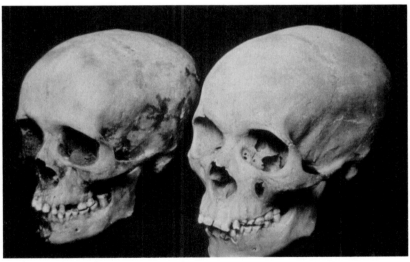

*The skulls of the two brothers Khnum-Nakht (right) and Nekht-Ankh (left). They were buried together in the rock-tombs of Der-Rifeh and date from the twelfth Dynasty.*

Within the coffins of the two brothers were two small statuettes, 15.5 and 25.4 cm high and carved from wood. These represented the

161

deceased as they had been when in full health and so skilfully were they carved that it was decided to make some comparisons of the heads of the statues, which were some 3 cm long, with the life-size reconstructions in the form of clay busts. The head of the statuette of Khnum-Nakht is powerfully built, with full lips and a broad nose and from the hieroglyphics on the coffins it appears that he was the younger of the two, a half-brother to Nekht-Ankh sharing the same mother but with a negro father. He is estimated to have been between 40 and 45 years old at death, and to have suffered from osteo-arthritis which seriously affected his back. After reconstruction, a sketch was made from the clay bust and although the similarity between the drawing and the statuette's head is not striking, it is unmistakable, in spite of the differences in size being so extreme.

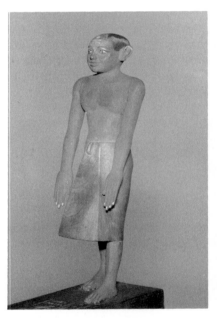

*The wooden figure of Khnum-Nakht which was found in the coffin.*

*The wooden figure of Nekht-Ankh which was found in the coffin.*

The head of the second statuette, Nekht-Ankh, is quite different from that of his half-brother, being more lightly formed and possessing a rather more delicate nose and a smaller mouth and chin. The similarity between statuette and reconstruction was again unmistakable. This elder brother was estimated to have been about 60 years old at death.

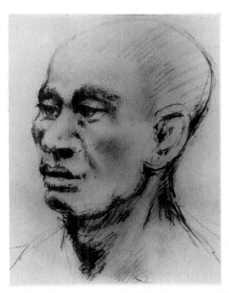

*Sketch of Knhum-Nakht made from the life-size clay model reconstruction.*

## '1770'—A 13-YEAR-OLD EGYPTIAN GIRL

The skulls of the *Two Brothers* were virtually unbroken, but the third mummy whose face was to be reconstructed by the Manchester group, labelled by them '1770' since no name was known and she was only an anonymous mummy No 1770 in their records, provided more of a problem. The skull had been broken into some 30–40 pieces but, fortunately, the majority of the bones which form the face were largely undamaged, although covered by a thick layer of mud and packing. The mandible was fractured to the left of the midline and both mandibular heads were broken, the vault was totally shattered with many pieces missing. The first stage in the facial reconstruction process was therefore, for 1770, to rebuild the skull in plastic.

The conservation of specimens of such antiquity as Egyptian mummies is of prime importance, and thus before any reconstruction work can be undertaken it is necessary to make accurate casts of a skull. The technique, using plaster of Paris and alginate, is similar to that used for manufacturing plastic face casts for certain radiotherapy patients so that they may be fixed into the correct treatment position before x-rays or gamma rays are delivered from a purpose-built machine such as a linear accelerator (for x-rays) or a cobalt-60 unit (for gamma rays). The reconstruction of the soft tissues on the prepared plaster skull was commenced by first *blocking-in* with soft model-

163

ling clay the head, neck and face, allowing the features to develop naturally and it is interesting to note how a skull will start to take on the character of a face at a very early stage.

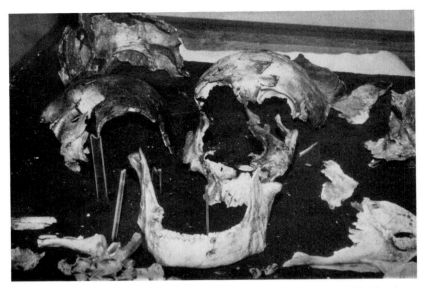

*Skull fragments of 1770 prior to reconstruction. Information provided by the remains of 1770 was somewhat limited at the time of making the reconstruction. Both legs appeared to have been amputated, one above and one below the knee, but whether before or after death was not known. There was evidence to indicate some intestinal infection, and an irregularity in the formation of the bone in the region of the nose suggests the probability of nasal congestion, which may have given rise to a slightly adenoidal appearance.*

The muscles of the face are so delicate and fine that they leave very few marks on bones to indicate their origins or insertions. Cartilage forms a large part of the nose, and fat is also an integral part of the soft tissue of the face. As the controlling factor in facial reconstruction was to be soft tissue, there seemed to the Manchester group little point in following anatomical structures too closely. Measurements were made, the thickness of the clay being increased or decreased as necessary until it corresponded to the mean of the maximum and minimum thicknesses for soft tissue of the face at 21 specific points as described by Kollmann and Büchly in 1898†. The measurements were taken by passing a thin steel probe through the clay at specific points. A slight variation of this system was adopted for 1770 in that

† 'Persistance of the strain and the reconstruction of the physiognomy of the pre-historic skull' *Arch. Anthropol.* **25** 345–59.

pegs cut to the appropriate length were fixed in position on the plaster skull. The clay was built up until it was of the correct thickness. Adjustments were made at the stage when the clay model was completed, which enabled minor ethnic considerations to be highlighted. These involved the eyes, which in ancient Egyptians were more almond shaped, and the tip of the nose. In addition, with 1770, some artistic licence was taken to avoid the somewhat cadaverous look of the *Two Brothers*, bearing in mind that 1770 was an adolescent.

The final reconstruction of 1770 was in many ways very satisfactory to the Manchester group because they were able to give some form of identity to a handful of broken bones, rather than reconstruct from almost perfect skulls, as in the case of the *Two Brothers*. 1770 was therefore taken a stage further than clay bust only. Glass eyes were fitted, eyelashes added, colour given to mouth and skin together with a limited amount of make-up which would almost certainly have been worn by a girl of this age. The hair (reputed to have been obtained from the theatrical props of the television series *Brideshead Revisited*!) was added in the form of a wig, but was only for exhibition purposes and was in no way judged to be part of the scientific reconstruction process. The result is startling and lifelike as can be seen from the photograph.

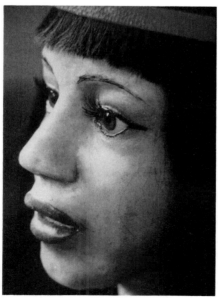

*The final bust of 1770.*

1770 contributed one final surprise to her reconstructors in that radiocarbon dating showed that although the bandages around the mummy dated from between AD 105 and AD 405, 1770 was in fact very much older than had been suspected, even though it had been thought that the mummy might have been re-wrapped at some stage. The bones dated from somewhere between 1250 BC and 1600 BC!

## ALEXANDER THE GREAT'S FATHER

The *Two Brothers* and *1770* are in no way famous historical figures with a reasonably well documented past. They were just three ordinary Egyptians of their time. There is, however, nothing ordinary about the last case history on reconstruction, one in which a member of the Manchester group was involved, Mr Richard Neave of the Department of Medical Illustration at Manchester University.

In late 1977, Professor Manolis Andronicos of the University of Thessaloniki discovered first two graves and later a third while excavating the Great Tumulus at Vergina in Macedonia: he has himself described the excitement of the discovery, the culmination of a long series of campaigns, on numerous occasions both in lectures around the world and in print. The principal burial was in a barrel-vaulted chamber, with a second burial in an antechamber, a cist tomb at the side contained the remains of a third body. There was little doubt that the burials were royal, not only because of the quality and wealth of the grave-goods—gold and silver, ivory, bronze vessels, armour of all kinds—but also because of the cremated bones in the antechamber and the main tombs were found wrapped in cloth of purple wool and gold thread placed inside magnificent gold *larnakes*, or chests, on each of which was emblazoned a starburst, the emblem of the Macedonian royal house. It confirmed the identification of Vergina with ancient Aigai, the first capital of Macedon and the traditional burial-place of its kings even after the capital was moved to Pella.

It was possible, to say the least, that the occupant of the principal tomb was the father of Alexander the Great, Philip II of Macedon, who was assassinated in 336 BC, and on many grounds he seemed a much more likely candidate than the other names put forward, such as Philip III Arrhidaeus, the mentaly ill half-brother and successor of Alexander who was murdered in 317 BC. However, apart from one name of indirect relevance engraved on a silver strainer found in the main tomb there were no inscriptions, and the dating evidence provided by the metalwork and the black-glazed pottery was inconclusive. Some scholars thought that it sat more comfortably at the end of the fourth century BC; it was also argued that the silver-gilt diadem found in the main tomb was the crown of Macedon adopted after Alexander's campaigns in Persia, and that the Macedonians could only

166

have learnt the technique of constructing a barrel vault during these campaigns.

Andronicos himself felt that Philip II was the most probable occupant, but he was not prepared to commit himself until proof could be found. Although excavations at Vergina are still continuing and further exciting discoveries are still being made, he felt that in view of their importance he should publish a preliminary account— by no means an automatic procedure in Greece. However, this only fuelled the controversy, which seemed unlikely to be settled before the eventual publication of the final report, inevitably still some years off.

The proof that Andronicos needed came from an unexpected quarter, by a series of happy coincidences. Richard Neave, who had been working on reconstruction of the heads of three Egyptian mummies, expressed an interest in working on some Greek skulls.

In 1981 Richard Neave went to Thessaloniki, where, with the ready cooperation of the staff of the Archaeological Museum, he was able to take casts of the bones of the skull. Although the body had been cremated, in general the bones were in remarkably good condition. Nevertheless, the skull was in fragments: the back had been blown out during the cremation, and as one tried to piece the fragments together it was clear that there was some distortion. At this point a two-pronged approach was adopted by showing casts to anatomists and to plastic surgeons.

The discoveries became more and more exciting. First of all it was found that although the heat of the cremation had made the bones shrink by around 10%, it had in general produced very little dis-tortion. In fact detailed study confirmed that the malformation that was causing problems in fitting the pieces together was largely the result of injury and of congenital deformity. This was particularly striking on the mandible and at the right eye-socket, although other anomalies were noted on the upper jaw and the right cheek-bone.

On the mandible the heights of the right condyle, coronoid process and incisura are much greater than those of the left, the right ramus is broader than the left, and there are some other anomalies: in layman's language, the right side of the jaw is higher and more developed than the left. At some stage too the chin and the dental midline have become realigned for reasons that are still unclear to us, so that the natural midline between the central incisors now lies visibly to the right. In short, the left side of the dead man's face was markedly under developed, and the right side over developed, to compensate for this. Yet such is the ability of the human body to make up for such defects that it was probably hardly noticeable to his contemporaries, particu-larly if he wore the short, thick beard typical of Macedonians before the reign of Alexander. In fact the mandible shows traces of such

167

compensation in additional bone growth at the new centrepoint; on the other hand very few faces are exactly symmetrical anyway.

The plastic surgeons who examined the casts also noted injuries to the right eye-socket: a notch on the upper margin of the right orbit and traces of a healed fracture on the zygomatic bone (the cheek-bone) below it. These they concluded could only have been caused by a missile coming from above. *The eye injury provided the conclusive proof of the identification of the dead man*, and eliminated the possibility that it was Philip III Arrhideaus.

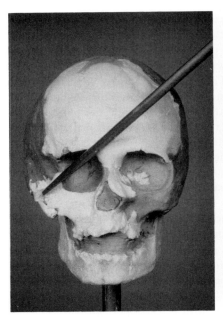 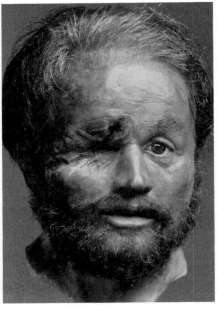

*(Left) The reconstructed skull of King Philip II of Macedon showing the line taken by the missile, which injured the right eye. The white parts are casts of the surviving bones and the grey area is the clay in which the bones are set. It also includes the missing areas.*

*(Right) Wax cast of the reconstructed head of King Philip II. The technique used was similar to that previously described for the Two Brothers and for 1770. It contains hair, beard and skin colouring, shows the scar to the right eye, and a prominent nose. This 'nasal decision' was made since a nose with a prominent bridge was typical of the Macedonian Royal Family, including Alexander the Great.*

The published portraits of Philip II of Macedon, the father of Alexander the Great are a gold medallion from Tarsus, now in Paris, made at the time of Caracalla; a marble head in Copenhagen, of

Trajanic date but based on an original of a very striking miniature ivory, one of a set found in the tomb at Vergina which must originally have decorated a bier or couch; and a small late Hellenistic marble head in Chicago. All give the impression of a bearded man with a square face having marked brow-ridges, a nose with a pronounced bridge, and an obstinate chin. The most striking feature, however, is that the Copenhagen and Vergina versions show an injury to the right eye, depicted as a nick to the eyebrow with an apparently sightless right eye, rendered on the Chicago marble as a hollow eye-socket.

The Athenian orator Demosthenes, an implacable opponent of Philip, lists the injuries that he was prepared to suffer for the sake of the empire, power, and glory—to have an eye cut out, his collar-bone broken, his hand and leg maimed. The grammarian Didymus Chalcenterus who wrote a commentary on Demosthenes in the first century BC, drawing on various fourth century sources now lost to us, enlarges on this and says quite categorically that 'he had his right eye cut out when he was struck by an arrow while inspecting the siege engines and the protective sheds at the siege of Methone'. This was in 354 BC, eighteen years before his assasination. Didymus' account is confirmed by other ancient authors—and it tallies exactly with the injury we noted on the skull from Vergina.

The proof was provided.

REFERENCES

David A R 1978 *Mysteries of the Mummies* (London: Book Club Associates)
David A R 1979 *Manchester Museum Mummy Project* (Manchester: Manchester University Press)
Marsh B 1984 'Faces from the past' *World Medicine* 21 Jan. 19–21
Neave R A H 1979 'Reconstruction of the heads of three ancient Egyptian mummies *J. Audiovisual Media in Medicine* **2** 156–64
Neave R A H 1979 'Mummies waren eens mensen' *Organorama* **4** 17–20
Neave R A H, Barson A J and Percy S 1976 'Models for medical teaching' *Medical and Biological Illustration* **26** 211–18
Prag J 1984 The flesh on the bones *Popular Archeology* **5** (March) 8–11

ACKNOWLEDGMENT

I am most grateful to Richard Neave for his first-hand story-telling of these most interesting and unusual events. For the section on Philip II, I have drawn closely on the paper by the archeologist John Prag. The illustrations are by courtesy of the Manchester University Department of Medical Illustration, Royal Infirmary, Manchester.

# Descartes and the Mind

It is not enough to have a good mind: the main thing is to use it well.

---

## The Irish Giant, The Sicilian Dwarf and the American Giant

The exhibit no visitor to the Hunterian Museum in the Royal College of Surgeons, Lincolns Inn Fields, London, should miss is the glass case containing the skeletons of two giants and one dwarf, together with a giant boot and glove; the death mask, shoes, ring and some wax anatomical impressions of the dwarf.

*John Hunter painted by Sir Joshua Reynolds. The feet of the Irish Giant can be seen in the top right-hand corner of the portrait.*

John Hunter (1728–93) was the founder of modern scientific surgery, and he investigated, both by dissection and experiment, human and comparative anatomy, and pathology. His museum differed from others in existence during the latter half of the eighteenth century since it was not only a collection of exhibits but also an illustration of his theories—in particular, of the constant adaption in living things of structure to function. Hunter also obviously realised the value of his unique collection and expressed a wish that it should not be sold piecemeal but preserved as an entity.

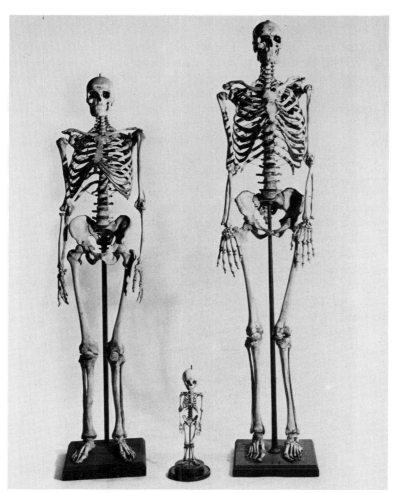

*The American Giant, the Sicilian Dwarf and the Irish Giant, in the Hunterian Museum in the Royal College of Surgeons, London. By kind permission of the President and Council of the Royal College of Surgeons.*

## THE IRISH GIANT

This was Charles Byrne, born in Ireland in 1761 and died in London 1783. When he was exhibited in London he was known as *O'Brian, The Irish Giant* and contemporary newspaper accounts gave his height variously as 8′, 8′ 2″ and 8′ 4″. These were all exaggerations as the actual height of the skeleton, with due allowance for intravertebral substances, is only 7′ 7″. There is no verified account of how Hunter came to possess the giant's body, and at the time it was legal to dissect only the bodies of executed criminals and of suicides. The popular version of events is that Hunter told the Irishman 'Giants are usually short lived. Your body has scientific value, so when you die, it should be left to me for dissection'. The terrified giant fled from Hunter, and on later falling ill made his undertaker promise that after death his body would be sunk in a lead-lined coffin off the Thames estuary. Hearing this, some medical students constructed a diving bell, so they could recover the body. However, this was not needed since Hunter bribed the undertakers with £500 to deliver the coffin to his house in Earls Court—where he prepared the skeleton in a large boiler.

## THE AMERICAN GIANT

On 17 March 1843, the American, Charles Freeman, appeared on stage in London at the Olympic Theatre with a dwarf named Signor Hervio Nano, in a piece specially written for the pair, entitled *The Son of the Desert and the Demon Changeling*. Ten years later, in 1853, Freeman was serving behind the bar of the Lion and Ball public house at 63 Red Lion Street, Holborn, claiming a height of 7′ 6″ and a weight of 21 stone. One of his handbills poetically invited the members of the general public to visit him.

> You need not unto Hyde Park go,
> For without imposition
> Smith's Bar Man is, and no mistake
> The true Great Exhibition.
>
> The proudest noble in the land,
> Despite caprice and whim,
> Though looking *down* on all the world,
> Must fain look *up* to him.
>
> His rest can never be disturbed,
> By chanticler in song,
> For though he early goes to bed,
> He sleeps so very *long*.

Though you may boast of many friends,
　　Look in and stand a pot,
You'll make a new acquaintanceship,
　　The *longest* you have got.

Then come and see the Giant Youth,
　　Give Edward Smith a call,
Remember in Red Lion Street,
　　The Lion and the Ball.

Freeman's skeleton was not as tall as 7′ 6″ but was only 6′ 8½″. He died in 1854, and in September of that year his bones were purchased for £2 and articulated for display by a Mr Willmost for the sum of £4 12s. 6d.

## THE SICILIAN DWARF

With a Sicilian father and an Italian mother, Caroline Crachami (1815–24) was brought to England at the end of the summer of 1823 and exhibited at Oxford, Birmingham, Liverpool and in Bond Street, London. Her parents and four brothers and sisters were all of normal size. Her skeleton is 19.8″ high.

REFERENCES
Silverberg R 1968 'Anatomy of a surgeon' *Reader's Digest* (December)
Allen E 1974 *Guide to the Hunterian Museum* (London: Royal College of Surgeons of England)
Exhibit captions in the Hunterian Museum

# The Wax Museum of Florence University

For many years the dissection of corpses for medical teaching in anatomy was dissaproved by church authorities and in most centres the educational material was limited to diagrams. However, in Florence in the eighteenth century the study of anatomy was greatly assisted by the use of wax models, culminating in the work of Clemente Susini (1754–1805) who was famous not only for the amazing number of waxworks he executed during his lifetime, but also for the high standard of his techniques. He was assisted by several first-class anatomists, including Paolo Mascagni (1755–1815) who was a specialist on the lymphatic system and the author of excellent anatomical drawings. There are now over a thousand models on exhibition at *La Specola* (so called, because in 1780 a telescope was placed on top of the building for astronomical observations and this event stirred the public imagination to such an extent that the whole museum—originally called the Royal and Imperial Museum of Physics and Natural History when it first opened in 1775—was called 'la Specola' from specula meaning look-out, observatory). This is the Zoological Museum of Florence University and is to be found in the Via Romana, the old road from Florence to Rome, and is open to the general public on only one half-day per week. The models on display are contained in their original eighteenth-century glass cases and one of the most impressive is a whole life-sized body illustrating the lymphatic system, and known as *the skinned man*. In addition to the work of Susini there are also some models from the seventeenth century, by the abbé Gaetano Giulio Zummo or Zumbo (1665–1701) who worked for several masters such as the Medici in Florence and Louis XIV in Paris. Only a few examples of his work have survived to the present day, and these include an anatomy of a man's head.

*The skinned man by Clemente Susini (1754–1805).*

Although dissection of corpses for medical purposes was supposed to be banned by the church, a large number of bodies for dissection were nevertheless obtained by the wax modellers of Florence, and according to papers in the State Archives these bodies had to be duly recorded by a doorman in a daily register. The burdensome task of

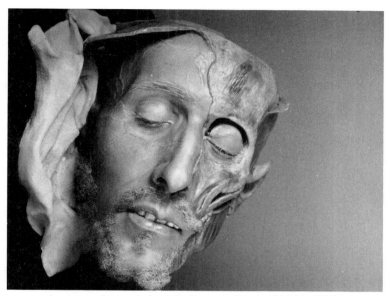

*Anatomy of a man's head by Gaetano Zumbo (1656–1701).*

transporting these corpses to and from the Santa Maria Nuova Hospital fell to one Giacinto Guidetti, the cleaner who each day made his gruesome rounds with a basket. The Florentine Archives contain many of his pleas, the most pathetic of which might be that dated 29 November 1792 in which he asked for . . .*a waxed hat, a coat and a pair of boots, so that in transporting corpses to and from the Hospital I am protected from the rain and cold, as it is not possible for me to stop anywhere along the way with the basket. . . .* The Archives also tell us that in January 1790 the Grand Duke of Florence approved of dividing the morning into two sessions, the first for the populace and the second for intelligent people of consequence!

The techniques followed by the wax modellers varied according to the size and nature of the piece to be reproduced; the smaller the model, the more delicate the process.

As far as the larger models are concerned, they were copied from the actual limb or corpse dissected expressly by an anatomist for the purpose; drawings from scientific books were of much aid (those of Paolo Mascagni were extremely useful for this purpose). A model in coarse wax or in clay was made and subsequently a plaster mould was made on it. Once the mould was ready, it was rubbed inside with soft soap in order to fill the small pores and to facilitate the subsequent detachment of the model. After this was dried, a first layer of soft wax was spread inside and given the general background colour, yellowish if it was a skin, dark red for the liver, and so on. The thickness of this layer varied according to the size of the piece to be reproduced. When

175

this had been cooled a second layer of wax of about the same thickness was added on the first one, and in the case of a large piece, a third thicker layer was added. It is likely that in the most carefully executed pieces, the first step before these rough layers were poured inside, was to cover the mould with a thin layer of almost fluid wax, applied with a paint-brush, having a colour somewhat deeper or paler than the underlying material, according to the effect which was desired.

For the largest pieces of work like thorax and abdomen with all the viscera, it was necessary to improve the solidity; in this case the cast was left open on the underside and, when the wax layers were cooled, the cavity was filled with tow impregnated with wax. In the case of organs such as the liver, kidneys, lungs and stomach, which appear in the more complex compositions, these were modelled separately, and by being warmed were welded to the common base with wax.

The full-bodied models, both the standing and the lying ones, bear inside an iron frame, accurately shaped in the attitude which the body is expected to show. This iron frame was well wrapped in tow and cloth soaked with wax in order to prevent the finished statue from sliding along it; the remaining cavities were filled in the same way. The whole figure was not cast in one piece; the head, bust, arms and legs were prepared separately and eventually joined together at a later stage.

The striations of the muscle fibres were made on the finished model, then the nerves, lymphatics and blood-vessels were added. The finer lymphatic fibres were given their knotty appearance by dripping white warm wax along a thin silk thread. the finer details were applied in paint. The bones were made of wax containing 'gilder's chalk' ground very finely and with colour added. When a transparent membrane was required, a marble table was kept at the right temperature with warm water; the wax was then poured and made extremely thin by means of a roll; several layers were thus applied, until the required transparency was obtained. When the model with all its final touches was completed, it was eventually covered with a film of copal for protection.

The workers were all specialists, in the sense that an expert on the circulatory system laid the veins and arteries, others the nerves, the lymphatic vessels, the viscera.

In the more delicate pieces, for example the small full-bodied models measuring about 60 cm, the method was slightly different. After making the model in clay, the mould was made in plaster in the usual way; then the interior of the mould was irrigated with a soft soap to block the pores, and was then smeared with a mixture of oil and lard. At this point, while still keeping the plaster mould opened in two halves, an extremely thin layer of wax, coloured like the surface, was spread inside, by means of a brush with long soft hairs; the wax had to be lukewarm (temperature tested on the cheek) and one continued in this way, with lukewarm wax, until the required thickness, 1 mm or less, had been obtained. To this first layer was added a second stronger layer, composed of a mixture of bee's wax and vegetable wax (Colophonian resin) at a slightly lower temperature and of a different

176

colour. According to the desired transparency, these thin layers were sometimes built up from as many as five or six separate applications of wax. At this point the mould was closed, leaving an outlet; then a third denser layer was trickled inside. When the whole had been well cooled, the mould was opened and the imperfections were removed with a lukewarm iron tool.

In conclusion, it is recorded that other small collections can be found elsewhere, in the Institutes of Human Anatomy of Padova, Cagliari and Bologna, of Pathological Anatomy at Florence and others. Some were ordered by Napoleon (1796) and sent to Paris but, because of some jealousy perhaps, they remained unpacked and were eventually sent to the School of Anatomy of Montpellier (1802). The Emperor of Austria Joseph II commissioned a great number of wax models for the Surgical Military Academy in Vienna. The order was too exacting for the Museum workshop and the Grand Duke refused it; Felice Fontana then took the appointment upon himself and gathered the necessary workers in his own house. The amazing work, over 1200 pieces, was completed in 5 years and in 1786 the great collection set off for Vienna through Italy and the Alps on mule back!

REFERENCES
Azzaroli M L 1975 *La Specola, The Zoological Museum of Florence University* (Firenze: Olschki) (Book prepared on the occasion of the 1st International Congress on Wax Modelling in Science and Art, Florence, June 1975.)
Berzi A, Cipriani C and Poggesi M 1980 'Florentine scientific museums' *J. Soc. Bibliogr. Natural History* **9** 413–25

# Martin Luther on Medicine, Mathematics and Theology

Medicine makes people ill, mathematics makes them sad and theology makes them sinful.

# A Plea for the Pigtail

The Chinese pigtail is an appendage which the progress of European civilisations now threatens with extinction. He maintains that the pigtail, like many other national customs, owed its origin to hygienic motives, and in one aspect formed the basis of China's ancient civilisation. He assures us that the effect of the pigtail is a more active circulation of the blood, which benefits the brain. He writes: 'The observations we hear now and then that Chinese without pigtails show less intelligence strikes me as not altogether unreasonable, as an active circulation of the blood will not fail to influence the nourishment and development of the brain.' He relates how the Chinese give special care to the head even of a newly born baby, and how it is shaved, and no cloth, cap, or soft pillow is allowed to interfere with the circulation of the scalp. When the child grows bigger, the hair is tied together in bunches, so as to expose the skin to the air, and thus promote perspiration. Later the hair is grown so as to form a pigtail, and superfluous hair is shaved away. The effect of the pigtail is a high and smooth forehead, and a face free from wrinkles; and so even old Chinese show smooth faces and a juvenile appearance. When rolled up on the top of the head, the pigtail acts as a substitute for a cap, and protects the head from the glare of the summer sun and the cold of winter. It also serves as a neckcloth and a pillow. As a cord, it is ever at hand to check haemorrhage. In addition to these virtues, the author attributes to it an inner and a moral meaning, for it is the symbol of the common nationality of 400 millions of people. Thus have Europeans, in dread of the power of China, adopted a method of protection by enticing the Chinese to become Europeans, and cut their pigtails off. This is the opinion of a European who, as we see, is more Chinese than John Chinaman himself. He neglects to remind us that the pigtail was only introduced by the Manchus somewhere about the middle of the seventeenth century of the Christian era, although it seems to be the fact that at an earlier date Chinese men let their hair grow long, and gathered it into a knot at the top. Still it remains for Dr Budberg to convince us that the European Delilah is really devising the ruin of the Chinese Samson by tonsorial methods.

REFERENCE
*British Medical Journal* 1912 **2** 1633–4

---

## Spanish Proverb

El Médico lleva la plata, pero Dios es que sana. (The physician takes the fee, but God sends the cure.)

# The Two-Headed Nightingale

*(Courtesy Wellcome Institute Library, London.)*

---

# Fire Damage

*World Medicine* of 21 September 1977 reported that the following information had been given in the *Daily Telegraph*.

> A Dutch vet was fined £140 for burning down a farm with a jet of flame from the rear end of a cow. He lit a match to test the gas and the flame set alight to bales of hay, causing damage estimated at £45,000. The cow escaped with shock.

## Barnum's Prize Ladies

*(Courtesy Wellcome Institute Library, London.)*

---

## Anecdote on Dr Cunningham

Contributed by Alan Frank after an interview on BBC Radio London about *Mould's Medical Anecdotes*. It relates to a story told to keep students awake during lectures in Cambridge.

> A woman patient told the eminent physician Dr Cunningham, author of the much-used manuals of dissection, that she had been given a digitalis examination by another doctor. 'I assume that he wore foxgloves, was Dr Cunningham's retort.

**ELISA & MARY CHULKHURST**

A 34 Y — IN 1100

BIDDENDEN.

A SHORT AND CONCISE ACCOUNT OF

## ELIZA & MARY CHULKHURST,

Who were born joined together by the Hips and Shoulders,

IN THE YEAR OF OUR LORD, 1100,

### At Biddenden, in the County of Kent,

COMMONLY CALLED

# THE BIDDENDEN MAIDS.

THE READER will observe by the Plate of them, that they lived together in the above state Thirty-four Years, at the expiration of which time one of them was taken ill and in a short time died; the surviving one was advised to be separated from the Body of her deceased Sister by dissection, but she absolutely refused the separation by saying these words,--"As we came together we will also go together," and in the space of about Six Hours after her Sister's decease, she was taken ill and died also.

By their Will, they bequeathed to the Churchwardens of the Parish of Biddenden and their Successors Churchwardens for ever, certain Pieces or Parcels of Land in the Parish of Biddenden, containing Twenty Acres, more or less, which now let at 40 Guineas per annum. There are usually made in commemoration of these wonderful Phænomena of Nature, about 1,000 Rolls with their Impressions printed on them, and given away to all Strangers on Easter Sunday after Divine Service in the afternoon: also about 300 Quartern Loaves and Cheese in proportion to all the Poor Inhabitants of the said Parish.

THOMSON, PRINTER, TENTERDEN.

*(Courtesy Wellcome Institute Library, London.)*

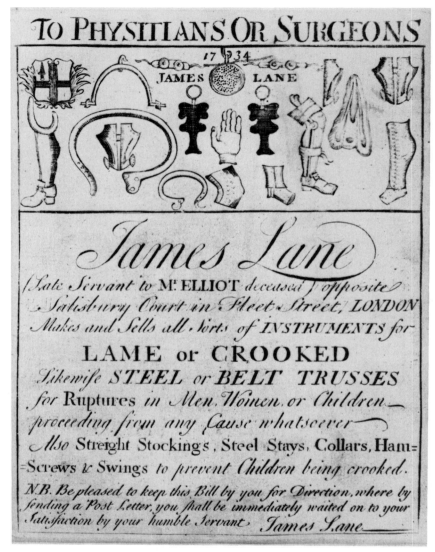

*(Courtesy Wellcome Institute Library, London.)*

# Hippocratic Hoax

There was published in 1958[1] in the student's club journal of the Welsh National School of Medicine, what purported to be a short Hippocratic text which had previously been overlooked by scholars and was stated to have been at one stage in the Library of the Church of Santa Eufrasia in Pisa. The journal editor described the author as a *special correspondent in New York* and the paper had all the outward signs of respectability with references, footnotes and, of course, the editor's backing. The text of this rare find dealt with many fields, including prognostics, epidemics, aphorisms, in the surgery, fractures and articulations and ancient medicine. The following are some of the words of wisdom from the pen of Hippocrates.

(*a*) Absence of respiration is a bad sign.

(*b*) It is unfavourable for the patient to be purple, especially if he is also cold. The physician should not promise a cure in such cases.

(*c*) Haemorrhoids are not improved by horseback riding or by riding on an ass.

(*d*) On the island of Tenebros, in the spring of the year, several maidens were attacked by swelling of the abdomen and absence of menstrual discharge. Giddiness, face flushed, absence of the membrane. A voyage to Asia was recommended. Complete crisis at the end of the ninth month.

(*e*) Life is short and the art is long; patients are inscrutable; their ignorance is impenetrable and their relatives impossible.

(*f*) Dislocation of the neck, which has been produced by a rope, cannot be treated with barley water.

(*g*) Where symptoms are severe and protracted, it is good if the patient's relatives are few, and best if they be absent altogether.

REFERENCE
[1] Gano S N 1958 'A gloss attributed to the Hippocratic School' *Leech* **4** 17–19

---

# Anecdote of George James Guthrie, FRS
# Army Surgeon to Wellington

Guthrie (1795–1856) was a famous surgeon of the nineteenth century who was President of the Royal College of Surgeons three times

(1833, 1842, 1854) and surgeon to the Westminster Hospital 1827–43. In addition he had a very eventful medical career from the age of 13 until the year 1814 when the last battle of the Peninsula War was fought.

*George James Guthrie, FRS.*

At the age of 13 he was invited by Mr Rush, the Inspector General of the Army Hospitals, to become a 'surgeon's mate', and in 1800 went to the Army Hospital, York Hospital in Euston Square. But in 1801, Mr Keate the Surgeon-General refused to employ unqualified hospital mates any longer. Thus, Guthrie and his three other 'mates' were faced with a dilemma. The other three resigned, but Guthrie, at two days' notice, took the MRCS examination and passed. He was 16 years old. The Royal College of Surgeons shortly afterwards passed a regulation to prevent candidates acquiring the diploma before the age of 21. Guthrie was appointed Regimental Assistant Surgeon to 29th Foot and after service in North America went to the Peninsula with Wellington in 1808 and was present at the battles of Rolica and Vimeiro (where he was wounded in both legs by a musket ball).

On recovery he was at Talavera, Albuera (where there were 3000 wounded) and the sieges of Cuidad Rodrigo and Badajoz. After the battle of Salamanca his masterly handling of the wounded won him the approval of Wellington, who later appointed him Deputy-Inspector of Hospitals. He was also present in 1814 at the last battle of the Peninsula War, Toulouse, where he made a collection of specimens of gunshot wounds to bone: no doubt from the victims of Sir Thomas Picton's rash attack, whose bodies lying in their red tunics in the green water meadows led them to be called the 'Flowers of Toulouse'. The casualties that day were 3200 French and 4500 Allied soldiers, and it was after this battle that Napoleon abdicated and went into exile in Elba. Napoleon had in fact tried to commit suicide after the battle was lost, but the poison, prepared two years earlier for an emergency in Russia, was stale.

Guthrie was not rewarded very well, and, because of his youth, the medical authorities refused to gazette his appointment as Deputy-Inspector and he was placed on half pay. To improve his financial situation he managed to acquire two private patients in 1815. However, he was recalled to the colours for Waterloo (arriving the day after the battle) and it is recorded that neither of his two private patients ever spoke to him again!

REFERENCES
Humble J G and Hansell P 1974 *Westminster Hospital, 1716–1974* (London: Pitman Medical)
Longford E 1971 *Wellington, The Years of the Sword* (London: Panther Books)

## Pliny on Doctors

Pliny the Elder (AD 23–79) commented on doctors when compiling his *Natural History*: 'There is no doubt that all these, in their hunt for popularity by means of some new gimmick, trafficked for it with our lives. This is the cause of those wretched, quarrelsome consultations at the bedside of the patients, when no doctor agrees with another, in case he may appear to acknowledge a superior. This is also the cause of that unfortunate epitaph on a tomb: 'He died from an overdose of doctors'. Medicine changes every day, again and again it is revamped, and we are swept along on the puffings of the best brains of Greece. It is obvious that anyone among them, who acquires the power of speaking, immediately assumes supreme control over our life and death, just as if thousands of people do not in fact live without doctors, though not without medicine, as the Roman people have done for more than six hundred years. The Roman people themselves were not slow in welcoming science—indeed, they were even greedy for medicine—until they tried it, and condemned it.'

## Why Choose Anaesthetics or Dermatology?

'Anaesthetics' because: you can sit down during all operations and you never meet a patient who is awake.

'Dermatology' because:
    (1) your patients never call you out at night;
    (2) they never die;
    (3) they keep on paying.

## Excuses, Excuses!

Some of the replies from patients in response to a circular asking if they required transport to and from the chiropody clinic. (Contributed by Mr J M McGarry, FRCOG, Barnstaple.)

I am under the doctor and cannot breathe.
I can't walk to the bus stop and my wife is bent.
I can't use my legs or my wife's.
I cannot wait for a bus as it might rain.
I can't breathe and haven't done so for many years.
I am blind in one eye and my leg.
I live five miles from the clinic and postman says I should have it.
I have got arthritis and heart failure in both feet and knees.
I am unable to walk as my dog has died.
I have arthritis of the spine and can hardly walk into doors.
I suffer with my chest and have a job getting breath at the local shops.
I cannot walk up a hill unless it is down and the hill to the clinic is up.
I suffer from thyroid and cannot climb as I don't have a car.
I have not got a bus or a husband and my home help comes.
I cannot walk far because my husband is bedridden.
I haven't had my feet done for months but they are no better when I have been.
My husband is dead and will not bring me.
My wife must have transport as she is over 80 and drives me mad.

186

I must have your transport. We have a car but my husband is 76 and I haven't had it for a long time.

I can't walk any distance because I have a hip.

I must have transport as I have funny feet.

I cannot drive a car because I haven't got one.

My mother is 96 and must have a car as she has got long fingernails.

I want transport as bus drivers do funny things to me and make me feel queer, I am nearly 86.

If mother goes out alone she gets into trouble.

I hope you will still send your driver as my husband is quite useless.

I must have your man as I cannot go out or even do up my suspenders.

When your man brings me back will you ask him to drop me off at the White Swan.

I don't have a bus 83 years old.

I can come any time to suit you but not mornings as I don't feel too good. I can't come Monday or Wednesday as home help comes, and not Friday as the baker calls for his money. I can't come on Tuesday as my sister visits.

---

## Surgeon's Hazards, Babylon, 1000 BC

As early as 1000 BC it has been recorded that successful operations for cataract took place in Babylon. The fees were fixed by law and the surgeon would receive 10 silver shekels for a patient who was a rich freeman but only 2 silver shekels for a slave. These fees, however, were for successful operations. Mistakes were more costly. If the freeman lost the sight of his eye, then one of the surgeon's hands would be cut off, whereas if it was the slave who was blinded, the doctor had to provide a new slave for the owner.

---

## Anecdote of Dr Gilmore and Dr Graham

Dr James Gilmore was the first patient to have one whole lung removed surgically at a single operation.[1] This treatment was for squamous cell carcinoma of the lung when he was 48 years old in 1933. The surgeon was Dr Evarts Graham who, when he was positive that the tumour was malignant, told Gilmore and described to him the planned operation. Gilmore asked for a few days to go home and

187

arrange his affairs and included in this was the purchase of a cemetary lot. Most unusually, the patient also paid a visit to his dentist to fill some cavities that had been bothering him! Graham later said[2] in 1957 that Gilmore had told him that the visit to the dentist was to give him great assurance! The pneumonectomy was successful and Dr Gilmore died in 1963 at the age of 78 years: cause of death was cardio-renal disease. He outlived his surgeon by six years since Dr Graham died in 1957 of inoperable cancer of the lung.[3]

REFERENCES
[1] Graham E A and Singer J J 1933 'Successful removal of an entire lung for carcinoma of the bronchus' *J. Am. Med. Assoc.* **101** 1371–74
[2] Graham E A 1957 'A brief account of the development of thoracic surgery and some of its consequences' *Surg. Gynaecol. Obstet.* **104** 241–50
[3] Eloesser L 1970 'Milestones in chest surgery' *J. Thorac. Cardiovasc. Surg.* **60** 157–65

## God and the Doctor

Contributed by Michael Aspel, 31 May 1984 on Capital Radio during an interview about *Mould's Medical Anecdotes*. Verse by John Owen (1560–1622)

God and the doctor we alike adore
But only when in danger and not before
The danger o'er, both are requited
God is forgotten and the doctor slighted

## Australian Medicine Man with Magic Healing Crystal

This is a type of medical magic, seriously applied, which many primitive races practise after the ordinary domestic remedies have been used without much success. The healing crystal, the Oruncha, is invisible and imaginary, but this does not detract from its magic powers. The body markings are magical and represent healing crystals.

REFERENCE
Stubbs S G B and Bligh E W 1931 *Sixty Centuries of Health and Physik*
(London: Sampson Low, Marston & Co.)

---

## Physician's Fees in St Louis, 1829

At a meeting of the medical faculty of the city of St Louis, held at the
City Hall on the 23rd day of November, 1829, the following
regulations for fees were unanimously entered into:

For the first visit in the city ...................................................... $1.00
For two or more visits to regular patients, per day ................... $2.00
For a whole day's medical attention ....................................... $10.00
For a night visit (expressly) after 9 o'clock .............................. $2.00
For a whole night's medical attention ..................................... $10.00

For visit in the country, per mile ................................................ $1.00
For consultation ................................................................... $5.00
For writing a prescription ....................................................... $1.00
For verbal prescription or advice ............................................. $1.00
For treating syphilis .............................................................. $20.00
For treating gonorrhea ........................................................... $10.00
For natural labors, from ............................................. $8.00 to $20.00
For preternatural, difficult, etc, labors ...................... $30.00 to $40.00
For amputating fingers, toes and other small members ............ $10.00
For amputating arm, leg or thigh ........................................... $50.00
For introducing catheter ......................................................... $5.00
For extracting tooth ............................................................... $1.00
For bleeding ........................................................................ $1.00
For opening abscess ............................................ From $1.00 to $2.00
For the operation for strangulated hernia ............................... $60.00
For operating for lithotomy ....................... From $100.00 to $200.00

REFERENCE
'One hundred years of medicine and surgery in Missouri' *St. Louis Star* 1900 pp 52–3

## Confucius on Medicine

When Chi-sun Fei sent him medicine, he received it with a deep bow but said, 'since I do not recognise it, I will not put it in my mouth'.

One of the 450 sayings of Confucius (551–479 BC)

## An Opinion on Quacks, 1714

Those who have little or no faith in the abilities of a quack will apply themselves to him, either because he is willing to sell health at a reasonable profit, or because the patient, like a drowning man, catches at every twig and hopes for relief from the most ignorant when the most able physicians give him none.

REFERENCE
*The Spectator* 26 July 1714

# CIGARS DE JOY

GIVE IMMEDIATE

RELIEF IN CASES

## OF ASTHMA, COUGH, BRONCHITIS, HAY-FEVER AND SHORTNESS OF BREATH.

One of these Cigarettes gives IMMEDIATE RE-LIEF in the worst attack of ASTHMA, HAY FEVER-CHRONIC BRONCHITIS, INFLUENZA, COUGH, and SHORTNESS OF BREATH, and their daily use effects a COMPLETE CURE. The contraction of the air tubes, which causes tightness of chest and difficulty of breathing, is at once lessened by inhaling the medicated smoke of the Cigarette, a free expectoration ensues, and the breathing organs resume their natural action. Persons who suffer at night with COUGHING, PHLEGM, and SHORT BREATH, find them invaluable, as they instantly check the spasm, promote sleep, and allow the patient to pass a good night.

These Cigarettes, invented by Mons. Joy, have been successfully tested and recommended by the Medical Profession for many years. They are perfectly harmless, and can be smoked by ladies, children, and the most delicate patients, as they are pleasant to use, and contain no substance capable of deranging the system.

Price 2s. 6d. per box of 35, and may be obtained of all Chemists, or, post free, from WILCOX & CO., 336, OXFORD STREET, LONDON, on receipt of Stamps or P.O.O. NONE GENUINE UNLESS SIGNED ON BOX, E. W. WILCOX.

*(Courtesy Mary Evans Picture Library.)*

# Take a Wine-Moistened Topaz

The writings of ancient alchemists and philosophers such as Paracelsus, Van Helmont, Cornelius Agrippa and other students of the Caballa and Gnostics indicate the importance of gems in diagnosis and therapy. They considered that there was nothing purer than a precious stone because it was the product of a gradual process of purification and crystallisation going on in the mineral kingdom through long stages of evolution.

The medicinal powers of diamond—for good and bad—were widely accepted. In 1532, his doctors prescribed costly doses of powdered precious stones to cure a lingering ailment of the pontiff,

Clement VII. Much of the mixture was diamond. Unfortunately, it is on record that the patient died after the fourteenth spoonful. The Spanish favoured diamond dust for treatment of bladder trouble and the plague, with entirely predictable results.

A well known medieval gemmologist, Marbodus Redonensis, twelfth century Bishop of Rennes, commended the healing powers of topaz and the virtues of a green jasper amulet against fever and dropsy. He suggested amethyst for warding off drunkenness and emerald as a cure for epilepsy and malaria. Later, Paracelsus used powdered jacinth mixed with an equal amount of laudanum as a remedy for fevers of various origins.

Ruby has been claimed to be the most precious of the twelve stones created by God and this 'Lord of Gems' was placed on Aaron's neck by His command. It has always been held in high esteem by the Hindu people who consider that the glowing hue of the gem is an inextinguishable fire. Ruby was said to preserve the mental and bodily health of the wearer; it removed evil thoughts, and, although considered to be associated with passion, it was also thought to control amorous desires and to dispel pestilential vapours and reconcile disputes. Rubies were considered to be a remedy for haemorrhage and inflammatory disease. They were believed to confer invulnerability from wounds, but the Burmese said that it was not sufficient to wear the stones, but that they must be inserted into the flesh, becoming part of the wearer's body.

## IN MEDICAL USE TODAY

Even in this day of men landing on the Moon, there is a hospital in India's state of Hyderabad which practises a Moslem system of medicine calling for the use of powdered gems. Entitled *Unani*, this system prescribes mixtures of oxidised jewels, dried fruit, honey, sugar and a variety of Indian spices. *Unani* practitioners claim that jewels provide chemicals in their purest form and have virtues beyond their chemical composition. Emeralds are recommended for ailments of the liver and kidneys, rubies for the heart, coral for asthma and the brain, diamonds for external use. Yet this is a modern, up-to-date hospital.

Among the *materia medica* of St Hildegard who lived in the tenth century was topaz to brighten dim vision. She advised her patients to rub their eyes with a wine-moistened topaz before retiring for the night. She also claimed that the same stone neutralised any liquid poison. The holy lady claimed that precious stones worked their medical wonders when merely held in the hand, grasped while making the Sign of the Cross, placed next to the skin, and breathed upon, held in the mouth (especially while fasting), or taken to bed to be warmed by the heat of the body. It was also the saint who suggested that epileptics cook their food only in water in which an agate had been soaking for three days.

Ancient physicians used gems both in diagnosis and healing paying particular regard to the 'dispositions' and 'moods' of the patient when tested against a gem.


## RELATION OF GEMS TO DISPOSITIONS AND MOODS

*Amethyst.* Regarded as the emblem of self-control and self-discipline. It was said to preserve the wearer from intoxication, both alcoholic and psychic, thus keeping the mind and body clear and well balanced.

*Agate.* Worn as a charm against scorpions and spiders. People who found the radiations of the stone disturbing were usually found to be prone to fits of anger, bad temper and jealousy.

*Turquoise.* Associated with health and the preservation and restoration of vital energy. It was also given as a symbol of conjugal fidelity and chastity. If this stone was worn by a person suffering from a chronic condition, malignant disease or infection of the blood, it was reputed to lose its lustre, the blue or blue-green colour changing to a muddy, dull brownish moss-green.

*Lapus-lazuli.* Used medicinally in a finely pulverised form, mixed with honey, as a remedy for sterility. The ancients considered that people who are emotionally unbalanced and very sensual will not be able to keep this stone long—always losing it.

*Opal.* Medieval philosophers praised this gem very highly, considering it conducive to idealism and impersonality, taking the wearer into abstract lands away from the material sphere of life.

*Coral.* Although not a true gem Paracelsus recommended red coral as a remedy against melancholy and evil influences. Brown coral, according to him, should not be worn as it attracts undesirable forces.

*Amber.* Fossilised resin of coniferous vegetation of the Tertiary period. With its high electrical properties, it was worn against the body to prevent the wearer from losing nervous energy.


## PREPARATION

For healing purposes, gems were used in the following ways:
    (*a*) The gem was soaked in pure water, kept in a glass or crystal

193

container and exposed to sunlight from the east. The irradiated water was then given as a drink or preserved with the admixture of a little wine and given in drops.

(b) Some of the gems were pulverised into a fine powder which was then mixed with honey and given internally, or mixed with tallow (mutton fat) and used as an ointment. Or it was mixed with a little water into a thin paste which was painted over the affected part. Pulverised jet mixed with honey was given as a remedy for flatulence.

## DIAGNOSIS

For diagnosis, certain gems were related to various parts of the body:

| | |
|---|---|
| Amber | Nervous system |
| Amethyst | Lymph and glands |
| Agate | Stomach and inflammations |
| Alabaster | Skin |
| Beryl | Eyes |
| Carbuncle | Blood stream |
| Cornelian | Chest and heart |
| Coral | Abdomen |
| Garnet | Liver and gall bladder |
| Jade | Kidneys |
| Jet | Intestines |
| Lapus lazuli | Reproductive organs |
| Malachite | Eye lids |
| Moonstone | Throat |
| Opal | Spleen |
| Rose quartz | Head—Brain—Epilepsy |
| Topaz | Rectum and bladder |
| Turquoise | Blood purifier and a tonic |
| Tourmaline | Spinal column |

The sixteenth century physician and pharmacist, Bulleyn, a cousin of Henry VIII's wife, Anne Boleyn, wrote a prescription containing 'two drachms of white perles; two little peeces of saphyre; jacinthe, corneline, emeraulds, granettes, of each an ounce; redded corrall, amber, shaving of ivory, of each two drachms; thin peeces of gold and sylver, of each half a scruple.' These very expensive prescriptions were in common use against disease and poison.

In today's world of wonder drugs and medical miracles, one wonders what the Ministry would reply to a doctor who recommended ruby for a National Health patient and how the charges would be met.

REFERENCE
Glogger T J 1971 *History of Medicine* **3** 29–30

# Toothache in the Third Century AD?

The illustration is of a funerary slab, probably of the third century AD in ancient Rome. It shows a doctor examining the face of a woman patient. The scene below may represent mourners at her deathbed.

*(Courtesy Mansell Collection.)*

---

# Oscar Wilde's Dying Words

I expect I shall have to die beyond my means (said after accepting a glass of champagne).

## Napoleon on Doctors

Doctors will have more lives to answer for in the next world than even we generals.

## Medical Ants

The closing of surgical wounds using catgut or thread sutures is widely known but a forerunner of these materials will come as a surprise to most people. In an early medical text by Albucasis[1] it is stated in Latin: 'Deinde aggrega duo labia vulneris et pone fornicam unam ex eis que habebant os apertum super duo labia vulneris'. This refers to sewing up the colon with a kind of ant with a big head. As soon as the ant has closed up its mouth, the head is cut off and it rests attached to the wound. Immediately a second ant is used for the following suture, and so on, until the entire wound is closed. The method had previously been described by the ancient Hindu surgeon Susruta who recommended the heads of red ants, and this practice was even followed in the nineteenth century[2] in the war between the Greeks and the Turks in 1821 when Turkish surgeons followed this method.

REFERENCES
[1] De Lint J G 1927 'The treatment of the wounds of the abdomen in ancient times' *Annals of Medical History* **7** 403–7
[2] Gurlt E J 1898 *Geschichteder Chirurgie, Berlin* **1** 97

## Pour the Medicine Over the Patient's Head!

In addition to poultices and drugs to cure irritated membranes, to reduce swellings, to open the bowels and to get rid of wind, the ancient Babylonians and Assyrians also recommended such curious remedies as pouring concoctions over a patient's head. An example is for a man with cramps: *either*

Let that man sit down, with his feet under him, pour boiled . . . and cassia juice over his head and he will recover.

*or*

Let him kneel and pour cold water on his head.

The choice was usually between a warm and a cold douche and if one afforded no relief then the other could be tried. A variation of the treatment for cramps included a massage to stimulate the circulation, but it sounded worse for the patient than the original cramps!

Place his head downwards and his feet under him, manipulate his back with the thumb, saying 'be good', manipulate his arms 14 times and his head 14 times, rolling him on the ground.

The address 'be good' was apparently directed at the demon causing the cramps. Why 14 was a magic number is not known.

REFERENCE
Jastrow M 1917 'Babylonian-Assyrian Medicine' *Annals of Medical History* **1** 241

---

## Remarkable Prevention of Poaching

A gentleman of Hampshire, who was in the habit of being robbed almost every night by poachers, adopted a novel and effectual mode of putting an end to this depredation. He went to London, purchased a man's leg at a hospital, and on his return had it hung up near the next place of public meeting, with a label attached to it, stating it had been caught on his grounds, and requesting the right owner would send for it. This had such an effect that he has not since been robbed.

REFERENCE
*John Bull Newspaper* No 10 18 February 1821 p 80

---

## Gravestone Inscription in America's Rural South

I told you I was sick.

# 'Sorry Sir, But What Your Wife Wants is Not Available on Prescription'

One of a series of drawings by Mr Harald Nymark for the posters of H Lundbeck A/S Copenhagen.

---

## An Experiment with Rats

The investigator reported that one third of the rats were improved on the experimental medication, one third remained the same, and the other one third could not be reported on because that rat got away.

Edwin Bidwell Wilson (1879–1964)

# 'In My Spare Time I Shoe Horses'

In my spare time I shoe horses.

*Another Nymark drawing for a Lundbeck poster.*

---

# Saint Who's?

By my last reckoning there were 160 hospitals in Britain named after a saint. St Mary has easily the largest number (26) with St John (13) in second place and St George and St Luke jointly third, with nine each. Others are St Andrew and St Michael (six each); Anne, Margaret, James, and Peter (five each); Paul and David (four each); Nicholas, Thomas, Clement, Lawrence, Leonard, and Catherine (three each); Matthew, Stephen, Bartholomew, Francis, Joseph and Martin (two each). Thirty two saints patronise just one hospital—Aldhelm, Anthony, Audrey, Augustine, Barnabus, Blazey, Cadoc, Chad, Charles, Christopher, Crispin, Ebba, Editha, Edmund, Edward, Faith, Giles, Helen, Helier, Hilda, Mark, Monica, Olave, Oswald, Pancras, Philip, Saviour, Theresa, Vincent, Wilfrid, Woolas and Wulstan.

These simple facts give us considerable opportunity for speculation. When there are several saints with the same name which is the one concerned? Why was he or she chosen? Does the saint have any special medical connection? And who were some of the saints on the list with unfamiliar names?

It seems a reasonable assumption that most if not all the St Marys refer to the Blessed Virgin, pre-eminent among the saints although with no medical association; St Mary Magdalene is a possibility for one or two but she has no medical association either.

There are three main contenders for the hospitals named St John. Either the Baptist or the Evangelist may be intended but St John of God is the strongest candidate although he is usually given his complete title to distinguish him from the other two. He was a fifteenth century Portugese who fought for years in the Spanish army before becoming fired with religious zeal. Unable to reach North Africa, where he hoped to die a martyr's death ransoming slaves, he became a highly successful seller of religious books and pictures. Then he went mad, gave away his books, and ran wildly through the streets. Eventually he recovered and devoted the rest of his life to helping the sick.

## PATRONESS OF CHILDBIRTH

Several St Margarets contest the five hospitals with this name but the favourite is St Margaret of Antioch. She was a mythical figure whose legend tells of her refusal to marry the pagan governor of Antioch since she was an avowed Christian virgin. She was tortured in a variety of ways and at one point was swallowed by a dragon which found her indigestible and burst asunder. She became the patroness of childbirth because of a promise made before she died that women who invoked her would be safe during pregnancy and labour. The Holy See supressed her cult in 1969 but her name lives on. One St Margaret's Hospital, however, in Auchterarder, Scotland, surely suggests a dedication to St Margaret of Scotland, grand daughter of Edmund Ironside and Queen to King Malcolm III. She was a woman of saintly behaviour and disposition who bore her husband eight children. One of them, King David of Scotland, also became a saint, but it seems unlikely that any of the four hospitals which bear that name refer to him. All are in Wales and are surely named after the

200

patron saint of that country who flourished in south Wales in the sixth century.

We have only one small clue to suggest which St James—James the Greater or James the Less—deserves the title to four hospitals. James the Greater used to be invoked for the cure of rheumatism although there is nothing in his life or legend to suggest that he suffered from it.

Three hospitals are named St Thomas but after whom—St Thomas the Apostle, Thomas Aquinas, Thomas of Canterbury, or even Thomas of Hales or Thomas of Hereford? If we are to be guided by medical considerations the apostle and Thomas Becket are the favourites. St Thomas, the twin, who refused to believe in the risen Lord until he could put his finger in the marks of the nails and his hand in Christ's side, used to be invoked for the cure of blindness and eye disorders because of the spiritual blindness he showed on that occasion. In this sense, therefore, we might regard him as a specialist but St Thomas of Canterbury was clearly a general practitioner since a prodigious number of cures from a wide variety of disorders were attributed to drinking the water of St Thomas—a large volume of water into which had been sprinkled tiny particles of the dried blood of the saint.

St Catherine of Alexandria is more likely to be intended as patron of three hospitals bearing her name than St Catherine of Genoa or of Siena. Like Margaret of Antioch, Catherine of Alexandria was a mythical figure whose cult was also suppressed by the Holy See in 1969. Her legend tells us that she, like Margaret, refused to marry a high-ranking pagan suitor since she had dedicated her life to Almighty God. Fifty philosophers were called in to show her the error of her Christian ways but she confounded them and many who heard the debate became Christian. She was condemned to die on a huge spiked wheel (the Catherine wheel of 5 November) but the wheel collapsed killing her torturers, and thousands more became Christians. Finally, she was beheaded when milk—the milk of human kindness—spouted from her body instead of blood, leading to her patronage of nurses.

The search for a specific medical connection in the many other hospital patrons reveals surprisingly few—Anthony, Chad, Christopher, Edmund, George, Giles, Leonard, Luke and Oswald.


## ST ANTHONY OF EGYPT OR OF PADUA?

Two St Anthonys, each with a medical association, contest one hospital at Cheam. St Anthony of Egypt was the archetypal desert hermit who gave away all his money and lived in solitude, enduring formidable privations and temptations until his death at the age of 99.

Ergot poisoning was once called St Anthony's fire, an association which came about because, in the Middle Ages, the monks of the Order of Hospitallers of St Anthony opened hospitals with red painted walls for sufferers from the disease, in which affected fingers and toes burn like fire. The original name was holy fire (*ignis sacer*) before St Anthony had his name attached. Representations of the saint show him with a bell, denoting his hermit status, and a pig since the monks of the order were allowed to let their pigs roam the streets. St Anthony of Padua, famous as a finder of lost articles, used to be invoked also for the cure of infertility, an association hard to understand. But which is patron of Cheam Hospital?

An equally puzzling link with infertility is found in St Edmund who was also invoked by barren women. He was King of the East Angles and was martyred by the Danes in 869. On 20 November, his feast day, it was at one time the custom for infertile women to walk alongside a white bull stroking its flanks until they reached the monastery gates when the women went inside to the saint's shrine to offer prayers that they might conceive. But this does not explain why St Edmund's Hospital is in Northampton.

More geographically appropriate is a hospital named after St Chad, the first Bishop of Lichfield, in Birmingham. It is also medically appropriate since, in former times, pilgrims visiting his shrine could put their hands through a little aperture in the wall, take some dust from within, sprinkle it into water, and give it to an invalid to drink with every expectation of cure.

We have seen a similar condition already with the water of St Thomas and we meet it again with St Oswald who has a hospital at Ashbourne. Oswald, King of Northumbria, was killed fighting the heathen King of Mercia, Penda, in 642. The Venerable Bede describes several miracles of healing near his grave and the cure of a man from the plague after he had drunk water into which had been placed a fragment of the stake which had impaled the severed head of the saint.

Plague also has an association with St Christopher, the patron of travellers who has a hospital at Fareham. He used to be invoked for protection against plague, tempest and flood, all hazards which a traveller in former times could expect to encounter.

It is not surprising that all nine hospitals named after St George are in England. Apart from being the national patron he is also patron of soldiers which explains why he was invoked for the cure of skin disease. Serious cutaneous sores and skin eruptions severely troubled soldiers in former times so the association is not surprising. All we know of St George is that he was a soldier killed in Palestine in the second century. The story of his encounter with the dragon is entirely mythical.

# LEPERS, NURSING MOTHERS AND CRIPPLES

St Giles has, nowadays, only one hospital dedicated to him and that is in London. At one time, however, when lepers were common, leper hospitals were named after him. He is patron saint of the unlikely combination of lepers, nursing mothers and cripples, perhaps one reason why his church in London is at Cripplegate. His legend tells, how, while living in the forest in the south of France, he was fed for a time by the milk of a hind—hence the nursing mothers. One day a hunting party pursued the hind and an arrow intended for it wounded the saint instead—hence the cripples. But we are in the dark about the lepers.

St Leonard, like St Giles, was a forest hermit, in which unlikely setting he became one of the patron saints of childbirth. The story goes that he was visited by King Clovis whose wife, heavy with child, was with him. She was safely delivered, thanks to St Leonard, but whether he acted as midwife or interceded with God on her account we do not know. He got a most original grateful patient present none the less—as much land as he could ride round in one night on a donkey. the animal covered so much ground that the saint was able to build the Abbey of Noblac on it.

We know from St Paul's Epistle to the Colossians that St Luke was a physician—the only one among the group of hospital patrons. Strangely he is almost always shown in hagiographic art in another guise—painting a picture of the Blessed Virgin with his emblem, an ox, close by. He is of course one of the patron saints of doctors but artists claim his as their patron also.

We will find few medical connections among the other patrons. Some have a simple geographical connection with a local saint. Frome, for example, has St Aldhelm's Hospital named for an abbot of Malmesbury and bishop of Wessex who died in 709; St Blazey, an early Cornish saint, has a hospital at Par bearing his name; St Cadoc, a sixth century saint from south Wales has a hospital at Caerleon, and his father, St Woolas, one at Newport; St Wulstan, the eleventh century Bishop of Worcester, has a hospital at Malvern, St Augustine, one at Canterbury, and St Hilda, abbess of a nunnery at Hartlepool in the seventh century has a hospital there; St Edith, or Editha, of Polesworth has given her name to a hospital at nearby Tamworth and St Wilfred to one at Ripon where he was abbot in the seventh century.

There are some patronages, however, which defy any such geographical explanation. Why does St Helier, the first martyr of Jersey, have a hospital in London; St Ebba, who was abbess of the nunnery of Coldingham on the remote Northumbrian coast, one at Epsom; or St Edmund of the East Angles one at Northampton?

Northampton, however, has one of the most delightfully original dedications in St Crispin's Hospital. The town is famous for its shoes and St Crispin and his brother St Crispinian are the patron saints of shoemakers since, to avoid taking alms from those to whom they preached, they supported themselves by their trade as cobblers.

With this ingenious dedication in mind is it too much to hope that future National Health Service hospitals—if any are built—might be more suitably named? Surely we need a children's hospital named after St Elmo, invoked for the cure of colic in children in the mistaken belief that he was martyred by having his intestines wound on to a windlass; and ear, nose and throat hospital named for St Blaise, the patron saint of throats; a mental hospital for St Vitus, invoked for the cure of mental disorder, epilepsy and possession by devils; a cancer hospital called after St Peregrine, a thirteenth century monk miraculously cured from a repulsive cancer on his foot the night before he was to have it amputated; and eye hospital called after St Lucy, the principal patron of the blind; and a dental hospital for St Apollonia, invoked for the cure of toothache.

And if the acquired immune deficiency syndrome (AIDS) ever becomes the scourge many fear and a hospital is built for its sufferers may it please be called St Fiacre's Hospital? St Fiacre was a sixth century Irish missionary who went to Gaul where he set up a hospice for the poor and sick but would never let a woman enter it. He was invoked for the cure of venereal disease.

REFERENCE
Dewhurst, Sir John 1986 *Br. Med. J.* **293** (December) 1618–21

---

# Caesarean Section by the Patient Herself

Cases of Caesarean section by the patient herself are most curious, but may be readily believed if there is any truth in the reports from savage tribes. In 1884 there was an account of a successful case performed, with preservation of the lives of both mother and child, by a native African in Kahura, Uganda. The young girl operated in the crudest manner, the haemorrhage being check by a hot iron. The sutures were made by means of seven thin, hot iron spikes, resembling acupressure-needles, closing the peritoneum and skin. The wound healed in eleven days and the mother made a complete recovery.

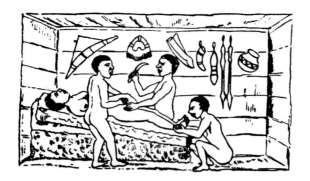

An earlier case, described in the *London Medical Journal* of 1785, was of a negro woman who, being unable to bear the pains of labour any longer, took a sharp knife and made a deep incision in her belly—deep enough to wound the buttocks of her child, and extracted the child, placenta and all. A negro horse-doctor was called, who sewed the wound up in a manner similar to the way dead bodies are closed.

REFERENCE
Gould G M and Pyle W L 1900 *Anomalies and Curiosities of Medicine* (Philadelphia: Saunders)

---

## John Shaw Billings on Statistics

Statistics are somewhat like old medical journals, or like revolvers in newly opened mining districts. Most men rarely use them, and find it troublesome to preserve them so as to have them easy of access; but when they do want them, they want them badly.

REFERENCE
*Medical Record* 1889 **36** 589

---

## Somerset Maugham on Training

I do not know a better training for a writer than to spend some years in the medical profession.

# Childbirth in Sicily

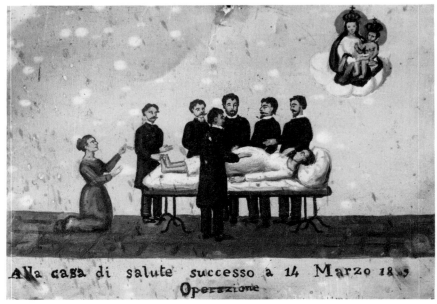

*Nineteenth-century votive pictures. Thanksgiving after a successful operation and safe delivery in childbirth, Ethnographic Museum, Palermo, Sicily. (Courtesy Wellcome Institute Library, London.)*

---

# Pompeian Surgery and Surgical Instruments

### POMPEII AND HERCULANEUM

The Romans occupied Campania in 88 BC, and thereafter Pompeii takes its place in Roman history, and is frequently mentioned by Seneca, Pliny and other contemporary writers. Toward the close of Nero's reign—that is to say, in the year AD 63—the whole region was visited by severe earthquakes, which made such a havoc that the cities were deserted for several years. The rebuilding of Pompeii appears to have been begun about AD 69, ten years before its final destruction, which took place on the 23 November AD 79 and appears to have commenced in the afternoon. Although Herculaneum and Pompeii were destroyed by the same eruption, they were destroyed in quite different ways. The former was filled up by a flow of warm muddy water, which filled it with a soft paste; and subsequent eruptions have covered it with molten lava no less than eleven times, rendering excavation exceedingly difficult and costly. Pompeii, on the other hand, was covered with loose ashes and pumice-stone, which were ejected from the volcano to a considerable height and blown into the city by the violent northwesterly gale which Pliny tells us was raging at that time. In short, Pompeii can be excavated with a trowel, but it takes a chisel to make an impression on Herculaneum.

The streets of the uncovered city of Pompeii bring vividly to the mind of the visitor the life, works, virtues, and vices of its former inhabitants. The old aqueduct that supplied the city with pure water from the mountains is well preserved and remains as one of the marvels of engineering of that time. The pavements of the streets can compare favorably with those of our day. The bare walls of public and private buildings testify to the unrivalled perfection masonry had attained at that day. The crude stone mills operated by human power furnished the city with flour, which in the adjacent bakery was converted into bread.

The enormous wine-jugs, so numerous in places where wine was sold and drunk, remain as lasting mementoes that the Pompeians were by no means prohibitionists. The numerous houses of prostitution, both public and private, remain as silent witnesses of a vice which appeared to have been unusually prevalent at that time. The capacious forum, amphitheatre, comic and tragic theatres that remain in a wonderful state of preservation, show that the people of that day—male and female, old and young—enjoyed the glittering stage and the bloody contests of the gladiators. The public bath-house is a marvel of its kind, and it is doubtful if in its artistic design and luxury it could be duplicated today. The private dwellings are all constructed on the same plan—masterpieces of comfort and sanitary construction. The numerous fountains furnished pure water for beast and man. The temple of Esculapius is one of the prominent landmarks of the former city, and fortunately time and the elements have dealt gently with its precious contents. In the centre of the capacious anteroom stands the altar of pure marble, beautifully carved, at which the priests of old worshipped in the interests of suffering humanity. It is here where the sick, the maimed and the injured sought relief.

A walk through the narrow, stone-paved streets of the uncovered part of the ruins of Pompeii is necessarily attended with serious thoughts of the past and present. The wider streets show deep grooves made by the chariot-wheels, while the narrower streets were reserved for pedestrians. The one-story buildings, both public and private, show a singular uniformity in their construction—evidence that the Pompeian architects and builders had in view more the comfort and health of their occupants than a desire to exhibit their talent. The many shops in the principal street were the homes and business places of merchants who supplied the citizens with the luxuries and necessities of life. A large building on the corner of two streets served as a drug-store, where crude drugs were dealt out to those in need of remedial agents. The proprietor of this primitive pharmacy—living, as he did, next door to a public house of prostitution—in order to protect himself and family against intrusion of an undesirable nature, found it necessary to place above the entrance a sign to indicate to the prospective customer the legitimate character of his business, and to direct him properly if he was in search of pleasure.

## SURGICAL INSTRUMENTS

These instruments were found in a house which has since been called the 'Surgeon's House'. They are made of bronze, and some of them show a high degree of artistic workmanship. Some of them show the destructive effect of heat and oxidation, while others are in a state of excellent preservation, as will be seen from the illustrations. The illustrations are taken from specimens from the Naples Museum.

(*a*) Actual cautery. Length 10 in.

(*b*) Bivalve speculum working on a central pivot. Length 6 in. Width, when open, 2½ in.

(*c*) Scissors with a spring-like shears. Length 4 in.

(*d*) A male catheter which is almost a facsimile of the one devised by J L Petit in the last century. At the closed end is an eye, as in the modern instrument. Length 10½ in.

(*e*) Hook. Length 6 in.

(*f*) Point of injection-syringe, with eight small perforations near the distal end. The other end was, no doubt, filled with a syringe. Length 6 in.

(*g*) Pompeian forceps, formed of two branches, crossing and working on a pivot. Each branch is fitted with with an engine-turned handle and a spoon shaped blade. A powerful forceps, undoubtedly used for the extraction of foreign bodies. Length 8 in.

(*h*) Forceps with serrated bite. Length 4½ in.

(*i*) Cupping-glass of bronze. Height 6 in, diameter 3 in.

(*j*) Medicine-box with medicines, 5 × 3 in.

(*k*) Spatula for mixing ointments. Length 7 in.

(*l*) Lancet for bleeding. Length 5 in.

(*m*) Fleam for bleeding horses. Length 5½ in.

(*n*) Forceps. Length 4½ in.

(*o*) Toothed dissecting-forceps with the engraved name, A. C. A: G. L. V. S. F. Length 7½ in.

(*p*) Trocar for tapping, with a hole at the end for the escape of the fluid. Length 5 in.

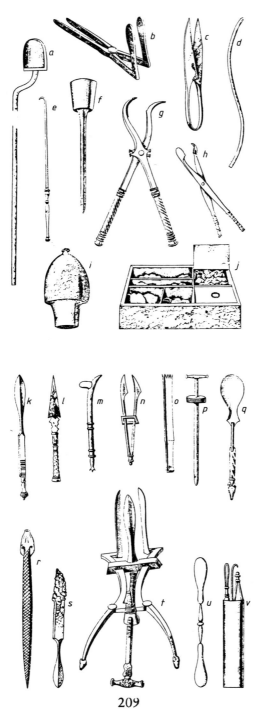

(*q*) Small spoon with bone handle, ending in the head of a ram. Length 5½ in.

(*r*) Female catheter. Length 4 in.

(*s*) Bistoury, the blade oxidised and the handle in bronze. Length 5¾ in.

(*t*) Trivalve speculum, an instrument which, like the bivalve and the quadrivalve, has been much discussed by archaeologists and physicians. It is composed of three valves standing at right angles to the rest of the instrument, and jointly dependent on one another in the expansion transmitted only to one of them. By turning the screw one valve is drawn nearer the operator, and this forces the other two to open in a sidelong direction. The instrument can be held by the two curved handles with the left hand, while the right hand turn the screw. Length 8¼ in.; widest expansion of the two valves 1½ in.

(*u*) Spatula. Length 7 in.

(*v*) A metallic case containing surgical instruments. Length 8 × ¾ in.

These are some of the most important instruments found in the ruins of Pompeii, and which were employed by our ancestors two thousand years ago in the practice of surgery. I searched carefully, but without avail, for traces of needles or something else which would indicate that at that time wounds were sutured. The collection contains no saws, trephines, chisels or any other instruments for operations upon bones. All of the instruments with the exception of the specula and catheters are diminutive in size as compared with the same instruments of less remote and modern times. The absence of saws and chisels is noteworthy, as among the agricultural instruments these tools are represented by specimens of a high degree of perfection.

In the writings of Hippocrates raspatories, mallet and trephine are mentioned, and consequently must have been used in operations upon bones other than those of the skull. Hippocrates gives very minute directions as to the use of the trephine in the treatment of fractures of the skull:

With regard to trepanning, when there is a necessity for it, the following particulars should be known: If you have had the management of the case from the first, you must not at once saw the bone down to the meninx, for it is not proper that the membrane should be laid bare and exposed to injuries for a length of time, as in the end it may become fungous. And there is another danger if you saw the bone down to the meninx and remove it at once, lest in the act of sawing you should wound the meninx. But in trepanning, when only a very little of the bone remains to be sawed through, and the bone can be moved, you must desist from sawing, and leave the bone to fall out of itself. For to a bone not sawed through, and where a portion is left of the

210

sawing, no mischief can happen; for the portion now left is sufficiently thin. In other respects you must conduct the treatment as may appear suitable to the wound, and in trepanning you must frequently remove the trepan, on account of the heat in the bone, and plunge it into cold water. For the trepan, being heated by running round, and heating and drying the bone, burns it and makes a larger piece of bone around the sawing to drop off than would otherwise do. And if you wish to saw at once down to the membrane and then remove the bone, you must also in like manner frequently take out the trepan and dip it into cold water. But if you have not charge of the treatment from the first, but undertake it from another after a time, you must saw the bone at once down to the meninx with a serrated trepan, and in doing so must frequently take out the trepan and examine with a sound, and otherwise along the track of the instrument. For the bone is much sooner sawn through, provided there be matter below it and in it, and it often happens that the bone is more superficial, especially if the wound is situated in that part of the head where the bone is rather thinner than in other places. But you must take care where you apply the trepan and see that you do so only where it appears to be particularly thick, and, having fixed the instrument there, that you frequently make examinations and endeavour by moving the bone to bring it up. Having removed it, you must apply the other suitable remedies to the wound. And if, when you have the management of the treatment from the first, you wish to saw through the bone at once and remove it from the membrane, you must in like manner examine the track of the instrument frequently with the sound, and see that it is fixed on the thickest part of the bone, and endeavour to remove the bone by moving it about. But if you use a perforator you must not penetrate to the membrane, if you operate on a case which you have had the charge of from the first, but must leave a thin scale of bone, as described in the process of sawing.

As Hippocrates at one time lived and practised in Athens during a great epidemic, it appears strange that his teachings in reference to the treatment of injuries of the skull should not have reached Pompeii, as evidenced by the absence of trepans and other bone-instruments in the 'House of the Surgeons'.

If we judge the worth of the Pompeian surgeon from the collection of instruments he left behind him, it is evident that bloody operations were confined to bleeding, cupping, extraction of foreign bodies and opening of abscesses. The metallic medicine-box, the spatula and spoon indicate that the surgeons of that time made free use of medicines and ointments in the treatment of injuries and disease. The instruments and implements of wood, splints, etc, were of course destroyed by fire and heat, and their absence in the collection leaves undoubtedly a large gap in the surgical resources of the Pompeian surgeon.

211

## SURGEON'S HOUSE

The surgeon's house does not differ from the private houses in its vicinity. It is roofless like the rest, all that remains being the bare walls. It is here that most of the surgical instruments were found. This house was undoubtedly occupied by the principal surgeon of Pompeii, who ministered to those in need of surgical aid. It is here that bleeding and cupping were practised for all kinds of ills, real and imaginary. It is difficult to imagine what transpired from day to day. That the surgeon was a busy man there can be but little doubt. Competition was then not as active and pressing as it is now, and it is therefore safe to assume that the capacious waiting-room was crowded day after day by patients anxious to be bled, cupped or burned. These bare walls, if they could talk, could tell of many sad and exciting scenes. Fainting from loss of blood and writhing under the actual cautery, must have been frequent and familiar sights. How often the neighbourhood must have been disturbed by the cries of the suffering and the shrieks of the tortured! How often the atmosphere and adjacent streets must have been stifled with the smell of burned human flesh! Let us hope that the master escaped, leaving in his haste his instruments of skill and torture as lasting mementoes of his so suddenly interrupted professional career. The house is deserted and silent now, a permanent reminder of the great antiquity of the art of surgery. If the last representative of Pompeian surgery could return today and behold the improvements in surgery which have been made since his time, he would indeed be astonished and amazed. What would be his surprise if he could visit one of our modern hospitals and inspect an aseptic operating-room. He would find his old occupation gone. No need now for lancet, cupping-glass and actual cautery. He would find the science of surgery developed to a wonderful degree of perfection and its practice in consonance with its principles. He could make use of anaesthesia to prevent pain, Esmarch's bandage to guard against haemorrhage and operate under aseptic precautions to protect accidental and intentional wounds against complications the treatment of which made up a large part of the work of the ancient surgeon. He would perhaps be astonished to learn that since Pompeii was buried surgery not only came to a standstill, but retrograded for centuries, and that its present state of perfection is owing largely to the improvements and advancements made during the present century. Let us not forget, however, that our colleagues of the distant past, possessed of a primtive knowledge of anatomy, physiology and pathology, and armed with few and imperfect instruments to practise their art, laboured faithfully in the interest of suffering humanity, and unquestionably did much toward prolonging the lives and adding to the comfort and happiness of those who were entrusted to their care.

*Bas relief from Pompeii of Nestor healing Machaon. (Courtesy Mansell Collection.)*

## POMPEIAN SURGERY

There can be but little doubt that the Pompeian surgeons practised surgery in accordance with the teaching of Hippocrates, who is justly entitled to be called the father of medicine, and was born on the island of Cos in 450 BC; hence his lifework was contemporaneous with the early history of Pompeii. It is not difficult to conceive that his teachings penetrated to this city, or that some of its surgeons might have been his pupils. In all probability, Pompeian surgery was Hippocratic surgery. As has been remarked before, the instruments which have been recovered from its ruins so far seem to indicate that no major operations were performed at that time, and that the surgeon's work was limited to cupping, bleeding, the treatment of injuries and the performance of minor operations. The discovery of a number of very ingenious specula in the 'House of the Surgeon' furnishes us with positive evidence that at that time gynaecology was not practised as a specialty, but constituted a legitimate part of the surgeon's work. Considering the character of the moral atmosphere of Pompeii, it is not astonishing to learn that diseases of the genito-urinary organs taxed the ingenuity and occupied much of the time and attention of its surgeons, as shown by the different kinds of specula then in use and the wonderfully perfect construction of the male and female catheter. The numerous wine-shops and houses of prostitution,

213

private and public, that constitute such a conspicuous part of the ruins of Pompeii, stand as lasting monuments of the debauchery and licentiousness of its former inhabitants, and furnish a satisfactory explanation of the prevalence of genito-urinary diseases among males and females, and which so often necessitated the services of a surgeon. The fleam for bleeding horses found in the instrumentarium of the Pompeian surgeon goes to show that he extended his sphere of usefulness to the domestic animals, which furnished him with an additional field for observation and undoubtedly added materially to his income. That the surgeon of Pompeii was a man of means and good social position is amply testified to by the size and location of his house. This house is capacious, and is located in the aristocratic part of the city. A liberal income undoubtedly rewarded his labours and placed him in a position to enjoy the luxuries of life, which seems to have been the main object in life of the mass of the people at that time. The existence of a separate house occupied as a pharmacy shows that the people then, as now, had great faith in the healing powers of herbs and drugs, and the medicine-box found in the 'Surgeon's House' was replenished from time to time from this source, and its contents were undoubtedly frequently made use of by the surgeon in the practice of his profession.

REFERENCE

Adapted from Senn N 1895 'Pompeiian surgery and surgical instruments' *Medical News* **67** (28 December) 701–8

## Cure for a Babylonian Hangover

The following is surely one of the most ancient remedies for the after-effects of a good night out on the bottle!

> If a man has drunk unmixed wine and has a severe headache, he forgets his words, his speech is heavy, his mind is clouded, his eyes are set, take (eleven plants are enumerated), mix them with oil and wine, let him drink before the approach of evening, and in the morning before anyone has kissed him.

REFERENCE

Jastrow M 1917 'Babylonian–Assyrian medicine' *Annals of Medical History* **1** 243

## FREE TRIALS
AT 27, PRINCES STREET, HANOVER SQUARE, LONDON, W.

THE

# CARBOLIC SMOKE BALL

WILL POSITIVELY CURE

For Inhalation only.

For Inhalation only.

COLDS

COLD IN THE HEAD

COLD ON THE CHEST

CATARRH

ASTHMA

BRONCHITIS

HOARSENESS

LOSS OF VOICE

INFLUENZA

HAY FEVER

THROAT DEAFNESS

SORE THROAT

SNORING

CROUP

WHOOPING COUGH

NEURALGIA

HEADACHE

*This Infallible Remedy is Used by*

Marchioness of Bath.
Marchioness of Conyngham.
Marchioness de Sain.
Countess of Dudley.
Countess Dowager of Meath.
Countess of Enniskillen.
Countess of Ravensworth.
Countess of Lanesborough.
Countess of Aberdeen.
Countess of Home.
Countess of Elgin.
Countess of Chichester.
Countess of Hardwicke.
Countess of Carnwath.
Countess Manvers.

Countess Ferrers.
Viscountess Cranbrook
Dowager Viscountess Downe.
Baroness de Linden.
Dowager Lady Garvagh.
Lady Elizabeth Home.
Lady Leucha Warner.
Lady Eleanor Harbord
Lady Florence Duncombe.
Lady Henrietta Pelham.
Lady Eva Wellesley.
Lady Algernon Percy.
Lady Aline Beaumont.
Lady Blanche Hozier.
Lady Frances Hawke.

Lady Alfred Paget.
Lady Campbell of Garscube.
Lady Erskine.
Lady Mostyn.
Lady Clavering.
Lady Borthwick.
Lady Annesley.
Lady Churchill.
Lady Cavendish.
Lady Wellesley.
The Lady Mayoress (Lady Isaacs).
Mrs. S. B. Bancroft.
Mrs. Bernard Beere.
Miss Ellen Terry.
Mrs. W. H. Kendal.

Earl Cadogan.
Earl of Leitrim.
Lord Rossmore.
Lord Montagu.
Lord Fitz-Gerald.
Rt. Hon. Sir John Saville Lumley, Bart.
Sir John Whittaker Ellis, Bart., M.P.
Sir Digby Murray, Bart.
Sir Barnes Peacock, Bart.

Sir Edward Colebrooke. Bart.
Sir Edward Birkbeck, Bart.
Sir Robert Cunliffe, Bart.
Dr. Russell Reynolds, F.R.S., Physician to Her Majesty the Queen's Household.
Sir John Banks, M.D., K.C.B., Physician to Her Majesty the Queen.
Henry Irving, Esq.
Leopold de Rothschild, Esq.

## New American Remedy.

*A handbill, c. 1890. (Courtesy Wellcome Institute Library, London.)*

**FOR THE CURE OF COUGHS, COLDS,**

CONSUMPTION, ASTHMAS, &c.

## LOTT'S
## LUNG PILLS

have justly obtained the unqualified approbation of all those who have tried them for speedily curing the Influenza and Bronchitis; also as a most valuable and efficacious Remedy for Colds, Catarrhs, Coughs, Hoarseness, Difficulty of Breathing, Asthma, Consumption, Spitting of Blood, and all disorders of the Lungs. In violent and distressing Coughs they never fail to afford immediate relief, by allaying the tickling and irritation of the Throat and Windpipe, which is the cause of frequent Coughing. They promote Expectoration, remove Fever, Lassitude, and Chilliness. Public Orators and the most Celebrated Vocalists have long held them in the highest estimation for their invaluable properties in removing Huskiness, Wheezing, and Oppression of the Chest, Strengthening the Lungs, and giving Power, Tone, and Clearness to the Voice. A Dose at bed-time seldom fails to cure a recent Cold, and procure sound and refreshing Sleep.

*Sold by Appointment of the Proprietor, by*

## J. W. STIRLING, CHEMIST,

86, HIGH STREET, WHITECHAPEL,

From whom a SINGLE DOSE can be had; or in Stamped Boxes at 13½d., 2s. 9d., 4s. 6d. & 11s. each.

*(Courtesy Wellcome Institute Library, London.)*

## Removal of Stones from the Head

A curious form of quackery which prevailed in the seventeenth century and which was taken as a subject for painting by several

Dutch artists (the illustration here is a seventeeth-century drawing on parchment by Pieter Jansz Quast of Amsterdam). It consisted of treating headache or supposed brain disease by extracting a stone from the head. An incision was made in the scalp and the quack took out a stone which he had ready for the purpose—and the patient was cured!

*(Courtesy Wellcome Institute Library, London.)*

REFERENCE
*Br. Med. J.* 1911 **1** 1262

---

# *A Yorkshire Water Doctor*

In the town of Leeds in Yorkshire, there once lived a quack who had received no professional instruction whatever, but was known far and wide for his wonderful cures, and especially for his power of diagnosing the diseases of patients whom he had never seen by simply examining their urine. A celebrated surgeon, Mr X, wishing to see his method of working, desired to be present one day, and the quack readily acceded to his request, feeling much flattered that so great a

The Sick Lady *by Caspar Netscher. (Courtesy Mansell Collection.)*

man should patronise him. Shortly after Mr X had taken his seat a woman came in with a bottle of urine, which she handed to the quack. He looked at her, then at the bottle, held it up between him and the light, shook it, and said 'Your husband's?'.

'Yes, Sir'.

'He is a good deal older than you?'.

'Yes, Sir'.

'He is a tailor?'.

'Yes, Sir'.

'Here' he said, handing her a box of pills, 'tell him to take one of these pills every night and a big drink of water every morning and he will soon be all right'.

No sooner had the woman gone out than Mr X turned to the quack, curious to know how he had made all this out.

'Well, you see', said the quack, 'she was a young woman, and looked well and strong, and I guessed that the water was not hers. I saw she had a wedding ring on her finger so I knew she was married, and I thought that the chances were that it was her husband's water. If he had been about the same age as she it was hardly likely that he was going to be ill either, so I guessed he was older. I knew he was a tailor because the bottle was stopped not with a cork but with a bit of paper rolled up and tied round with a thread in a way no one but a tailor could have done it. Tailors get no exercise and consequently they are all very apt to be constipated. I was quite sure that he would be no exception to the rule and so I gave him opening pills'.

'But how did you know that she came from S?'

'Oh, Mr X have you lived so long in Leeds and don't know the colour of S clay? It was the first thing I saw on her boots the moment she came in'.

REFERENCE

*Br. Med. J.* 1911 **1** 1261

---

## *Eleven Blue Men*

This story by Berton Roueché is taken from *The Medical Detectives* (Times Books). It was originally published in the *New Yorker*.

At about eight o'clock on Monday morning, September 25, 1944, a ragged, aimless old man of eighty-two collapsed on the sidewalk on Dey Street, near the Hudson Terminal. Innumerable people must have noticed him, but he lay there alone for several minutes, dazed, doubled up with abdominal cramps, and in an agony of retching. Then a policeman came along. Until the policeman bent over the old man, he may have supposed that he had just a sick drunk on his hands; wanderers dropped by drink are common in that part of town in the early morning. It was not an opinion that he could have held for long. The old man's nose, lips, ears, and fingers were sky-blue. The policeman went to a telephone and put in an ambulance call to Beekman-Downtown Hospital, half a dozen blocks away. The old man was carried into the emergency room there at eight-thirty. By that time, he was unconscious and the blueness had spread over a large part of his body. The examining physician attributed the old man's morbid colour to cyanosis, a condition that usually results from an insufficient supply of oxygen in the blood, and also noted that he was diarrheic and in a severe state of shock. The course of treatment prescribed by the doctor was conventional. It included an instant gastric lavage, heart stimulants, bed rest, and oxygen therapy. Presently, the old man recovered an encouraging, if painful, consciousness and demanded, irascibly and in the name of God, to know what had happened to him. It was a question that, at the moment, nobody could answer with much confidence.

For the immediate record, the doctor made a free-hand diagnosis of carbon-monoxide poisoning—from what source, whether an automobile or a gas pipe, it was, of course, pointless to guess. Then, because an isolated instance of gas poisoning is something of a rarity in a section of the city as crammed with human beings as downtown Manhattan, he and his colleages in the emergency room braced themselves for at least a couple more victims. Their foresight was promptly and generously rewarded. A second man was rolled in at ten-twenty-five. Forty minutes later, an ambulance drove up with three more men. At eleven-twenty, two others were brought in. An additional two arrived during the next fifteen minutes. Around noon, still another was admitted. All of these nine men were also elderly and dilapidated, all had been in misery for at least an hour, and all were rigid, cyanotic, and in a state of shock. The entire body of one, a bony, seventy-three-year-old consumptive named John Mitchell, was blue. Five of the nine, including Mitchell, had been stricken in the Globe Hotel, a sunless, upstairs flophouse at 190 Park Row, and two in a similar place, called the Star Hotel at 3 James Street. Another had been found slumped in the doorway of a condemned building on Park Row, not far from City Hall Park, by a policeman. The ninth had keeled over in front of the Eclipse Cafeteria, at 6 Chatham Square. At a quarter-to-seven that evening, one more aged blue man was brought in. He had been lying, too sick to ask for help, on his cot in a cubicle in the Lion Hotel, another flophouse, at 26 Bowery, since ten o'clock that morning. A clerk had finally looked in and seen him.

By the time this last blue man arrived at the hospital, an investigation of the case by the Department of Health to which all outbreaks of an epidemiological nature must be reported, had been under way for five hours. Its findings thus far had not been illuminating. The investigation was conducted by two men. One was the Health Department's chief epidemiologist, Dr Morris Greenberg, a small, fragile, reflective man of fifty-seven, who is now acting director of the Bureau of Preventable Diseases; the other was Dr Ottavio Pellitteri, a field epidemiologist, who, since 1946, has been administrative medical inspector for the Bureau. He is thirty-six years old, pale, and stocky, and has a bristling black moustache. One day, when I was in Dr Greenberg's office, he and Dr Pellitteri told me about the case. Their recollection of it is, understandably, vivid. The derelicts were the victims of a type of poisoning so rare that only ten previous outbreaks of it had been recorded in medical literature. Of these, two were in the United States and two in Germany; the others had been reported in France, England, Switzerland, Algeria, Australia, and India. Up to September 25, 1944, the largest number of people stricken in a single outbreak was four. That was in Algeria, in 1926.

The Beekman-Downtown Hospital telephoned a report of the occurrence to the Health Department just before noon. As is customary, copies of the report were sent to all the Department's administrative officers. 'Mine was on my desk when I got back from lunch,' Dr Greenberg said to me. 'It didn't sound like much. Nine persons believed to be suffering from carbon-monoxide poisoning had been admitted during the morning, and all of them said they had eaten breakfast at the Eclipse Cafeteria, at 6 Chatham Square. Still, it was a job for us. I checked with the clerk who handles assignments and found that Pellitteri had gone out on it. That was all I wanted to know. If it amounted to anything, I knew he'd phone me before making a written report. That's an arrangement we have here. Well, a couple of hours later I got a call from him. My interest perked right up.'

'I was at the hospital,' Dr Pellitteri told me, 'and I'd talked to the staff and most of the men. There were ten of them by then, of course. They were sick as dogs, but only one was in really bad shape.'

'That was John Mitchell,' Dr Greenberg put in. 'He died the next night. I understand his condition was hopeless from the start. The others, including the old boy who came in last, pulled through all right. Excuse me, Ottavio, but I just thought I'd get that out of the way. Go on.'

Dr Pellitteri nodded. 'I wasn't at all convinced that it was gas poisoning,' he continued. 'The staff was beginning to doubt it, too. The symptoms weren't quite right. There didn't seem to be any of the headache and general dopiness that you get with gas. What really made me suspicious was this: Only two or three of the men had eaten breakfast in the cafeteria at the same time. They had straggled in all the way from seven o'clock to ten. That meant that the place would have had to be full of gas for at least three hours, which is preposterous. It also indicated that we ought to have had a lot more sick people than we did. Those Chatham Square eating places have a big turnover. Well, to make sure, I checked with Bellevue, Gouverneur, St. Vincent's and the other downtown hospitals. None of them had seen a trace of cyanosis. Then I talked to the sick men some more. I learned two interesting things. One was that they had all got sick right after eating. Within thirty minutes. The other was that all but one had eaten oatmeal, rolls, and coffee. He ate just oatmeal. When ten

221

men eat the same thing in the same place on the same day and then all come down with the same illness . . . I told Greenberg that my hunch was food poisoning.'

'I was willing to rule out gas,' Dr Greenberg said. A folder containing data on the case lay on the desk before him. He lifted the cover thoughtfully, then let it drop. 'And I agreed that the oatmeal sounded pretty suspicious. That was as far as I was willing to go. Common, ordinary, everyday food poisoning—I gathered that was what Pellitteri had in mind—wasn't a very satisfying answer. For one thing, cyanosis is hardly symptomatic of that. On the other hand, diarrhoea and severe vomiting are, almost invariably. But they weren't in the clinical picture, I found, except in two or three of the cases. Moreover, the incubation periods—the time lapse between eating and illness—were extremely short. As you probably know, most food poisoning is caused by eating something that has been contaminated by bacteria. The usual offenders are the staphylococci—they're mostly responsible for boils and skin infections and so on—and the salmonella. The latter are related to the typhoid organism. In a staphylococcus case, the first symptoms rarely develop in under two hours. Often, it's closer to five. The incubation period in the other ranges from twelve to thirty-six hours. But here we were with something that hit in thirty minutes or less. Why, one of the men had got only as far as the sidewalk in front of the cafeteria before he was knocked out. Another fact that Pellitteri had dug up struck me as very significant. All of the men told him that the illness had come on with extraordinary suddenness. One minute they were feeling fine, and the next minute they were practically helpless. That was another point against the ordinary food-poisoning theory. Its onset is never that fast. Well, that suddenness began to look like a lead. It led me to suspect that some drug might be to blame. A quick and sudden reaction is characteristic of a great many drugs. So is the combination of cyanosis and shock.'

'None of the men were on dope,' Dr Pellitteri said. 'I told Greenberg I was sure of that. Their pleasure was booze.'

'That was OK,' Dr Greenberg said. 'They could have got a toxic dose of some drug by accident. In the oatmeal, most likely. I couldn't help thinking that the oatmeal was relevant to our problem. At any rate, the drug idea was very persuasive.'

'So was Greenberg,' Dr Pellitteri remarked with a smile. 'Actually, it was the only explanation in sight that seemed to account for everything we knew about the clinical and environmental picture.'

'All we had to do now was prove it,' Dr Greenberg went on mildly. 'I asked Pellitteri to get a blood sample from each of the men before leaving the hospital for a look at the cafeteria. We agreed he would

222

send the specimens to the city toxicologist, Dr Alexander O. Gettler, for an overnight analysis. I wanted to know if the blood contained methemoglobin. Methemoglobin is a compound that's formed only when any one of several drugs enters the blood. Gettler's report would tell us if we were at least on the right track. That is, it would give us a yes-or-no answer on drugs. If the answer was yes, then we could go on from there to identify the particular drug. How we could go about that would depend on what Pellitteri was able to turn up at the cafeteria. In the meantime, there was nothing for me to do but wait for their reports. I'd theorized myself hoarse.'

Dr Pellitteri, having attended to his bloodletting with reasonable dispatch, reached the Eclipse Cafeteria at around five o'clock. 'It was about what I'd expected,' he told me. 'Strictly a horse market, and dirtier than most. The sort of place where you can get a full meal for fifteen cents. There was a grind house on one side, a cigar store on the other, and the 'L' overhead. Incidentally, the Eclipse went out of business a year or so after I was there, but that had nothing to do with us. It was just a coincidence. Well, the place looked deserted and the door was locked. I knocked, and a man came out of the back and let me in. He was one of our people, a health inspector for the Bureau of Food and Drugs, named Weinberg. His bureau had stepped into the case as a matter of routine, because of the reference to a restaurant in the notification report. I was glad to see him and to have his help. For one thing, he had put a temporary embargo on everything in the cafeteria. That's why it was closed up. His main job, though, was to check the place for violations of the sanitation code. He was finding plenty.'

'Let me read you a few of Weinberg's findings,' Dr Greenberg said, extracting a paper from the folder on his desk. 'None of them had any direct bearing on our problem, but I think they'll give you a good idea of what the Eclipse was like—what too many restaurants are like. This copy of his report lists fifteen specific violations. Here they are: "Premises heavily infested with roaches. Fly infestation throughout premises. Floor defective in rear part of dining room. Kitchen walls and ceiling encrusted with grease and soot. Kitchen floor encrusted with dirt. Refuse under kitchen fixtures. Sterilizing facilities inadequate. Sink defective. Floor and walls at serving tables and coffee urns encrusted with dirt. Kitchen utensils encrusted with dirt and grease. Storage-cellar walls, ceiling, and floor encrusted with dirt. Floor and shelves in cellar covered with refuse and useless material. Cellar ceiling defective. Sewer pipe leaking. Open sewer line in cellar." Well . . .' He gave me a squeamish smile and stuck the paper back in the folder.

'I can see it now,' Dr Pellitteri said. 'And smell it. Especially the kitchen, where I spent most of my time. Weinberg had the proprietor and the cook out there, and I talked to them while he prowled around. They were very co-operative. Naturally. They were scared to death. They knew nothing about gas in the place and there was no sign of any, so I went to work on the food. None of what had been prepared for breakfast that morning was left. That, of course, would have been too much to hope for. But I was able to get together some of the kind of stuff that had gone into the men's breakfast, so that we could make a chemical determination at the Department. What I took was ground coffee, sugar, a mixture of evaporated milk and water that passed for cream, some bakery rolls, a five-pound carton of dry oatmeal, and some salt. The salt had been used in preparing the oatmeal. That morning, like every morning, the cook told me, he had prepared six gallons of oatmeal, enough to serve around a hundred and twenty-five people. To make it, he used five pounds of dry cereal, four gallons of water—regular city water—and a handful of salt. That was his term— a handful. There was an open gallon can of salt standing on the stove. He said the handful he'd put in that morning's otameal had come from that. He refilled the can on the stove every morning from a big supply can. He pointed out the big can—it was up on a shelf—and as I was getting it down to take with me, I saw another can, just like it, nearby. I took that one down, too. It was also full of salt, or, rather, some-thing that looked like salt. The proprietor said it wasn't salt. He said it was saltpetre—sodium nitrate—that he used in corning beef and in making pastrami. Well, there isn't any harm in saltpetre; it doesn't even act as an anti-aphrodisiac, as a lot of people seem to think. But I wrapped it up with the other loot and took it along, just for fun. The fact is, I guess, everything in that damn place looked like poison.'

After Dr Pellitteri had deposited his loot with a Health Department chemist, Andrew J. Pensa, who promised to have a report ready by the following afternoon, he dined hurriedly at a restaurant in which he had confidence and returned to Chatham Square. There he spent the evening making the rounds of the lodging houses in the neighbour-hood. He had heard at Mr Pensa's office that an eleventh blue man had been admitted to the hospital, and before going home he wanted to make sure that no other victims had been overlooked. By midnight, having covered all the likely places and having rechecked the down-town hospitals, he was satisfied. He repaired to his office and composed a formal progress report for Dr Greenberg. Then he went home and to bed.

The next morning, Tuesday, Dr Pellitteri dropped by the Eclipse, which was still closed but whose proprietor and staff he had told to return for questioning. Dr Pellitteri had another talk with the proprietor and the cook. He also had a few inconclusive words with

224

the rest of the cafeteria's employees—two dishwashers, a busboy, and a counterman. As he was leaving, the cook, who had apparently passed an uneasy night with his conscience, remarked that it was possible that he had absent-mindedly refilled the salt can on the stove from the one that contained saltpetre. 'That was interesting,' Dr Pellitteri told me, 'even though such a possibility had already occurred to me, and even though I didn't know whether it was important or not. I assured him that he had nothing to worry about. We had been certain all along that nobody had deliberately poisoned the old men.' From the Eclipse, Dr Pellitteri went on to Dr Greenberg's office, where Dr Gettler's report was waiting.

'Gettler's test for methemoglobin was positive,' Dr Greenberg said. 'It had to be a drug now. Well, so far so good. Then we heard from Pensa.'

'Greenberg almost fell out of his chair when he read Pensa's report,' Dr Pellitteri observed cheerfully.

'That's an exaggeration,' Dr Greenberg said. 'I'm not easily dumbfounded. We're inured to the incredible around here. Why, a few years ago we had a case involving some numbskull who stuck a fistful of potassium-thiocyanate crystals, a very nasty poison, in the coils of an office water cooler, just for a practical joke. However, I can't deny that Pensa rather taxed our credulity. What he had found was that the small salt can and the one that was supposed to be full of sodium nitrate both contained sodium nitrite. The other food samples, incidentally, were OK.'

'That also taxed my credulity,' Dr Pellitteri said.

Dr. Greenberg smiled. 'There's a great deal of difference between nitrate and nitrite,' he continued. 'Their only similarity, which is an unfortunate one, is that they both look and taste more or less like ordinary table salt. Sodium nitrite isn't the most powerful poison in the world, but a little of it will do a lot of harm. If you remember, I said before that this case was almost without precedent—only ten outbreaks like it on record. Ten is practically none. In fact, sodium-nitrite poisoning is so unusual that some of the standard texts on toxicology don't even mention it. So Pensa's report was pretty startling. But we accepted it, of course, without question or hesitation. Facts are facts. And we were glad to. It seemed to explain everything very nicely. What I've been saying about sodium-nitrite poisoning doesn't mean that sodium nitrite itself is rare. Actually, it's fairly common. It's used in the manufacture of dyes and as a medical drug. We use it in treating certain heart conditions and for high blood pressure. But it also has another important use, one that made its presence at the Eclipse sound plausible. In recent years, and particularly during the war, sodium nitrite has been used as a substitute for

225

sodium nitrate in preserving meat. The government permits it but stipulates that the finished meat must not contain more than one part of sodium nitrite per five thousand parts of meat. Cooking will safely destroy enough of that small quantity of the drug.' Dr Greenberg shrugged. 'Well, Pellitteri had had the cook pick up a handful of salt—the same amount, as nearly as possible, as went into the oatmeal—and then had taken this to his office and found that it weighed approximately a hundred grams. So we didn't have to think twice to realize that the proportion of nitrite in that batch of cereal was considerably higher than one to five thousand. Roughly, it must have been around one to about eighty before cooking destroyed part of the nitrite. It certainly looked as though Gettler, Pensa, and the cafeteria cook between them had given us our answer. I called up Gettler and told him what Pensa had discovered and asked him to run a specific test for nitrites on his blood samples. He had, as a matter of course, held some blood back for later examination. His confirmation came through in a couple of hours. I went home that night feeling pretty good.'

Dr Greenberg's serenity was a fugitive one. He awoke on Wednesday morning troubled in mind. A question had occurred to him that he was unable to ignore. 'Something like a hundred and twenty-five people ate oatmeal at the Eclipse that morning,' he said to me, 'but only eleven of them got sick. Why? The undeniable fact that those eleven old men were made sick by the ingestion of a toxic dose of sodium nitrite wasn't enough to rest on. I wanted to know exactly how much sodium nitrite each portion of that cooked oatmeal had contained. With Pensa's help again, I found out. We prepared a batch just like the one the cook had made on Monday. Then Pensa measured out six ounces, the size of the average portion served at the Eclipse, and analyzed it. It contained two and a half grams of sodium nitrite. That explained why the hundred and fourteen other people did not become ill. The toxic dose of sodium nitrite is three grains. But it didn't explain how each of our eleven old men had received an additional half grain. It seemed extremely unlikely that the extra touch of nitrite had been in the oatmeal when it was served. It had to come in later. Then I began to get a glimmer. Some people sprinkle a little salt, instead of sugar, on hot cereal. Suppose, I thought, that the busboy, or whoever had the job of keeping the table salt shakers filled, had made the same mistake that the cook had. It seemed plausible. Pellitteri was out of the office—I've forgotten where—so I got Food and Drugs to step over to the Eclipse, which was still under embargo, and bring back the shakers for Pensa to work on. There were seventeen of them, all good-sized, one for each table. Sixteen contained either pure sodium chloride or just a few inconsequential traces of sodium nitrite mixed in with the real salt, but the other was point thirty-seven per cent nitrite. That one was enough. A spoonful of that salt contained a bit more than half a grain.'

'I went over to the hospital Thursday morning,' Dr Pellitteri said. 'Greenberg wanted me to check the table-salt angle with the men. They could tie the case up neatly for us. I drew a blank. They'd been discharged the night before, and God only knew where they were.'

'Naturally,' Dr Greenberg said, 'It would have been nice to know for a fact that the old boys all sat at a certain table and that all of them put about a spoonful of salt from that particular shaker on their oatmeal, but it wasn't essential. I was morally certain that they had. There just wasn't any other explanation. There was one other question, however. Why did they use so *much* salt? For my own peace of mind, I wanted to know. All of a sudden, I remembered Pellitteri had said they were all heavy drinkers. Well, several recent clinical studies have demonstrated that there is usually a subnormal concentration of sodium chloride in the blood of alcoholics. Either they don't eat enough to get sufficient salt or they lose it more rapidly than other people do, or both. Whatever the reasons are, the conclusion was all I needed. Any animal, you know, whether a mouse or a man, tends to try to obtain a necessary substance that his body lacks. The final question had been answered.'

© Berton Roueché 1948.

---

## Old Dog, New Tricks

Contributed by Monica Harding (who was the lady from Liverpool) and extracted from *World Medicine* 11 February 1976, p. 102.

A lady from Liverpool related how something rather bizarre happened when she took the occupants of an old folk's home on a river cruise. Down in the loo, an octogenarian lady called Mildred suddenly decided she needed a tampon and put a coin in the machine . . . which promptly stuck.

Spellbound at the thought of medical history in the making, my correspondent handed her another coin; out popped the object in question. The old lady heaved a sigh of relief and gasped:

'Thanks, luv. I'm dying for a fag.'

Her friend, slightly older, interposed:

'They're not ciggies, Mildred. They're sweets—and you're on a diet so give them to me.'

Alas, history does not record the effect of this unusual confection on her gastrointestinal tract!

# A Barber Surgeon Operating in His Office

The plate with the indented rim, resting on the window ledge to the right, was held under the chin of his customers while he shaved them. The stuffed animal suspended from the ceiling was intended to keep away evil spirits. A stuffed alligator was usually sufficient for this purpose.

REFERENCE
Haggard H W 1929 *Devils, Dugs and Doctors* (London: Heinemann)

# The Indian Manner of Blood-letting

*A Central American native letting the blood by piercing the patient's arm with an arrow. From 'The new voyage and description of the isthmus of America' by Lionel Wafer, published in London, 1699. The illustration relates to Panama in 1681.*

*(Courtesy Wellcome Institute Library, London.)*

## Breathing a Vein

*1804 engraving by H Humphrey, London. (Courtesy Wellcome Institute Library, London.)*

---

## Rejects

The Chinese are certainly a sweet and sour lot. An English writer has just received quite the most elegant letter of rejection, from a Chinese economics magazine: 'We have read your manuscript with boundless delight. If we were to publish your paper, it would be impossible for us to publish any work of a lower standard;

'And as it is unthinkable that, in the next 1000 years we shall see its equal, we are, to our regret, compelled to return your divine composition and beg you a 1000 times to overlook our short sight and timidity.' Oh well, all right, since you put it like that.

REFERENCE
*Observer* 25 September 1983

## Setting Limbs

*Illustration from a thirteenth-century treatise on surgery which was translated from the Latin of Roger of Salerno. (Courtesy Mansell Collection.)*

# Physician's Tombstone

*The carving of a doctor about to make his diagnosis is from the family tomb of the famous Greek physician, Jason. (Courtesy Mansell Collection.)*

# French Proverb

Forty is the old age of youth but fifty is the youth of old age.

# Consultation with a Thirteenth-century Arab Physician

Medical learning was highly developed in all the chief cities of Moslem power, and in the eleventh century Arabian medicine began to be

*(Courtesy Mansell Collection.)*

*(Courtesy Mansell Collection.)*

233

known through translations in the western world. For example, in Sicily it was encouraged by King Frederick II, who as a youth had several Saracen teachers.

The two pictures are from fragments of an Arabic translation, dated 1222, of the *Materia Medica* of Dioscorides, a Greek physician who lived *c*. 90–40 BC. One of the Arabic paintings is of a consultation between doctor and patient whereas the other is of a clinical consultation involving several doctors.

---

## The Earliest Recorded Clinical Trial

Nebuchadnezzar II, having invested Jerusalem and defeated the Israelis in 600 BC, took several youths back to his own country for indoctrination and training. They were carefully selected. All were of royal or princely blood, social class 1 in fact. All were physically fit with a high IQ. They were to be put on a rigid diet of meat and wine for three years with one of the eunuchs acting as the monitor. Daniel persuaded the monitor to give him and three others a diet of pulse and water for ten days, when, it is recorded, they were fairer in countenance and fatter in body than the other subjects who were given meat and wine. Daniel had ruined the trial, the eunuch had defied the King, and the trial had become uncontrolled. It was not recorded what Nebuchadnezzar did to the eunuch.

REFERENCE
Book of Daniel ch. 1

---

## Malingering

In the following article, Richard Asher[1,2] has provided what is probably the best published essay on malingering:

I do not want my contribution to consist only of anecdotes about malingerers, so I shall not spend long recounting stories of ingenious patients who fooled the doctors or tales of clever doctors and how they caught the patient in the act.

The pride of a doctor who has caught a malingerer is akin to that of a fisherman who has landed an enormous fish; and his stories (like

234

those of fisherman) may become somewhat exaggerated in the telling.

Nor will I spend long on the details of technique: where the pins were hidden, how the blood was obtained, how the food was disposed of and how the thermometer was manipulated. These are matters of prestidigitation and are more worthy of consideration by the Magic Circle than by the Medical Society of London.

I am not going to confine myself to pure malingering, but I shall also discuss the borderland of malingering which provides even more difficult and obscure cases for study.

True malingering is defined in the *Concise Oxford Dictionary* as the pretending production or protraction of illness in order to escape duty, especially in soldiers and sailors. I allow a wider meaning to the term. To start with, I do not exclude the Air Force, especially as I remember an aircraftsman treated by Squadron Leader Harry Gosset who was a genuine haemophiliac and required over twenty pints of blood to stop the bleeding from his appendicectomy wound. This poor fellow, believing he was dying and wishing to confess his sins, admitted that he had simulated his appendicitis to get home for Christmas. He said he would never have done this had he known there would be so much bleeding trouble.

I define malingering as the imitation, production or encouragement of illness for a deliberate end. The patient is quite conscious of what he is doing and quite cognisant of why he is doing it. With that definition, pure malingering—the planned fraudulent faking of illness —is, in my experience, a very rare condition. Either that, or else I am a very gullible physician. I know I have been mistaken before now and it is possible that many malingerers have deceived me without being suspected. I shall show you the portrait of a case where I was grossly deceived. It was over a matter of malingering associated with hysterical ataxia, scissor gait and aggressive outbursts.

One Sunday, when my daughter was two years old, I promised my wife that, if it was not taken as a precedent, I would myself get her up from her afternoon rest, dress her and take her for a walk. I performed these duties without difficulty or loss of dignity, until the walk started. Then there was trouble. The child kept falling to the left; she walked with a ridiculous scissor gait and she frequently fell to the ground. She cried and said she had a pain. She behaved abominably, and I spent a wretched afternoon in Park Square West attempting to coax her into good behaviour. I knew this was sheer devilment, a malignant aggressive demonstration against the father figure; I would not submit.

At last my wife returned and undressed her for her evening bath. There was a sudden cry:—'Do you realise you've put both her legs through the same hole in her knickers?' I can still remember, after those tortured limbs had been freed from the crippling garments, how that gay, naked figure raced unrestrictedly to the bathroom without a trace of malingering. That incident taught me to be cautious about diagnosing malingering or hysteria. These diagnoses must not be made for the sole reason that the clinical picture is not yet hung in the clinical picture gallery of the doctor in charge. It may be something he hasn't heard of. There are too many examples of apparent malingering turning out to be cases of organic disease, and jokes about the high mortality of malingering or hysteria are commonplace.

I could show you a picture of a lady who had a lesion on her face and other lesions on her fingers and elsewhere. For a long time they were thought to be self inflicted, but biopsy studies have left no doubt that she is suffering from Phagaedena Geometricum of Von Brocq, and the control of the lesions by cortisone has brought great relief to the patient and to her doctor.

A young National Serviceman, who complained for over a year of peculiar abdominal pains, was thought to be malingering, but in the end he was found to have Von Recklinghausen's neurofibromatosis of the intestine, and several members of his family were discovered to be afflicted by the same rare condition.

Nevertheless, genuine authenticated malingering does occur. True malingering is best classified by motives rather than by techniques—the principal prime movers being Fear, Desire and Escape.

*Fear:* fear of call-up, fear of overseas duty, fear of warfare.

*Desire:* desire for compensation, desire for a comfortable pension, desire for revenge against a surgeon for some (usually imagined) wrong. Desire for the comforts of hospital life. Desire to stay in the ward longer because one has fallen in love with the staff nurse.

*Escape:* escape from a prisoner-of-war camp by uncurable disability, escape from prison by transfer to hospital, escape from an impending court case, escape from battle.

Malingering by prisoners of war has evolved a variety of ingenious techniques, even to the extent of passing borrowed albuminuric urine, secreted in a false bladder and passed in the presence of a suspicious German doctor, through a hand-carved and hand-painted penis of life-like verisimilitude. The technical term for such realistic art is *trompe l'oeil,* an expression usually applied to oil painting but presumably correct for sculpture and wood carving.

It is well known that opposing forces try to weaken the enemy's army by dropping pamphlets persuading them to malinger. A particular pamphlet dropped in large numbers early in 1945 on English troops in Italy is worthy of your attention. Neatly produced in bookmatch form (to make it easy to hide), it opens with Three Golden Rules for Malingering which I do not think could be beaten:

236

1. You must make the impression you hate to be ill.
2. Make up your mind for one disease and stick to it.
3. Don't tell the doctor too much.

This pamphlet catered for all tastes and resembles the 'Do it Yourself' guides so popular today.

There is only time to give details about one of those. Here is how to have tuberculosis—according to the instruction book:

'First you must smoke excessively to acquire a cough. Then tell the doctor that you have lost weight, you do not feel well and that you cough a great deal. Say that sometimes you cough up streaks of blood. Sometimes you wake drenched with sweat. Stick to these symptoms, do not invent any new ones.

'Mix a very little blood with your sputum; suck it from your gums or a prick on your finger, one or two drops are sufficient.' Then there follow certain supplementary techniques of which I propose to omit the details. They concern the correct use of the smegma bacillus to enrich the bacteriology of the sputum. They are not attractive. The instructions finish with these encouraging words: 'All the above directions are based on scientific observations in well-known University Hospitals.'

The pamphlet was printed in the *Lancet* who gave it rather a discouraging review as follows: The authors would be disappointed to see how lightly their elaborate propaganda is received by our soldiers. They say they know far better tricks than these!

At Cardiff last week an ex-prisoner-of-war, Dr Archie Cochrane, told me a technique (he called it 'Fudging' X-rays) whereby he annually returned three healthy prisoners to England totally disabled with radiological tuberculosis.

Painting in barium requires delicate brushwork very different from painting in oils. The technique of Van Gogh is useless for painting tuberculosis cavities however suitable it may be for sunflowers.

The general run of malingerers are of poor intelligence, and naivety rather than ingenuity is the rule. Among ingenious and distinguished malingerers Sherlock Holmes is worth recalling. In 'The Case of the Dying Detective' Holmes had to convince Watson that he was dying of an obscure tropical disease. He started by reducing Watson's limited faculties to their lowest ebb:

'Facts are facts, Watson. You are only a general practitioner with very limited experience and mediocre qualifications. It is painful for me to have to say these things but you give me no choice. Shall I demonstrate your own ignorance?

'What do you know, pray, of Tapanuli Fever? What do you know of the Black Formosa Corruption?'

Mr. President, I do not wish to discomfort you in front of this

237

distinguished audience, but can you improve on Watson's answer: 'I have never heard of either'?

After this preliminary humiliation Holmes had little difficulty in convincing Watson, who in his turn convinced the villain—Culverton Smith—that he, Holmes, was dying. At the end of the case when Culverton Smith was in handcuffs and Holmes had revived himself with claret and biscuits, he described his methods: Three days of fast does not improve one's beauty. Then, with Vaseline upon one's forehead, belladonna in one's eyes, rouge over the cheek-bones and crusts of beeswax round one's lips, a very satisfying effect can be produced. A little occasional talk about half-crowns, oysters, or any other extraneous subjects produces a pleasing effect of delirium.

Mention of the imitation of delirium reminds one that mental as well as physical disease can be simulated. Those with little experience of mental disease may learn with surprise that it is very hard to pretend to be mad. For instance, the peculiar distorted thinking of the schizophrenic is something a sane person cannot manage. An experiment was done in which twenty normal people were asked to feign insanity; they, and twenty genuinely psychotic patients, were interviewed by psychiatrists (who had no other means of telling which was which). The psychiatrists were able to pick out the malingerers in nearly ninety per cent of the cases. Yet the most remarkable feat of feigning insanity was carried out by Lieut. Jones and Lieut. Hill who were prisoners in Yozgad, Turkey, in 1917. They were carefully coached by a fellow-prisoner who was a doctor and they went through almost unendurable privations in their successful pretence of madness. Their book, *The Road to Endor*, is still a best seller and a classic.

I now pass to those cases of illness which, though self produced or prolonged, do not constitute malingering. They do not have so definite a purpose. As you see, I have grouped them together as The Borderland of Malingering.

*The Advantages of Illness*

1. Illness as a Purpose (True Malingering)
2. Illness as a Comfort (Hysteria)
3. Illness as a Hobby (Hypochondriasis)
4. Illness as a Profession (Psychopaths)

} The Borderland of Malingering

Hysterics differ from malingerers because, although they may produce illness and enjoy it, they are unaware of what they are doing. They possess a capacity for self-deception; they can wall off part of their mind so that it is impervious to self scrutiny. This process, dignified by psychiatrists with the term 'dissociation', is coloquially called 'kidding yourself'. Some cases of hysteria are very close to malingering. Others start as malingerers, and as they become better at kidding

238

others, they finally succeed in kidding themselves, and become hysterics.

### The Borderland of Malingering

*A. Illness as a comfort*
- (*a*) Hysteria.
- (*b*) The Proud Lonely Person (Lucy's disease).

*B. Illness as a hobby*
- (*a*) The Grand Tour Type (rich hypochondriac).
- (*b*) The Chronic Out-Patient (poor hypochondriac).
- (*c*) The Eccentric Hypochondriac (faddist).
- (*d*) The Chronic Convalescent (daren't recover).

*C. Illness as a profession*
- (*a*) Anorexia Nervosa.
- (*b*) The Chronic Artefactualists.
- (*c*) Munchausen's Syndrome.

I have seen very little proven hysteria and I diagnose the condition only with diffidence. A fair proportion of 'hysterics' turn out to have organic disease, as many of us know to our cost.

Notice that among illness as a comfort I have put the proud, lonely person. Allow me to explain this. This pathetic type of case usually occurs in later life when praise and companionship are hard to come by. To lonely people a medical consultation may represent an event of great importance. It supplies that need to be noticed that exists in all human beings. A child cries: 'Look at my sandcastle!' A lonely old person cries: 'Look at my stomach.' The child says: 'I got two goals this afternoon!' The lonely old person says: 'I got two giddy turns this afternoon.'

I have heard from more than one practitioner of elderly people who have a weekly or fortnightly consultation of this kind. A visit from the doctor allows them the illusion of seeking medical advice rather than companionship. A patient may be too proud to complain of loneliness, but there is no loss of pride in complaining of symptoms. Lonely people miss, not only companionship, but also the advice and criticism that go with it. Under the guise of seeking advice about health, a lonely lady may be seeking advice about family affairs. Ostensibly she is asking for advice about her bad heart, but *au fond* she seeks advice about her bad nephew. This mild and most benign form of malingering, in which solitary elderly people look for company or counsel under the thin disguise of chronic ill health, is understandably tolerated or even enjoyed by quite a lot of practitioners. It could well be called Lucy's disease after Wordsworth's· Lucy:

> She dwelt among the untrodden ways
> Beside the springs of Dove,
> A maid whom there were none to praise
> And very few to love.

Turning to the hypochondriacs, first we have the rich hypochondriac. I call this one The Grand Tour Type, because she spends much of her time touring the larger cities in Europe visiting consultants. She always carries a large dossier, opinions from consultants, X-rays and laboratory reports; and usually a list of her own symptoms which she has carefully written out. During her tour she may have persuaded surgeons to remove some of her less essential organs. She has usually had her gall bladder and a quota of her pelvic organs removed by the time she reaches one's consulting room. Such women discuss consultants with their friends as one recommends a fashionable tailor or a smart hairdresser, and rival each other in the grandeur of the consultants they have visited as well as in the complexities of their investigations and treatments. They are usually married to rich but unattentive husbands who supply them with money for their hobby as a substitute for affection. To the consultant in private practice they are a familiar, tedious and lucrative burden.

The poor hypochondriac (or perpetual out-patient). Every hospital has a number of out-patients who have attended for many years. Whenever they are discharged from one department they turn up in another, thus acquiring a very large collection of documents, rivalling that of the rich hypochondriac although written on less luxurious writing paper and penned by less illustrious names. Though some have genuine chronic illness, many of them attend because they like the companionship of hospital; instead of going to the local public house for a glass of beer and a chat with the landlord, they go to the hospital for a bottle of medicine and a chat with the other patients. One enlightened doctor tried the experiment of arranging out-patient sessions where the patients did not see the doctor at all unless they asked for him. It was a great success.

The eccentric hypochondriacs. These people like peculiar or unorthodox treatments. They like so-called 'natural' things and belong to the homespun, wholemeal, handicraft school of living. They know a clever little man who can find out what is wrong with you by holding a pendulum over you or using some form of magic box. They believe with apostolic fervour in nature cures, osteopaths, astrologers and herbalists. Their preoccupation is more with treatment than with illness and they are harmless and often entertaining.

The chronic convalescent. When a patient has had longstanding organic disease for many years it becomes so familiar to him that it is almost a friend. If the illness is suddenly cured, he may feel deserted

240

and friendless. He misses the familiar pain, the sympathetic enquiries of his friends and the security of his medical routine. He feels like a man who has worn shoes so long that he can no longer go barefoot. He does not really want to get well; he has become a hypochondriac.

Now the last group: Illness as a profession. First, I consider anorexia nervosa. The reason why these people go to such lengths to avoid eating is rarely clear. They will resort to a variety of artifices to avoid food. They will hide food in their bed lockers, pour milk into their hot-water bottles and insist on starving in the midst of plenty. As a result of the starvation, amenorrhoea is an early symptom and this formerly led to the illness being confused with Simmonds disease, with which it has nothing to do.

The next group is that of the chronic artefactualists, who may spend years in self-mutilation or the production of spurious fevers. Skin diseases are most favoured by the sufferers from chronic autogenous disease, but various other forms of self-damage are reported. A group of nineteen patients was reported by workers in the Psychological Medicine Department of the Birmingham Hospitals under the title 'Deliberate Disability'. Their most surprising finding was that most of these patients are highly intelligent. One possibility is that their intelligence prevents them from learning the art of hysterical dissociation. One young woman was referred to me by Dr Harold Wilson. She was a bit of a problem child; she had been more or less bald for several years and he suspected the condition was due to trichotilomania (dermatologese for hair-pulling). He suggested that hypnosis, which is a part-time activity of mine, might stop her. At our first interview I asked her if it was possible she was occasionally pulling hairs out of her scalp. She said, yes, she thought she was. I then said as it seemed to be making her bald it might be a good idea to stop pulling her hair out. She said, all right, she would try. That was the only treatment given. I never used hypnosis on her at all and yet she had a normal head of hair within three months.

Lastly, Munchausen's syndrome. This is the strangest and rarest form of chronic autogenous disease. I described and named the syndrome in a paper in the *Lancet* seven years ago, but it had been recognised for generations; all I did was to provide a convenient label.

*Munchausen's Syndrome*
Three main types

A. Laparotomophilia migrans.
   With mainly abdominal symptoms.
B. Haemorrhagica histrionica.
   Colloquially—'haemoptysis merchant,' 'haematemesis merchant.'
   Largely specialising in haemorrhage.
C. Neurologica diabolica.
   Specialising in faints, fits, stupor, etc.

The patient with this syndrome is nearly always brought into hospital by police or bystanders, having collapsed in the street or on a bus with an apparently acute illness, supported by a plausible and yet dramatic history. Though his history seems most convincing at first, later his story is found to be largely false, and his symptoms and signs mostly spurious. He is discovered to have attended and deceived an astounding number of other hospitals. At several of them he may have been operated upon, and a large number of abdominal scars is often found. So skilfully do these people imitate acute illness that the diagnosis may be quite unsuspected until a passing doctor, ward sister or hospital porter says 'I know that man—we had him in St Quinidines last September. He says he's an ex-fighter pilot shot in the chest, in the last war, and he coughs up blood; or sometimes he's been shot through the head in the last war and has fits.'

These people differ from other chronic artefactualists in their constant progression from one hospital to another, often under a variety of false names, but nearly always telling the same story. They rarely stay at one hospital for long, but, if not already ejected, discharge themselves within a few days and off they go on their wanderings, deceiving one hospital after another, telling the same false story, faking the same fictitious symptoms and submitting to innumerable operations and investigations.

Almost all of you have encountered examples of this, but I still say it is rare. It seems common because the afflicted (or afflicting) patients never stop working for a moment. One patient may visit over fifty hospitals in a year. With Munchausen's syndrome a little goes a long way, or, as Churchill might put it: Never in the history of medicine have so many been so much annoyed by so few.

The term Munchausen's syndrome is not beyond criticism and various other names, such as Hospital Hoboes, have been suggested, but the term is convenient and seems to stick.

The gentleman to whom I respectfully dedicated this syndrome originated as a real person—Hieronymus Karl Frederick Von Munchausen of Brunswick—who was wont to entertain his friends by recounting fabulous adventures of palpable absurdity. When Rudloph Raspe of Hanover anonymously published these stories without his permission, a gentlemanly old fellow with a gift for tall stories became the laughing stock of Europe and the eponymous representation of an incorrigible liar. It is to Raspe's fictitious Baron that my term refers; the patients resemble him in the dramatic nature of their stories, the wide extent of their travels and the untruthfulness of their tales.

I must tell you one of Baron Munchausen's stories, for they are truly splendid. The Baron tells how he was out hunting one day and, just when he had used up all his shot, he met a huge stag. In this emergency he charged his gun with cherry stones and let fly, hitting the animal in the forehead. The stag was momentarily stunned, staggered a few paces and made off into the forest. Two years later he was out hunting again when he encountered a particularly noble stag. Luckily on this occasion he was better equipped with ammunition and, after despatching the beast, he lunched off haunch of venison with cherry sauce, for there was a flourishing cherry tree growing from the beast's head.

The stories told by Munchausen patients are equally untrue, but less inherently unlikely; indeed it is the credibility of their stories that makes these patients such a perpetual and tedious problem. It seems that nothing can be done to prevent their continuing clinical depredations. Most doctors are so pleased if they succeed with detection and ejection they never think about protection. Though serious psychiatric studies of these people have been made, nobody can yet answer the two fundamental questions:

(*a*) Why do they do it?
(*b*) How can we stop them doing it?

It is lucky there are so few of them.

All that can be said about them, and indeed about the whole subject of malingering, are these words of Robert Burns:

But human bodies are such fools
For all their colleges and schools
That when no real ills do perplex them
They make enough themselves to vex them.

Shall we leave it at that then? The best explanation for all this self-manufactured disease is simply this: That human beings are such fools.

REFERENCES
[1] Asher R 1958–9 *Trans. Med. Soc.* **75**
[2] Asher R 1972 *Richard Asher Talking Sense* pp 145–55 (London: Pitman Medical). By permission of Pitman Publishing Ltd, London

---

# Milk Transfusion

An alternative to blood transfusion was tried in Toronto in 1854 during a cholera epidemic: transfusion of cow's milk. The underlying rationale was that milk would turn into white blood cells. Of the seven patients treated by intravenous transfusion of 12 ounces of milk, five died but two allegedly got better. Twenty years later in New York, intravenous sheep milk infusion was used as a treatment for terminal tuberculosis for two patients. One died in four hours and the other in 24 hours. Seven more patients were treated in New York between 1875 and 1879, by a Dr T G Thomas who claimed that fever, tachycardia and headache were all improved after a while, and predicted a 'brilliant and advantageous future' for intravenous milk infusion. Some enthusiasts even went so far as to predict that milk would eventually replace blood completely in transfusion work.

REFERENCE

Sterpellone L 1986 *Instruments for Health* Farmitalia (Carlo Erba: Freiburg i. Br.) pp 90–2

---

## Cow Pearl Jam

From the Khan el Khalili bazaar, Cairo, 1983.

Some 50 years ago, Egyptian ladies were only considered to be beautiful if they were fat and to enhance this process they were given Cow Pearl jam whose ingredients are given below, together with the Arabic marketing slogan.

COW PEARL JAM

Honey
Aleppo pistachios
Iraqi dates
Carbohydrates (probably flour!)
Shelled almonds
Moroccan sage
Gall-stones from the cow (this is the active ingredient!)

For vivaciousness, beauty and anti-slimming and anti-cachexia for women, men, girls and boys supposed to give you an appetite.

---

## Gin Mania in London

Gin was first developed in 1650 by Franciscus Sylvius, professor of medicine at the State University of Leyden in Holland. To produce a medicine with diuretic properties, he distilled the juniper berry with spirits derived from fermented barley. The French name for juniper, *genièvre*, was the origin of the English name, gin, for the resultant alcoholic refreshment which also became known by the term Mother's Ruin. It was introduced in England early in the eighteenth century by soldiers returning from the Low Countries, and it rapidly become a national addiction. This was not because of its taste, though, but because of vested interests. Beer and ale had been taxed since 1643, but gin was not because its production gave farmers a market for cereal at a time when prices were low and because of the political power of the distillers.

Between 1715 and 1750, gin mania reached epidemic proportions in London, a city which in 1736 contained more than 7000 gin houses, representing one house in every six. During this period it was blamed for the occurrences of more deaths than births in London, with the greatest mortality among children. Burials of those younger than five years represented some three-quarters of all children who had been christened.

It was in London that a strong counter-reaction began to the gin mania, primarily in the arts and letters of the time. This alcoholic way of life and its incongruities were held up for ridicule by writers such as Henry Fielding in his novels *Joseph Andrews* and *The History of Tom Jones*, by composers such as John Gay in *The Beggar's Opera* and by artists such as William Hogarth in engravings. Fielding, who became a magistrate in Westminster, quoted:

> What must become of the infant who is conceived in gin with the poisonous distillations of which it is nourished both in the womb and at the breast?

and a statement of the overall problem was made by the Middlesex Sessions Committee in 1735:

> Unhappy mothers habituate themselves to these distilled liquors, whose children are born weak and sickly, and often looked shrivel's and old.

William Hogarth (1697–1764) attacked the social evils of his day with graphic depiction of the reality of London society in the first half of the eighteenth century. One of his most effective and best known satirical engravings is *Gin Lane*, published in 1751. It is one of his many allusions to the detrimental effects of alcohol on society and certainly his most comprehensive and biting. According to Hogarth himself:

> In Gin Lane every circumstance of its horrid effects are brought in view, in teorem nothing but Poverty misery and ruin are to be seen Distress even to madness and death, and not a house in tolerable condition but the Pawnbrokers and the Gin Shop.

The overall impression in *Gin Lane* is that of total disintegration of society through the depiction of general tumult, disorderliness, collapse of buildings, and agitated persons and groups. The period of 1700–50 was a time of severe hardship for the poor in England and of disregard to public health measures. Prostitution and venereal disease were rife. Robbery, brutality and murder were common and hanging was an acceptable punishment for theft of as little as a shilling. These elements of dissipation and callousness were heightened in London and alcohol became almost a necessity for lower-class Londoners of the eighteenth century to tolerate the cold, the abject poverty, the filth and the hopelessness of their environment. Gin, because it was cheap, became the opiate of the massess who spent their meagre income for solace rather than for staple food.

More specific effects of alcohol are depicted in *Gin Lane*. On the left a carpenter is pawning the tools of his trade for gin. In the right upper corner, the suicide by hanging is that of a barber, as indicated by the adjacent barber's pole, In the right lower corner, an emaciated and dying man is still clutching his bottle of gin. The centre background of the engraving is highlighted by the sign of an undertaker, a hanging coffin. Also depicted are several detrimental effects of alcohol on infants of gin addiction. Close to the undertaker's sign, a female body is being lowered into a coffin. Besides the coffin is a neglected infant. To the right of this pitiful tableau is a more macabre event. A mad-

246

*Gin Lane by William Hogarth. (Courtesy Wellcome Institute Library, London.)*

dened alcoholic has skewered a live child on his staff. A less violent event, but still child abuse, is in the right central margin of the engraving. A woman is forcing gin into an unwilling infant. Immediately above this, a consequence is seen in the form of two schoolgirls who are voluntarily drinking gin. The most dominant figure is in the foreground of the engraving—an unkempt, stupified woman whose unattended child is falling to his death. The facial appearance of the child is worthy of note. The eyes have a shorter than normal palpebral fissure, resulting in relatively round 'Orphan Annie' eyes. These are one of the effects of maternal alcohol on fetal morphogenesis as described in a paper in the *Lancet* of 1973[1].

REFERENCES

Text adapted from Rodin A E 1981 'Infants and gin mania in 18th century London' *J. Am. Med. Assoc.* **245** 1237–9

[1] Jones K L *et al* 1973 'Pattern of malformation in offspring of chronic alcoholic mothers' *Lancet* **1** 1267–71

# Nicholas Copernicus and the Inception of Bread-Buttering

During the first quarter of the sixteenth century, the small principality of Ermland was the scene of frequent and terrible devastation. It bordered the Gulf of Danzig and was one of four dioceses into which Prussia had been divided for purposes of ecclesiastical government. The bishop of Ermland, Copernicus' uncle, was frequently engaged in conflict with the Order of Teutonic Knights which had formerly ruled Ermland and they periodically launched military expeditions to regain supremacy. There were three fortified towns in the bishopric and one of these was Allenstein, on the River Alle. Copernicus, appointed to the sinecure of canon of Frauenberg Cathedral by his uncle, was also commandant of Allenstein Castle.

The final attempt of the Teutonic Knights to reassert suzerainty occurred between the years 1519 and 1521, and in 1520 they beseiged Allenstein Castle which was considered to be a strategic enclave impregnable to any army with such limited resources as those possessed by the Teutonic Knights.

Six months prior to the lifting of the siege, plague struck within the walls of Allenstein Castle.

Much of the renowned astronomer's reputation at the time rested upon his skill as a physician. As a youth he had studied medicine at the University of Padua, and, although never obtaining a medical degree, he did acquire a professional proficiency reputed to be well above the level of his contempories. The specific nature of the Allenstein plague cannot now be identified, but it was a moderately serious affair with much morbidity, although only two fatalities.

Copernicus, who at first contented himself with prescribing routine treatments, became disquieted when in many instances the disease either persisted in spite of his therapy, or reappeared in cases which he regarded as cured. Rather than limit himself to ineffective treatment he decided to search out and define its aetiology and pathogenesis. News of this rather basic approach soon spread beyond Ermland and travelled as far west as Leipzig where it reached the ears of Adolph Buttenadt, who had known the young Copernicus in Padua and who had served as Copernicus' sponsor for memberhsip of the Guild of Apothecaries and Physicians.

Buttenadt, unlike Copernicus, did earn a degree in medicine but hardly ever practised. Instead he dedicated himself to informational and organisational tasks in the very powerful physicians' guild and became, in the highest sense of the word, a professional propagandist.

NICOLAS COPERNIC
Né a Thorn le 19 Février 1473. Mort a Warmie
le 24 Mai 1543.

Paris chés Odieuvre m.d d'Estampes quai del'Ecole vis à vis la Samarit à la belle Image c.p.

*Before Copernicus the Ptolemaic concept of the universe was that all the planets orbited in perfect circles around the Earth. With Copernican theory, though, the Sun was at the centre of the universe, and not the Earth. He published this idea at the end of his life and received the first copy of his book* De Revolutionibus Orbicum Coelestium *(On the Revolutions of the Heavenly Spheres) on his deathbed. Legend has it that when Copernicus read the Preface, he collapsed and died. This had been written by one of his disciples, Andreas Osiander, whom he had not bothered to mention in his own Introduction to the book. Osiander repaid this by including an unauthorised Preface explaining that what followed was pure hypothesis and need not be taken too seriously. To add insult to injury, Osiander had not even signed the Preface and so it looked as if it had been written by Copernicus himself! (Courtesy Wellcome Institute Library, London.)*

249

By 1517 he had become Grand Master of the Apothecaries of Medieval Allegmagne, which was the official name of the Northern European chapter of the Guild of Apothecaries and Physicians.

The news of Copernicus' research activities disconcerted the Grand Master and resulted in his visit to Allenstein to express his concern about events. He hoped that the reports had no basis in fact, but not only did Copernicus confess that they were true, but also revealed that he had actually discovered the cause of the Allenstein plague and had instituted measures to prevent its recurrence. Buttenadt was astonished and bewildered by this new development and contended that such activities could undermine the entire contemporary ethos of medical science. The Grand Master did not allow that in attempting good works man necessarily served as a beneficient instrument. Although he could appreciate Copernicus' noble intentions, he also suspected that these might have inadvertently altered Nature's balance and thus redirected the will of God. The latter possibility was the chief basis for his concern, which if true, he feared could occasion divine retribution. Buttenadt admitted that his fears, if put into operation, might act to suppress originality in medical science but he was only an advocate of cautious innovation.

Copernicus protested that this affair hardly justified such apprehension and to assure Buttenadt that his ministerings could have no significant impact beyond the castle, he described the activities leading to his discovery.

Suspecting some correlation between the plague and the food available at the castle, he divided the inhabitants into separate groups each with special diets. Before long it had become apparent that the small group denied bread was the only one free from plague. If it had been plausible, Copernicus would have dispensed with all use of bread, but this was difficult in a besieged castle. Christian charity had required that only those Allenstein inhabitants too old or infirm to provide for themselves be domiciled within the castle and that all able-bodied non-military residents were to withdraw from the vicinity. As a result the castle population consisted only of military personnel directly involved with defence or enfeebled peasants. The labour shortage meant that even the latter were pressed into some kind of service.

The most difficult function for this elderly and infirm group was that of occasional waiter! The military were not permitted to leave their posts during their 12-hour tours of duty and their meals during this period were brought to them by the peasants. They lacked the stamina for such work, climbing from the kitchens to the high turrets where the guards were stationed, up narrow stairways for almost 200 feet above ground level. Items of food were often dropped and though probably brushed free of the more obvious contamination were nevertheless delivered for consumption by the guards. The food most frequently dropped were the Allenstein baked bread, which were large, coarse, black loaves, which could collect substantial amounts of foreign debris without noticeable discolouration. The diet of all castle inhabitants consisted in large part of this bread.

It was then suggested to Copernicus as commander of the castle, by one Gerhard Glickselig, that the bread loaves should be coated with a thin layer of an edible light-coloured spread. After such treatment, any foreign matter adhering to fallen loaves would then be readily visible and potential plague victims would be able to protect themselves by removing the contaminated material. A decision was made to coat all bread with a spread of churned cream. This subsequently resulted in the elimination of all plague from the castle.

Buttenadt regarded the localised nature of Copernicus' clinical success as sufficiently mitigating to dispense with any necessity of discipline at his unprofessional behaviour (in the sixteenth century!). However, it was agreed that the incident should remain unpublicised.

Copernicus died in 1543, and two years after his death the Schmalkaldic War erupted and Buttenadt, who was still alive and by then executive secretary of the Guild of Apothecaries and Physicians, was appalled at the backwash of disease and pestilence which appeared in the wake of the wars throughout Europe at that time. He was desperate to prevent the rapid spread of plagues, particularly those of the Schmalkaldic War, and now broadcasted the details of the late Copernicus' Allenstein cure. As a result, the system of coating bread was rapidly adopted throughout Europe. There is no evidence that this worked in any plague other than at Allenstein but nevertheless, once adopted, the coating practice proved to be surprisingly durable and was eventually integrated into the dietary habits of the vast majority of Europeans. As an apparent tribute to its most effective proponent, it was described as *Buttenadting* which today is called *buttering*.

REFERENCES
Adapted from Hand S B and Kunin A S 1970 *J. Am. Med. Assoc.* **214** 2312–15. Additional biographical information on Copernicus from Wilson C 1980 *Starseekers* (New York: Doubleday).

---

## Irish Proverb

The beginning of health is sleep.

# Medicinal Uses of Wine

The oldest known record of the medicinal use of wine is an inscribed clay tablet excavated from the ancient Sumerian city of Nippur, south of the supposed site of Babylon. Now stored in the University of Pennsylvania Museum at Philadelphia, the tablet describes drugs made with wine. Archaeologists believe the tablet dates back to about 3000 BC.

Later, wine was a principal medicine in Greece. Homer's *Iliad* says that Machaon and Podalirius (physician sons of Asklepios, the Greek god-physician) administered wine to the wounded heroes of the Trojan War. Hippocrates of Cos (460–370 BC) also made extensive use of wine, prescribing it as a wound dressing, a cooling agent for fevers, a purgative, and a diuretic.

© American Medical Association 1977

REFERENCE
*J. Am. Med. Assoc.* 1977 **238** 464

---

# Onion Incantation to Exorcise Sickness

The tablets of the Babylonians and Assyrians have preserved for posterity a long list of exorcising formulae in which various substances like onions, dates, palm clusters, bits of sheep's hide and goat skin and coloured wool are introduced to be peeled or torn to pieces, and as each bit is thrown into the fire an incantation is recited. The rite is purely symbolical and in the case of the onion is the following:

> As this onion is peeled and thrown into the fire,
> Consumed by the glowing fire-god,
> Never to be planted again in a garden,
> Never to be harrowed, never to take root,
> Will never again be placed in the ground,
> Its stalk will never grow, will never see the sunlight again,
> Will never come onto the table of a God or King,
> So may the ban curse, pain and woe,
> Sickness, groans, injury, sin, misdeed and transgression,
> So may the sickness in my body, in my flesh, in my limbs
> As this peeled onion, be consumed by the glowing fire-god.

REFERENCE
Jastrow M 1917 'Babylonian-Assyrian Medicine' *Annuals of Medical History* **1**
    236

## *The Oyster's Consolation*

(Lines suggested by an antivivisectionist's appeal against the cruelty
of eating oysters alive.)

I cannot laugh, I cannot wail, I cannot even droop a tail;
I cannot say how much I feel, I cannot squirm, I cannot
    squeal.
No eye have I for mute appeal, no tears a-down a cheek to
    steal;
No brow to dew with anguished sweat, no ear to hear, no mind
    to fret.

Sprinkled with pepper black or red, I calmly lie as if in bed;
I cannot even raise a quiver when forks are bedded in my liver.
But ah! I have a dainty taste, a palate all around my waist;
So all day long to feed I strive, and swallow all my food alive.

Can I complain if hungry man adopts this very simple plan,
Meting to me the self-same measure by bolting me alive with
    pleasure?
Yet, pity me not with frantic screech, the opening knife my
    vitals reach—
Swallowed by epicure or glutton, I am stone dead, as dead as
    mutton.

REFERENCE
Rayner H 1920 *Br. Med. J.* **2** 262

## *Biology Exam Howlers*

Edward Jenner was alive when smallpox was a popular disease.

Fertilization takes place in the Eustachian tube.

Sewage has been in existence since Roman times.

The female uterus is hidden by the pelvic gurgle.

Flies cannot live at a height of more than six feet underground.

Unwanted pregnancy is a sexually transmitted disease.

The function of the ciliary muscles is to focus rays of light onto the rectum.

TB patients should be given special care. His mistress and pillows should be put into the sun and washed with boiling water.

REFERENCE
*Biologist* 1982, Volume 29, p 203

---

# Illustrations from a Twelfth-Century Medical Treatise

Latin translations are courtesy of Dorothy Mould.

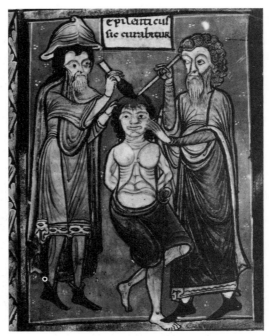

*Trepanning the skull, possibly for epilepsy. (Courtesy Mansell Collection.)*

*Diagnosis of stomach pains. In the twelfth century the anatomical position of the stomach would not necessarily be known. (Courtesy Mansell Collection.)*

*The language is medieval Latin and this illustration seems to refer to pain related to swellings in the armpits and possibly to breast cancer. (Courtesy Mansell Collection.)*

255

emoroida inci
ditur sic.

fungus
denare
sic inci
ditur.

The top part of the illustration demonstrates the method of surgical treatment of haemorrhoids, whereas the bottom part obviously refers to nosebleeds. (Courtesy Mansell Collection.)

albule oculorum sic extu cuciuntur.

Annointing the whites of the eyes. (Courtesy Mansell Collection.)

# Anecdote of an Australian Professor of Physiology

Before the Second World War, all students had to pay their lecture fees to attend the university and if they failed to attend lectures, they were plainly stupid. But after the war, when the Commonwealth Reconstruction and Training Scheme was paying the fees for returned ex-servicemen and not only paying their fees, but living expenses in addition, many students took second jobs in the city and did not attend their lectures. So, the Commonwealth Reconstruction and Training Scheme offices requested the university to get the various schools at the university to call the roll to ascertain which students were not attending.

The university then got in touch with the various professors, who chaired the schools and they all agreed to call the roll. Reports then came into the university from all the various schools, with the exception of the school of physiology. After about a month, the university sent a polite note to Professor Wright, saying that he might have forgotten.

After another fortnight had gone by and still no report from the school of physiology, the university sent over a rather stronger note, saying to the Professor that he had always cooperated in the past and what was wrong now. So, on the following Monday morning, he arrived at the lecture theatre with a big roll book.

He said, 'From now on, I call the roll and any student who does not answer runs the risk of losing his CRTS allowances. He will never become a doctor, so it is absolutely essential that every student attends every lecture. Do you all understand that?

'Yes, sir.'

He then commenced to call the roll: 'Aarons K.'

'Sir.'

'Akron D.'

'Sir.'

'Arkwright J.'

'Sir.'

And he continued to call the roll, until he came to Colwell J. He said, 'Colwell J'. there was no answer. He repeated 'Colwell J' in a louder voice. Still there was no answer.

Professor Wright called out, 'Can everyone hear me in the theatre, even those right at the back?'

There was a chorus of 'Yes, Sir'.

He called out again 'Colwell J'. Still there was no answer.

Then he said, 'Hasn't the poor bugger got any friends?'

REFERENCE

Gillespie-Jones A S 1978 *The Lawyer who Laughed* (Hutchinson of Australia: Richmond, Australia) p. 44

## Anecdote of a Roman Medical School, 80 AD

Asclepiades (*c*.124–40 BC) was the first Roman doctor to achieve high fame. He came to Rome from Bithynia and founded a medical school which continued after his death. At first, such schools were the personal following of the doctor and he would take all his students with him on visits. The writer Martial (40–102 AD) commented on this practice in the following manner:

> I was lying ill; but you immediately came to me, Symmacchus, accompanied by a hundred students. A hundred hands, frozen by the north wind, examined me. I did not have a fever, Symmachus, but I've got one now.

## Anecdote of Nansen

During the preparations for one of his polar expeditions, Fridtjof Nansen (1861–1930) was visited by a stout man who wanted to sign on for Nansen's future expedition. The man took a seat in front of the polar investigator with his bowler bravely on top of his head. Nansen had to get good qualities in the men he hired and to test this man, he reached for a revolver, aimed at the man and pulled off a bullet through the man's hat. The latter did not react at all.

'Well, please take this money for a new hat', Nansen said.

'Thank you very much' the man answered 'but what about the trousers'.

'I have not done anything to them' said Nansen.

'Well, I have' said the man.

REFERENCE

Contributed by Arne Skretting who translated from the Norwegian the anecdote in *Alverdens Anekdoter II, 1000 Person-Anekdoter* (Politikens: Copenhagen) 1964.

# Fission of Sir Arthur Conan Doyle into Sherlock Holmes and Doctor Watson

My favourite review of a TV programme is that by Nancy Banks-Smith in the *Guardian* of 11 December 1987. The TV programme was on the BBC earlier that week and was called *The Case of Sherlock Holmes.*

Conan Doyle is the visible proof that inside every fat man there is a thin one biding his time. Doyle looked a strapping tweedy chap with a moustache at the top and a dog at the bottom as you could see in the precious film clip in *The Case of Sherlock Holmes.* The very image of Dr Watson.

But what is that, round and glittering rising and falling on his substantial waistcoat? Not a fob, not a monocle. Eliminate the impossible and what remains, however improbable, must be the truth. A magnifying glass.

A hundred years ago, in a historic bit of fission, Conan Doyle split into good, old Dr Watson and Sherlock Holmes, 'Tall and slender, given to long bouts of depression relieved only by cocaine in a 7 per cent solution. A tortured as well as a lonely man.' The fall-out can still be measured.

I might mention that Dr Jekyll and Mr Hyde had been published the year before. These are deep waters, Watson.

Doyle was an Irishman whose father drank himself into an asylum, a background not dissimilar to Shaw's or Chaplin's. Geniuses are not mad but they probably know someone who is.

The case of Sherlock Holmes was a 40 Minutes Special, you could tell it was special because it lasted 70 minutes. The idea was to investigate the Holmes phenomenon and make a bit of money for the BBC. To this end it turned into a fair old international rag-bag of films and fans.

My favourite, I confess was a Czechoslovakian film in which a voluptuous woman in a sort of corset gripped the hand of Sherlock Holmes and said vibrantly, 'I hunger for you.' Her hand moved up his pale blue sleeve towards his velvet collar. 'Have you,' she added hungrily, 'your violin with you?'

'Of course.' said Holmes, 'I never go anywhere without it.'

'I love a man who gets hold of a violin by its neck,' she cried holding his hands to her pounding heart.

In America John Bennet Shaw, an undertaker who, I suspect, found his sense of humour too well developed for his job, has devoted his life to Holmes. He has the largest collection of Holmes memorabilia in the world including three chocolate rabbits called Inspector Hector ('Terrible chocolate but artistic of course') and ladies' panties with a picture of Holmes, his magnifying glass and 'It must be in here somewhere.' He is a collector, he cheerfully agrees, with the sensitivity of a vacuum cleaner.

Once a year he leads a pilgrimage to the confused hamlet of Moriarty in New Mexico. Wearing deerstalkers, they sing 'unhappy birthday to you, you bastard' (Bastard, Shaw alleges, was Moriarty's middle name) and deposit ceremonial dung. Moriarty considers it the highlight of the year but, then, there's not a helluva lot going on there the other 364 days.

At the end of a windy English pier, Paul Sparks, who has modelled his life on Holmes's methods, was wearing a helmet sprouting wire coat hangers and demonstrating a device for tracking down stolen bicycles which, one fears, is still in its infancy. Mr Sparks explained that he was 'Trying to see patterns in the chaos of life.'

When he finds one I hope he will let me know. Likewise if he finds my bicycle.

---

## Zip Answer to Stomach Problem

People with severe internal abdominal infections run a high risk of dying if their abdominal cavities are not regularly drained. However, this is easier said than done, because draining may involve constantly opening and closing the wound.

Spanish surgeons have come up with a neat solution to this problem: they close the access with a nylon/polythene zip which can be undone every day.

A report in the *Archives of Surgery* says that doctors at the Provincial Hospital, Madrid, have been using zippers since 1982 and have successfully treated more than 60 patients either with an infected pancreas or infections following perforations of the intestine.

REFERENCE
The *Independent* 1 March 1988

---

## Prisoner Surprises the Police Taking Him to Jail

'I have swallowed a spoon' was the amazing statement made to police who were taking Joseph Tyson, an electrician, to Preston Prison. He was removed to Preston Infirmary and examined under the x-rays, when a metallic substance was located near the stomach. An operation for its removal was performed, but Tyson still remains in the Infirmary in a serious condition. It appears that he broke off the bowl of the spoon and swallowed the shank end. He was under remand on a charge of obtaining £120 by false pretences from a woman to whom he had proposed marriage.

[X-rays were discovered in November 1895 but from the amount of press coverage of this incident it would appear that this was the first case, at least in the United Kingdom, of spoon-swallowing diagnosed by x-rays. Earlier reported swallowing incidents tended to refer to coins or false teeth.]

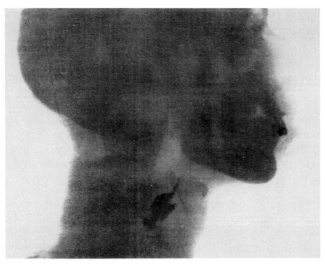

*A set of false teeth lodged in the throat, 1902 case, 20 years before the episode with the spoon.*

REFERENCE
*Hull Daily News* 16 March 1922

---

## $10 000 Radium Used to Cure Sing Sing Convict

Radium worth $10 000 was taken inside the walls of Sing Sing Prison and is being used today by Dr Amos O Squire, head prison physician, to treat John Duffy, a cancer patient in the prison hospital. The radium, which is the personal property of Dr Squire, was used to try and cure Duffy, who was operated on for cancer of the lip last Saturday. Duffy was committed to Sing Sing for assault in Brooklyn. Doctors also reported that John Amishoki, the ex-convict who voluntarily returned to Sing Sing in order to continue the treatment he was receiving for a tumour of the brain, was much improved and will probably leave the prison in a few days.

REFERENCE
*Brooklyn Daily Eagle* 11 October 1921

---

## The Medico-legal Significance of X-rays in the First Year after their Discovery

### 1 THE ACTRESS WHO FELL DOWN THE STAIRS (England)

One of the first lawsuits in which x-ray picture evidence became of great importance was heard in Nottingham in England. It was described in a number of British journals in March, 1896, and shortly thereafter was commented upon by several American journals. This case was described as follows in the *London Hospital* under the title, 'The New Photography in Court.'

> An interesting and novel case, in which the 'X-rays' practically decided the point, was tried by Mr. Justice Hawkins and a special jury at Nottingham the other day. Miss Ffolliott, a burlesque and comedy actress, while carrying out an engagement at a Nottingham theatre early in September last, was the subject of an accident. After the first act, having to go and change her dress, she fell on the staircase leading to the dressing-room and injured her foot. Miss Ffolliott remained in bed for nearly a month, and at the end of that time was still unable to resume her vocation. Then, by the advice of Dr. Frankish, she was sent to the University College Hospital, where both her feet were photographed by the 'X-rays.' The negatives taken were shown in court, and the difference between the two was convincingly demonstrated to the judge and jury. There was a definite displacement of the cuboid bone of the left foot, which showed at once both the nature and the measure of the injury. No further argument on the point was needed on either side, and the only defence, therefore, was a charge of contributory careless-ness against Miss Ffolliott. Those medical men who are accustomed to dealing with 'accident claims'—and such claims are now very numerous—will perceive how great a service the new photography may render to truth and right in difficult and doubtful cases. If the whole osseous system, including the spine, can be portrayed distinctly

262

on the negative, much shameful perjury on the part of a certain class of claimants, and many discreditable contradictions among medical experts, will be avoided. The case is a distinct triumph for science, and shows how plain fact is now furnished with a novel and successful means of vindicating itself with unerring certainty against opponents of every class.

The presentation of the new 'wonder pictures' before a solemn court had varying effects. Not everyone was readily convinced of the value of such pictures for court procedures, and 'some amusing remarks were made,' said the *Journal of Photography*:

On the defendant's counsel telling one of the witnesses that he ought to have scientific evidence as to the value of the rays, Mr. Justice Hawkins remarked, 'You might send a man to the lunatic asylum, you know, by photographing his head.' One of the barristers, looking at the photographs, asked, 'Is this the Trilby?' &c. Evidently the 'New Photography' was treated with a certain degree of levity on its first appearance as a witness, by the gentlemen of the long robe.'

## 2 BULLET IN THE LEG (Canada)

In April the *Medical Record* wrote under the title, 'Roentgen Rays in Court':

The *Canadian Medical Review*, quoting from a Montreal daily paper, said that the court of Queen's Bench of that city will be the first court on record where Roentgen's new photographic discovery will be used in evidence. A photograph of the leg of Tolson Cumming, who was shot on Christmas Eve, will be filed, showing the location of the bullet. The photograph was taken by Professor Cox, of McGill University.'

## 3 DOCK LABOURER'S BROKEN ARM (England)

Another English case, in which x-ray plates were used as evidence, was described in July 1896, in the *British Journal of Photography* under the title, 'A County Court Judge and the Roentgen Rays.'

At the Liverpool County Court on Monday, during the hearing of an action under the Employers' Liability Act, in which a dock labourer claimed £150 damages for personal injuries, the plaintiff's counsel produced two photographs of the injured arm taken by means of the roentgen rays. He proposed to use these photographs as evidence. The defendant's counsel objected, stating that he has no reason to believe that Dr. Buchanan, of the University College, by whom the photographs had been taken, was competent to produce reliable radiographs by the new process. Judge Shand replied that, from all he had read concerning the process, he believed that he himself should be able to take photographs by this method, providing he had the necessary apparatus. Dr. Buchanan agreed, stating that the process was perfectly simple, and could be carried out by anyone. Judge Shand felt compelled to admit the authenticity of the photographs, but stated that the question as to their value must be discussed by the defendant's counsel and the jury. When the photographs were produced, they showed clearly the injury which had been done to the bone of plaintiff's arm and the jury awarded the plaintiff £60 damages.

In June a detailed discussion of the legal significance of Roentgen pictures appeared in the *British Medical Journal*.


## 4 WIFE MURDER (England)

The radiograph is thought to be of one of the bullets fired by Hargreaves Hartley into the brain of his wife Elizabeth Ann Hartley at Nelson, Lancashire, on 23 April 1896, and 'photographed with x-rays' by the then Professor of Physics at Manchester University, Sir Arthur Schuster, on 2 May 1896.

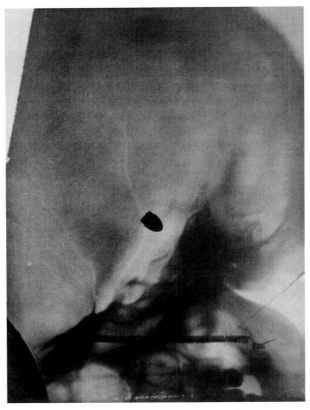

# 5 TROLLEY CAR ACCIDENT (USA)

W J Morton, in his book *The X-ray*, stated in September 1896:

> A very important application of the x-ray will be in connection with expert testimony in the court. Court records contain numerous cases in which the x-ray would have been of great service. Already it has been used for this purpose. I have a picture of the knees of a person which is likely to find its way into court. The patient was thrown down with violence in a trolley-car accident more than a year ago, and has suffered more or less ever since. An exposure was first made of the injured knee only and afforded no positive evidence of the seat or degree of the injury. By resorting to the comparative method, a picture of both knees was obtained which showed that the upper portion of the large bone of the leg below the knee was nearly three-quarters of an inch wider in the injured knee than in the normal one. This was doubtless due to fracture and subsequent growth of bone. Such a picture is very convincing and would be sure to have great weight with a jury.

Note: The original x-ray negative is now a court record and it is impossible to obtain possession of the same to make a half-tone reproduction until the case is settled. The picture will be reproduced in future editions of this work as soon as the original negative can be obtained.

# 6 SURGEON LOSES MALPRACTICE CASE (France)

At about the same time [as case 5 above] the *Journal of the American Medical Association* reported the use of x-ray plates in a trial in France as follows:

> We note that at the trial of an action for damages at Nancy, in France, the surgeon who had charge of the injured plaintiff was accused of having caused the damage by mistaking a dislocation for a fracture. The accusation was sustained by producing in court a roentgen photograph which showed clearly the bones in the dislocated position without a fracture.

The physician lost the suit.

# 7 ATTEMPTED MURDER (England)

An interesting use of x-rays in a case of attempted murder was also described in the English *Journal of Photography*.

> Henry Goodwin was again brought up last week at the Salford Police Court, and charged before Mr R Hankinson and Mr Alderman Jenkins with burglary and attempting to murder Mr Israel Rosenblum,

merchant, Northumberland Street, Higher Broughton. Prisoner was remanded for another week on the application of Chief Detective Inspector Lyogue, who said that there was now a likelihood of Mr Rosemblum's recovery, but he would not be in a fit state to attend court for two or three weeks. In the course of an interview, Dr Walmsley, the medical attendant of Mr Rosenblum, stated that his patient had been photographed by roentgen rays by Mr Chadwick, of St Mary's Street, and, as the result of the process, the bullet was discovered in the chest. An operation with a view to its extraction will be made in the course of a few days.

## 8 TERRORIST BOMBS (France)

The etching shows Post Office or Customs staff in the surroundings of a railway station examining a package. This use of x-rays had been described by Snowden Ward in 1896 to 'reveal the contents of certain suspected packets, which have proved to be infernal machines'. It is not known if this led to cases in the law courts.

## 9 THE LAW STUDENT WHO FELL OFF A LADDER (USA)

Towards the end of the year a law suit came up in Denver, Colorado, in which, again, roentgen-ray pictures were admitted as evidence. 'The suit was brought April 14, 1896, by one, James Smith, a poor boy who was reading law and doing odd jobs to see his expenses. He was injured in a fall from a ladder and after some time he consulted a well-known surgeon, who made no attempt at immobilization of the thigh, but advised exercise of various kinds as though treating a contusion.' The trial came up on December 2, 1896; the Judge was Owen E LeFevre, the Prosecutors for J Smith, Plaintiff, were B B Lindsey and N F W Parks. They qualified one H H Buckwalter as an expert in photography and in the use of roentgen rays because he had been making x-ray shadow-graphs for the past eight months. Plates of J Smith were made November 7, 11, 21 and 28, 1896. The most satisfactory one required an exposure of eighty minutes.

Judge Lindsey said that he had been in personal communication with Judges in the East who had refused to accept x-ray plates in evidence, and that one of them had stated to him that he had sustained an objection to the offering of x-ray plates in evidence, 'because,' said he, 'there is no proof that such a thing is possible. It is like offering the photograph of a ghost,' continued the judge 'when there is no proof that there is any such thing as a ghost.' In order to convey the idea of radiographic shadows to the Judge and the jury, Buckwalter, Lindsey and Parks contrived a shadow box and showed an ordinary shadow picture and then an x-ray shadowgraph of a hand. Next the shadow of a normal femur was shown and then the roentgen shadowgraph of such a femur and finally there was shown the x-ray plate of a picture of the plaintiff's left femur. This radiograph showed that the head of the bone was not in normal relationship to the great trochanter and the shaft, and it was proposed by the prosecutor that this radiograph be submitted to the jury as evidence that there had been a fracture of the femur in the region of the great trochanter with impaction of the fragments. C J Hughes, a Denver lawyer, argued for the defence for more than three hours against the admission of such evidence, stating that 'X-ray photographs' are not admissable under the law and past decisions of the court bore this out. He contended, furthermore, that even should it be admitted this was a photograph of J Smith's femur, so it could not be used as competent testimony under the broad principle of the law upon the matter of photographs as testimony because witnesses must testify to having seen the object which had been photographed and to having identified the photograph as a good likeness of the object—then only may any photograph be admitted as evidence.

On the following morning Judge LeFevre expressed his opinion on the exhibits consisting of $4 \times 6$ plates. (*Denver Republican*, Thursday, December 2, 1896.)

'The defendant's counsel objected to the admission in evidence of exhibits, the same being photographs produced by means of X-ray process, on the ground that, being photographs of an object unseen by the human eye, there is no evidence that the photograph accurately portrays and represents the object so photographed. This rule of law is well settled by a long line of authorities and we do not dissent therefrom as applied to photographs which may be seen by the human eye. The reason of this salutary rule is so apparent to the profession that as a rule of evidence we will not discuss it.

We, however, have been presented with a photograph taken by means of a new scientific discovery, the same being acknowledged in the arts and in science. It knocks for admission at the temple of learning; what shall we do or say? Close fast the doors or open wide the portals?

These photographs are offered in evidence to show the present condition of the head and neck of the femur bone which is entirely hidden from the eye of the surgeon. Nature has surrounded it with tissues for its protection and there it is hidden; it cannot by any possibility be removed nor exposed that it may be compared with its shadow as developed by means of this new scientific process.

In addition to these exhibits in evidence, we have nothing to do or say as to what they purport to represent; that will, without doubt, be explained by eminent surgeons. These exhibits are only pictures or maps to be used in explanation of a present condition, and therefore are secondary evidence, and not primary. They may be shown to the jury as illustrating or making clear the testimony of experts.

The law is the acme of learning throughout all ages. It is the essence of wisdom, reason and experience. Learned priests have interpreted the law, have classified reasons for certain opinions which, in time, have become precedents, and these ordinarily guide and control especially trial courts. We must not, however, hedge ourselves round about with rule, precept and precedent until we can advance no farther; our field must ever grow, as trade, the arts and science seek to enter it.

During the last decade, at least, no science has made such mighty strides forward as surgery. It is eminently a scientific profession, alike interesting to the learned and unlearned. It makes use of all science and learning. It has been of inestimable service to mankind. It must not be said of the law that it is wedded to precedent; that it will not lend a helping hand. Rather, let the courts throw open the door to all well-considered scientific discoveries. Modern science has made it possible to look beneath the tissues of the human body, and has aided surgery in telling of the hidden mysteries. We believe it to be our duty in this case to be the first, if you please to so consider it, in admitting in evidence a process known and acknowledged as a determinate science. The exhibits will be admitted in evidence.

In telling of the trial, Judge Lindsey stated: 'The electrical apparatus, batteries, Crookes tube, etc., were all in the court-room. We offered to show the jury the bones in their hands, which created such terrific excitement about the court-house that extra bailiffs were called in to keep the court in order during the argument. The excitement was intense, the 'gallery' on my side restrained from breaking into applause on several occasions because of their anxiety to have this 'miracle' demonstrated and actually recognized by a court.'

## 10 X-RAYS AND DIVORCE CASES (England)

The alacrity with which the medical and the judicial professions took up the new discovery of Röntgen and made use of it as evidence in court trials is remarkable. It is not surprising then that some persons less concerned with truth looked upon the use of the x-ray for more or less fraudulent purposes. On August 8, 1896, the London *Standard* carried the following advertisement: 'The New Photography—Owing to the success Mr Henry Slater has personally achieved with the New Photography, he is prepared to introduce same in divorce matters free of charge. Offices, No. 1, Basinghall Street, City.'

How the roentgen rays could be used for the purposes of clearing up divorce matters was not exactly explained, but a theory was propounded by the editor of the *Electrical Engineer*, who wrote on September 8, 1896:

> Mr. Slater, as a detective, is evidently up to date. We presume he uses the X-ray to discover the skeleton which every closet is said to contain. Ability to do the detective act without squinting through a keyhole is regarded evidently as one of the recommendations of the rays, which themselves prefer darkness rather than light.

REFERENCES
Case histories 1–3, 5–7, 9–10: Glasser O 1933 *Wilhelm Conrad Röntgen and the Early History of Roentgen Rays* (John Bale, Sons & Danielsson Ltd: London)
Case histories 4, 8: Mould R F 1980 *A History of X-rays and Radium* (IPC Business Press: Sutton)

---

# The Duc de Richelieu on Handwriting

If you gave me six sentences written by the most innocent of men, I will find something in them with which to hang him.

---

# Lord Palmerston's Last Words

Die, my dear doctor. That's the last thing I shall do.

# Radiation Protection in 1910

REFERENCE
X-ray equipment catalogue of Newton & Company, London

---

# A Specialist

Choose your specialist and you choose your disease.

REFERENCE
*The Westminster Review* 18 May 1906

# Persian Anatomy

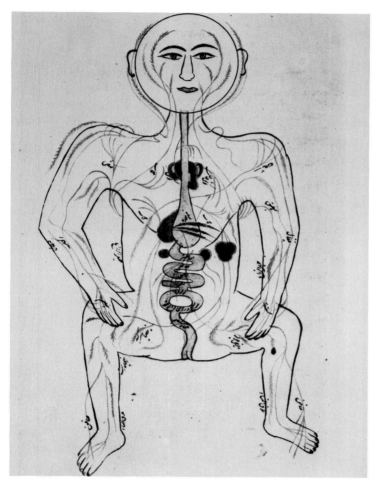

*Persian anatomical drawing showing the arterial system. (Courtesy Well-come Institute Library, London.)*

---

# Philosophy

Philosophy like medicine, has plenty of drugs, few good remedies and hardly any specific cures.
*Nicholas Chamfort* (1741–1794)

# Varicose Veins in Ancient Greece

*Votive relief of a man holding a huge leg showing varicose veins. From the Asklepion, Athens and now in the National Museum, Athens. (Courtesy Wellcome Institute Library, London.)*

---

## An Adult

An adult is one who has ceased to grow vertically but not horizontally.

# George Washington's Medical Supplies

In the bicentennial issue of the *Journal of the American Medical Association* many fascinating historical articles were published including one about an American physician in 1776. Within this text it was stated:

> The following order for medical supplies for his family and slaves at Mount Vernon was placed by George Washington with his London factor in 1767
>
> 2 best Lancets in one case
> 6 Common Do. [lancets] each in sepe. [separate case?]
> 25 lb. Antimony [used as an emetic]
> 10 lb. flour of Sulphur [an aperient]
> 2 Oz. Honey Water. [pectoral]
> 3 Quarts Spirits of Turpentine. [stimulant]
> 2 lb. best Jesuit's Bark, powdered. [quinine, used as a tonic and antiseptic]
> 3 Oz. Rhubarb Do. [powdered] and put into a bottle. [cathartic]
> 1 pint Spirit of Hartshorn [ammonia, a stimulant]
> 6 Do. Do. Nitre [diuretic and febrifuge]
> 1 lb. Blistering Plaister [made from Cantharides]
> 4 Oz. Tincture of Castor. [purgative]
> 8 Do. Balsam Capivi [copaiba, a corroborant, stimulant and diaphoretic]
> 1–4 lb. Termerick [Turmeric, a condiment used in curry; substitute for saffron]

© American Medical Association 1976

REFERENCE
*Journal of the American Medical Association* 1976 **236** 28

---

# Derbes's Law

(Contributed by Dr Vincent J Derbes of New Orleans.)

Most of our pleasures come from filling or emptying cavities, and vice versa.

# THE BUCK-PASSING MACHINE

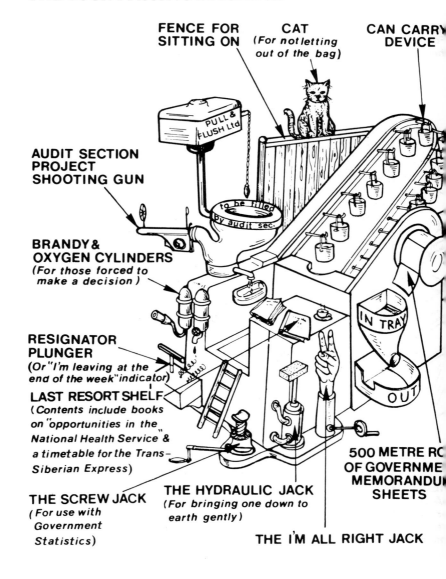

**FENCE FOR SITTING ON**

**CAT** (*For not letting out of the bag*)

**CAN CARRY DEVICE**

**AUDIT SECTION PROJECT SHOOTING GUN**

PULL & FLUSH Ltd

to be filled by audit sec.

**BRANDY & OXYGEN CYLINDERS** (*For those forced to make a decision*)

**RESIGNATOR PLUNGER** (*Or "I'm leaving at the end of the week" indicator*)

IN TRAY

OUT

**LAST RESORT SHELF** (*Contents include books on "opportunities in the National Health Service" & a timetable for the Trans-Siberian Express*)

**THE SCREW JACK** (*For use with Government Statistics*)

**THE HYDRAULIC JACK** (*For bringing one down to earth gently*)

**500 METRE RO OF GOVERNME MEMORANDU SHEETS**

**THE I'M ALL RIGHT JACK**

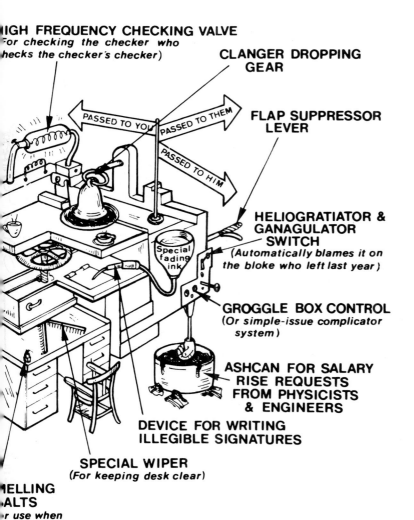

**IGH FREQUENCY CHECKING VALVE**
*(For checking the checker who checks the checker's checker)*

**CLANGER DROPPING GEAR**

**FLAP SUPPRESSOR LEVER**

PASSED TO YOU

PASSED TO THEM

PASSED TO HIM

**HELIOGRATIATOR & GANAGULATOR SWITCH**
*(Automatically blames it on the bloke who left last year)*

Special fading ink

**GROGGLE BOX CONTROL**
*(Or simple-issue complicator system)*

**ASHCAN FOR SALARY RISE REQUESTS FROM PHYSICISTS & ENGINEERS**

**DEVICE FOR WRITING ILLEGIBLE SIGNATURES**

**SPECIAL WIPER**
*(For keeping desk clear)*

**MELLING ALTS**
*(r use when ed to sign nething)*

# Seeking Health in Babylon

*'Seeking Health'. Supplicants approaching a shrine, the door of which stands open. Sketch of an impression from an early dynastic Assyro-Babylonian cylinder seal,* c. 2800 BC. *(Courtesy Wellcome Institute Library, London.)*

---

# The Pigtail of Li-Fang-Fu

The two British Music Hall comic monologues about doctors and hospitals which were included in *Mould's Medical Anecdotes* were well received and consequently a further two have been placed in this volume. This time the monologues concern that favourite topic of comedians—Death. Many are set in exotic places such as India and the Far East, such as this example of 'The Pigtail of Li-Fang-Fu' which involves opium addiction. The second example (p. 155) is just called 'Suicide', but it is far from morbid! 'The Pigtail' was written by Sax Rohmer and T W Thurban in 1919.

> They speak of a dead man's vengeance; they whisper a
>    deed of hell
>    'Neath the Mosque of Mohammed Ali.
>    And this is the thing they tell.
> In a deep and a midnight gully, by the street where the
>    goldsmiths are,
>    'Neath the Mosque of Mohammed Ali, at the back of
>    the Scent Bazaar,

Was the House of a Hundred Raptures, the tomb of a
    thousand sighs;
    Where the sleepers lay in that living death which the
        opium-smoker dies.
At the House of a Hundred Raptures, where the reek
    of the joss-stick rose
    From the knees of the golden idol to the tip of his
        gilded nose,
Through the billowing oily vapour, the smoke of the
    black *chandu*,
    There a lantern green cast a serpent sheen on the
        pigtail of Li-Fang-Fu.
There was Ramsa Lal of Bhiwâni, who could smoke
    more than any three,
    A pair of Kashmiri dancing girls and Ameer Khân
        Môtee;
And there was a grey-haired soldier too, the wreck of a
    splendid man;
    When the place was still I've heard mounted drill
        being muttered by 'Captain Dan'.
Then, one night as I lay a-dreaming, there was
    shuddering, frenzied screams;
    But the smoke had a spell upon me; I was chained to
        that couch of dreams.
All my strength, all my will had left me, because of the
    black *chandu*,
    And upon the floor, by the close-barred door, lay the
        daughter of Li-Fang-Fu.
'Twas the first time I ever saw her, but often I dream
    of her now;
    For she was as sweet as a lotus, with the grace of a
        willow bough.
The daintiest ivory maiden that ever a man called fair,
    And I saw blood drip where Li-Fang-Fu's whip had
        tattered her shoulders bare!
I fought for the power to curse him—and never a
    word would come!
    To reach him—to kill him—but opium had
        stricken me helpless—dumb.
He lashed her again and again, until she uttered a
    moaning prayer,
    And as he whipped so the red blood dripped from
        those ivory shoulders bare.
When crash! went the window behind me, and in leapt
    a greyhaired man,
    As he tore the whip from that devil's grip, I knew
        him: 'twas Captain Dan!
Ne'er a word spoke he, but remorseless, grim, his brow

with anger black.
  He lashed and lashed till the shirt was slashed from
    the Chinaman's writhing back.
And when in his grasp the whip broke short, he cut
    with a long keen knife.
  The pigtail, for which a Chinaman would barter his
    gold, his life—
He cut the pig-tail from Li-Fang-Fu. And this is the
    thing they tell
  By the Mosque of Mohammed Ali—for it led to a
    deed of hell.
In his terrible icy passsion, Captain Dan that pig-tail
    plied,
  And with it he thrashed the Chinaman, until any but
    he had died—
Until Li-Fang-Fu dropped limply down too feeble, it
    seemed, to stand.
  But swift to arise, with death in his eyes—and the
    long keen knife in his hand!
Like fiends of an opium vision they closed in a fight
    for life,
  And nearer the breast of the Captain crept the blade
    of the gleaming knife.
Then a shot! a groan—and a wisp of smoke. I
    swooned and knew no more—
  Save that Li-Fang-Fu lay silent and still in a red pool
    near the door.
But never shall I remember how that curtain of sleep
    was drawn
  And I woke, 'mid a deathly silence, in the darkness
    before the dawn.
There was blood on the golden idol! My God! that
    dream was true!
  For there, like a slumbering serpent, lay the pigtail
    of Li-Fang-Fu.
From the House of a Hundred Raptures I crept ere the
    news should spread
  That the Devil's due had claimed Li-Fang-Fu, and
    that Li-Fang-Fu was dead.
'Twas the end of that Indian summer, when Fate—or
    the ancient ties—
  Drew my steps again to the gully, to the Tomb of a
    Thousand Sighs;
And the door of the house was open! All the blood in
    my heart grew cold.
  For within sat the golden idol, and he leered as he
    leered of old!

And I thought that his eyes were moving in a sinister
     vile grimace
    When suddenly, there at his feet I saw a staring and
     well-known face!
With the shriek of a soul in torment, I turned like a
     frenzied man,
    Falling back from the spot where the moonlight
     poured down upon 'Captain Dan'!
He was dead, and in death was fearful; with features of
     ghastly hue—
    And snakelike around his throat was wound the
     pigtail of Li-Fang-Fu!

REFERENCE

Marshall M (ed) 1981 *The Book of Comic and Dramatic Monologues* (London: Elm Tree Books/EMI Publishing). Reproduced by permission of EMI Music Publishing Ltd.

## Camel Ambulance in Afghanistan

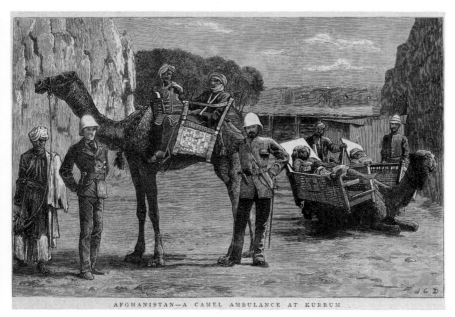

AFGHANISTAN—A CAMEL AMBULANCE AT KURRUM

*This engraving is circa 1880 and shows a camel ambulance at Kurrum during the Afghan War. (Courtesy Mary Evans Picture Library.)*

## Koko for the Hair

*(Courtesy Mary Evans Picture Library.)*

This advertisment appeared in the *Penny Pictorial Magazine*, *c*. 1900, and included a special offer for those who have not yet tried Koko—a 4/6 Trial Bottle for 2/-. The makers, The Koko-Maricopas Co of Bevis Marks, London, stated that 'We find it better to thus practically give away one bottle to make a customer than to spend large amounts on advertising'. However, they did issue another advertisement with complimentary remarks from: one John Strange Winter, the now long-forgotten author of 'Bootle's Baby', who claims it stopped his hair falling out (maybe due to worry of finding publishers for his stories!); from someone purporting to act for HRH Princess Victoria begging to send six bottle of Koko for this grandaughter of Queen Victoria; from HRH Princess Marie of Greece at the Palais Royal, Athens, to say she is very pleased with her hair; and finally, the most intriguing testimonial, from a Rear-Admiral A. Tinklar, at the address of Governor's House, H.M. Prison, Birmingham, who was very pleased with his bottle of Koko, claiming it to be a most excellent preparation. It is not stated whether this was first-hand knowledge, or whether it was somehow tested on the prisoners in Birmingham Jail!

---

## It's not What you Write—It's the Way that you Write it

It was a lazy July morning—bang in the middle of what people in the media call the 'silly' season.

Sifting through his morning mail, the Editor of the newspaper I worked for picked up three confidential letters unopened by his secretary and, knowing my penchant for analysing people's hand-writing, gave them to me.

'Here' he said. 'I haven't the vaguest idea who these are from, and I don't want to open them for a minute or two; so what can you tell me about them first?'

I examined the three envelopes briefly. Then I put my head on the chopping block.

The first was an intelligent hand; educated; from someone of high morals and principles. The writer was articulate, generous and just a shade aloof. Health was definitely not top line. I detected signs of a chronic heart condition. It was difficult to say much more from the few words on the envelope.

The second sample disturbed me. It was blatantly obvious that here was a very disturbed person. The writing crawled all over the place. The more I studied it, the more uneasy I became.

Schizoid tendencies were apparent and, I hesitated to say so, but felt instinctively that here was a potential suicide who needed professional surveillance.

Number three was neat and small. A tidy hand, from a tidy mind. Definitely from someone with keen powers of concentration, analytical, with good organisational ability.

But why so utterly frustrated in every way . . . even in relation to sex?

I handed back to my Editor his three unopened envelopes and, as he ripped them apart and read the contents, I watched his expression change from one of amazement to sheer incredulity.

He placed the first on his desk, gasping 'Good God!'—which was rather apt in view of the fact that it was from the Bishop.

'You're right about his being intelligent, educated and a man of high morals and principles. Aloof? Yes, I suppose he is that, too. A heart condition, you say? It's the first I've heard. To the best of my knowledge, he's perfectly healthy.'

One down, two to go.

As he read the contents of the second, he blanched. It was a patient in the local psychiatric hospital: a hopeless schizophrenic. He was writing to say that since he had recently developed a ray of gold light around his body, he could only assume that he was 'the long-awaited Messiah.'

And so to number three, for a little light relief:

My wicked Editor was rather tickled to announce that it was from a nun; a maths mistress at a grammar school in the next county.

About to renounce her vows, she was seeking some advice about accommodation and employment. She wondered if there were any openings for trainee journalists on our paper . . . while I wondered if there any chance of having my salary doubled.

The answer in each case was, unfortunately, no.

A month passed and, in the hustle and bustle of writing copy, I forgot about the saga of the three letters. Then, out of the blue, a newsflash came in. The Bishop had had a coronary.

A week or so later, the analysis of the hospital patient was also confirmed. Deeply upset because none of those in his immediate environment afforded him the deification he felt was his right, he hanged himself from a tree in the hospital grounds.

Can the future be foretold by graphology then?

No, although it may look like it from this little sequence of events, that conclusion must be entirely ruled out. The answer is much more simple than prophecy, precognition, clairvoyance or anything else you wish to call it.

Dominant factors in the Bishop's writing showed that all was not well in the region of his heart. The break in his script at the same point was repeated again and again.

It was obviously not a slip of the pen, because there were no breaks anywhere else.

The irregularity of his heart beat produced that unmistakeable 'jump' in the writing which most graphologists would recognise for what it was.

And the mentally disturbed man?

The downward slope of his writing was a clear indication of depression. Almost every symbol in that writing spelt despair. There wasn't a glimmer of hope to be seen anywhere.

Handwriting is not merely a case of the hand moving the pen. It is directly related to the motor mechanisms of the brain.

Suppose, for the sake of argument, you were unfortunate enough to lose the use of your right arm (or the left, if that's the hand you write with) and had to make its opposite partner take over.

You would find that, once you'd got the hang of it, your style of writing would not have changed very much.

So, having established that the way you write is connected with the way your unconscious nervous system functions: one can safely assume that a slip of the pen can be as revealing as a slip of the tongue.

General mentality, long-standing illness, physical and mental handicap show up easily. Sex doesn't, though sexuality—or the lack of it—usually does.

Depressives give themselves away easily. The way in which that psychiatric patient scored through his name was almost an unconscious desire to score through himself—which, predictably, he did.

Age is difficult as a rule, unless the writing is that of a young child, or of someone very old.

So what do we look for in someone's handwriting, in order to see how they function as individuals?

Various signs and symbols—the slope of the writing, its pressure, how large (or small) are the capitals in relation to the rest of the text, and whether they are entangled with the letters above, or below.

What is the spacing like between letters, words, lines?

Are the t-strokes and i-dots all in the right place? Are they there at all?

Corroborative evidence is essential.

Pencils and felt tips can be messy. Ballpoints are better, but there is nothing like a good fountain pen to produce a clean, legible script.

Small writing shows keen powers of concentration; the large variety can be from someone who is broadminded, self-assured. Equally, a small hand could come from an obsessively tidy housewife; a large one from a flamboyant megalomaniac.

One must never jump to conclusions. There are so many factors to take into account. Does the signature differ in style from the rest of the text? Is the surname the same size as the Christian name? Are some of its letters larger than others? Does it trail off to a line, or become enlarged and have a full stop after it?

Self-importance, pride in one's spouse, strange eccentricities, diplomacy could all be evident here.

Where did the art (or science) of graphology begin? It is generally believed that Dr Camillo Baldi, Professor of Theoretical Medicine at the University of Bologna in the eighteenth century, was the man who set the wheels—or was it the quills?—in motion.

During the next 100 years or so, the concept of drawing connections between personality and handwriting was taken up by a variety of scientists throughout Europe.

Towards the end of the nineteenth century, graphology was shown to have a possible relationship with psychology and physiology.

Beavering away in Germany, Doctors William Preyer and Hans Busse produced many fascinating periodicals on the subject and were helped by two more medical men—George Meyer and Ludwig Klages; who showed that it was possible, through handwriting, to pick out indications of ill-health.

Klages became the Freud of handwriting analysis and today is either regarded as the best thing since ballpoints, or his theories are thrown out with the (inky?) bathwater.

Today, mainly in America and Scandinavia, research is being conducted into how certain illnesses manifest themselves in one's writing. While in Britain, medical scientists are also beginning to accept graphology as an aid to conventional diagnosis.

On a personal level, I use any talent I might have, to assist my husband, a busy hypnotherapist. I can tell at a glance which of his letters are from patients in real need of help. Between us, we can establish a priority list and slot in the urgent cases quickly.

And the strange thing is that patients, once they have been successfully treated, change their handwriting accordingly. Nervous, spikey scripts become more rounded as repressions are cleared and emotions brought to the fore. The backward slant of the shy introvert straightens and leans forward as the writer becomes more confident and begins to mix socially. A whole host of other changes take place until the hand is that of a healthy, well-adjusted individual.

Though maybe that's not so strange, when you consider that the purpose of hypnotherapy is to put you in tune with your own subconscious, to iron out frustrations and phobias.

Ah yes, the nun.

I hadn't forgotten, but forgive me if I don't publicise details of how

284

I detected her problem. Otherwise, you'll all be examining each other's writing and having little private chuckles if you find those tell-tale signs.

REFERENCE

Harding M 1982 *Choice Magazine* (May) pp 36–7

## Is Graphology Accurate? Ask Lord Nelson and the Duke of Wellington

by Harold Winter, Monica Harding and R F Mould

### INTRODUCTION

When I was interviewed about *Mould's Medical Anecdotes* for the *Liverpool Echo*, the journalist, Monica Harding, turned out to also be a graphologist (see the previous anecdote) and unexpectedly gave me a reading when I inscribed my book! However, I was able to turn the tables by obtaining her agreement to a *blind trial* of her graphology prowess using some handwriting samples that one of us (HW) had recently obtained when cataloguing some hundreds of documents in an unpublished collection containing papers relating to the political life of the early nineteenth century and to the East India Company.

The conditions of the trial were:

(1) Glossy photographs of samples of handwriting, such as a few lines of a letter or an addressed envelope, to be provided with the signature and the address of the writer deleted.

(2) Any words in the handwriting sample which might provide a clue to the writer shall be deleted.

(3) The graphology reading for each sample shall include, when possible, a description of the attributes and health of the writer together with any general comments.

(4) It is agreed that the age and sex of the writer cannot be determined from graphology.

(5) The material for the study shall be posted and returned by recorded delivery, arriving in Liverpool on 3 August and being returned to London by 6 August 1984.

All conditions were fulfilled, and in addition Monica Harding was not told that the samples were from the early nineteenth century.

LORD NELSON, 1758–1805

*Engraving of Lord Nelson from a portrait by J Hoppner. (Courtesy Mansell Collection.)*

It was by chance that I saw the signature on the envelope and realised that this almost illegible signature *Nelson & Bronte* was in fact that of Lord Nelson, who was also created Duke of Bronte in Sicily. The address *Merton* is Nelson's country home, Merton Place, near London, which had been bought by Emma, Lady Hamilton on Nelson's instruction. The Dr Fisher of Christ's College was a Fellow of the College at that time and he appears to have been a lawyer who died in

1814 at the age of 65. The reason for Nelson's communication is not known, but Fisher was quite prominent in College affairs and the Nelson family had a long connection with Christ's. Nelson's brother, the second Lord Nelson, Nelson's younger brother Suckling and his son Horatio were all members of the College.

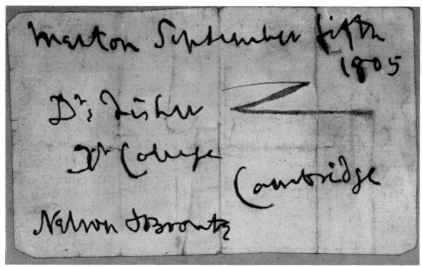

*An example (actual size) of the handwriting of Admiral Lord Nelson. If compared with the style of writing in his last letter to Lady Hamilton, written off Cadiz in October 1805 before the Battle of Trafalgar, it is at once obvious by comparing individual letters that the above sample is an original. His last letter to Lady Hamilton is in the British Museum and has been reproduced in Hattersley's biography[1] of Nelson.*

From engravings in the Mansell Collection it is seen that Nelson's signature was variously Nelson, Horatio Nelson or Nelson & Bronte. These signatures with his left hand, though, are more common than those with his right hand before it was shot off by the Spaniards at Tenerife in 1797. Indeed, at a recent sale of signatures at an autograph hunter's fair in London,[2] Horatio Nelson in right-hand signature was sold for £1250. This compared with £200 for a signed photograph of Winston Churchill and £180 for a cheque signed by Charles Dickens.

For the graphology trial the word Merton, the date 1805 and the signature were omitted from the handwriting sample.

The symbol Z on the envelope is thought to indicate that the writer had not paid the postage and that the recipient, Dr Fisher, was expected to pay. This was before the days of postage stamps, the penny black and penny red, and if the letter was prepaid then a symbol was printed on the envelope which consisted of a circle surmounted by a crown and three lines of writing, for example

FREE
3 AP 3
1832

The colour of the printing was red/brown.

## GRAPHOLOGY READING for Lord Nelson

The writer is proud and fiercely independent, to his own detriment. Moody—one has the impression, metaphorically speaking, that he is 'banging his head against a brick wall'. A divergent thinker who tends to live in 'cloud cuckoo land'. Totally apart from others. For this man, time is precious and not to be wasted on non-essentials. He considers there is important work to be done and believes he is the one to do it. Nor will he let go until he has completed his task: that for which he believes he was put on this earth. All of which makes life rather difficult for those around him. Bureaucracy be blowed, the 'people who matter' must take up where he leaves off. Note the determination expressed in his prolonged strokes. These are extended just enough to make their presence felt, but not so far as to become superfluous, so as to impress the masses. The final e, for example, ended rather more sharply than the writer intended, so he has added an extension, to ensure that those who read his words are quite clear about their meaning. Likewise, the capital C reinforces his innate desire to be understood by the intelligentsia.

It is my considered opinion that Alexander Pope must have known someone like this when he penned those immortal lines in his Essay on Criticism: 'Great wits are sure to madness near allied, and thin partitions do their bounds divide'. Here indeed is a *great wit*.

On health: 'Muddy' look could indicate changes in the writer's emotional state. However, I would need to see considerably more of it to determine whether this is due to epilepsy, alcoholism or some other serious cause of mental disturbance. It is also possible that the writer could be somewhat myopic. The grotesque formation of the second capital D certainly suggests some impairment. [The graphologist did not realise that the second D was the shorthand notation for Christ's.] Other interpretations of that thick, heavy stroke—a sensous nature; a strong, energetic drive in youth. Sexual activities appear normal, if spasmodic.

## COMMENTARY ON THE READING

The graphology reading for Nelson is remarkable. 'Bureaucracy be blowed' recalls the incident of the telescope to the blind eye, and the

Nelsonian remark 'I see no ships'. That Nelson was proud, independent and difficult to work with is not open to doubt. The comment about the one-eyed admiral being myopic is very good at first glance, although it seems that this was prompted by a misreading of nineteenth-century shorthand notation.

A graphology reading is relevant to the date at which the handwriting sample was written. In Nelson's case this was 5 September 1805, which according to the *Encyclopaedia Britannica* must only have been a mere 10 days before he sailed from Spithead for the Battle of Trafalgar:

> When his orders came, Nelson made his emotional farewell to Emma and Horatia, his daughter by her, and on September 15 sailed from Spithead in the Victory.[3]

What price the graphologist's comment about a muddy look to the writing that could indicate changes in the writer's emotional state? Finally, her comment that Nelson's sex life was normal, if spasmodic. As an eighteenth-century sailor, it would have had to be spasmodic, at least with Lady Hamilton!

## THE DUKE OF WELLINGTON, 1769–1852

Arthur Wellesley, fourth son of the first Earl of Mornington, an Irish peer, was created Earl of Wellington in 1812, the year of the battle of Badajoz at which this sketch was drawn. Later that year, after the victory of Salamanca and entry into Madrid, he was elevated to Marquis. His most famous success was, of course, Waterloo in 1815, after which he took to politics. He was Prime Minister from 1828–30, when he resigned office rather than accept parliamentary reform. He was Lord Warden of the Cinque Ports from 1829–52 and the letter used in this graphology trial, written from Walmer Castle, was therefore penned at the very start of his Wardenship.

The complete Wellington letter is shown (actual size), but for the trial, the text supplied was only:

presents his Compliments

and begs to acknowledge
the receipt of this memorial
to which
return an Answer after
reference to some Papers in
London—

On the folded piece of writing paper which formed the envelope there was written the words *From the Duke of Wellington 1829*. The handwriting in this letter is consistent with that of Wellington's letter of 9 August 1815, written in Paris to Lord Beresford less than two months after Waterloo, in which he states: 'The Battle of Waterloo was certainly the hardest fought that has been for many years I believe, and has placed in the power of the Allies the most important results'. He continues with a statement that the advantages which could stem from these results are now being wasted. This is not the only time in history that claims are made that the fruits of war are being squandered in the subsequent peace, but it also shows Wellington as other than a bloodthirsty soldier and several of his biographies emphasise the humane side to the man. There is, for example, the Duke's often quoted remark after the battle when he was given the long list of casualties:

Well, thank God I don't know what it is to lose a battle; but certainly nothing can be more painful than to gain one with the loss of so many of one's friends.

## GRAPHOLOGY READING FOR THE DUKE OF WELLINGTON

The pronounced rightward slant shows someone with a lively, outgoing personality. Tightened upper loops suggest inhibition, allied to bowed t strokes, one senses an ability on the part of the writer to keep his basic urges under control. Naked rompings and orgies taking place all around him would leave him unmoved. Mental exhaustion shows itself in those line droppings, *acknowledge* in line 2 and *memorial*

291

in line 3. Lower zones don't figure very prominently, indicating that appetites are small, both sexual and food. He eats to live rather than lives to eat, which is all very honourable and worthy. Considering the d in line 2, the capital letter M in line 3 and the capital letter A in line 5, all have an historic feel. This, like the other samples in the trial, is probably the hand of someone long dead. Upper zones are dominant throughout showing the writer to have been cerebral. Looped capital L in *London* hints at some personal vanity but not enough to have the writer awfully concerned about his personal image. In the words compliments, line 1, papers, line 6 and London, line 7, the upper loop of the lower case letters rises higher than the top of the capitals which in my experience inevitably means subjugation by siblings/parents in childhood. That is, the writer's confidence and abilities were undermined as a child, resulting in an extra special effort to make a success of life later. This is simply to prove those siblings/parents wrong, or maybe simply to gain their respect.

Yes, there is real repression here, of the 21 letter e's in the sample, all except two are blackened. These are memorial in line 3 and reference in line 6.

The writer has a habit of tying himself up in knots (see the following words, at, line 3, answer, line 5, to, lines 4 and 6) but gains a certain amount of pleasure in extricating himseslf from the same knots. It is almost as if it were deliberate, to illustrate his own cleverness to lesser mortals, or to those who once undermined him.

The occasional backward loops or letters already mentioned, and, in line 2, at, in line 3, suggest mother-influence or even mother-dominance. (I've seen a similar phenomenon in the youthful hand of Prince Charles.) All of this makes me wonder if this writer was very heavily influenced by his mother. The very high capital letter M in line 3 seems to suggest it. The little tick before he goes into the letter proper is also revealing and is seen in only two places, capitals M and A. Mother-influence is strong indeed if he has to take an unconscious breather before he even pens the letter. What *did* she do to him when he was young? Now let us turn to letter A. Was there some ghastly maiden aunt also wielding her influence over him? Let us just say the lady had such an effect on him that he subsequently found it very difficult to express, indeed, feel, emotions. If he ever got around to marrying and fathering children, it would have been out of a sense of duty.

This reading should provide a salutary lesson for all overpowering mothers and maiden aunts!

## COMMENTARY ON THE READING

The emphasis on an inhibiting mother–son relationship was something of a mystery and certainly the *Dictionary of National Biography*[4] gives no clues—only:

1769 Born
1784 Education at Eton, Brussels (15 years old)
1789 Angers Military Academy (17 years old)
1787 Lieutenant of Foot (18 years old)

It is known that from an early age he was taunted by his older brother as having a different father from the elder brothers, and also that he had a stern and unloving governess. However, if this is the female dominance in his early life, graphology makes too much of it. Comments about his integrity are obviously true and a successful Commander in Chief of a field army must have been able to keep his feelings under control, especially in the heat of battle. He may well have been self-opinionated and his marriage was not a great success. He died aged 81 years old from heart failure.

REFERENCES
[1] Hattersley R 1974 *Nelson* (London: Weidenfeld & Nicholson)
[2] The *Times* 1 October 1984 'Shakespeare tops the autograph price list'
[3] *Encyclopaedia Britannica*
[4] *Dictionary of National Biography,* The Concise Dictionary, Part 1, from the beginning to 1900 (Oxford: Oxford University Press)

# Finger Reckoning in the Middle Ages

One of the few Europeans who did enough in mathematics during the Middle Ages to be remembered today was Leonardo of Pisa (1175–1230) also known as Fibonacci, who wrote *A Book of Counting* in 1202. He described three methods of solving problems, one involving Arabic numerals, one the abacus and the third, finger reckoning. The people of the Middle Ages went in for the latter in a big way, and indeed, solving problems with fingers goes back far beyond the Middle Ages and it is not now generally realised that big numbers could be represented on the fingers. There was a system of finger signs in the same way that there is a system using hand signals for the deaf.

A person could also multiply with his or her fingers, and with finger reckoning one would only need to know one's multiplication tables up to 5 × 5.

Here is how one would do 6 × 9 on one's fingers: On one hand hold up one finger (the difference between the five fingers on that hand and

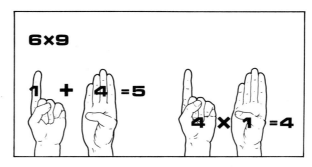

**6×9**

$1 + 4 = 5$   $4 × 1 = 4$

Add *the* standing fingers *to get the* tens *and* multiply *the* closed fingers *to get the* units.

the first of the numbers to be multiplied, i.e. six). On the other hand hold up four fingers, the difference between five and nine. Then *add* the standing fingers $(1 + 4 = 5)$ to get the tens of the answer. In this case there would be five tens, so one would know that the answer would be 50-something. To find the 'something' *multiply* the closed fingers $(4 × 1 = 4)$, thus reaching the answer of 54. A similar system could be used to multiply numbers between 10 and 15. Take the problem 14 × 12. Put up four fingers on one hand and two on the other. Add them to get six; then add 10 times that sum to 100, which would give you 160; finally add the product of the fingers, 4 × 2, to reach the answer: 168. If you know algebra the equation: $xy = 10 \ [(x - 10) + (y - 10)] + 100 + (x - 10)(y - 10)$ will show you why this system works.

REFERENCE
Adapted from Rogers J T 1968 *The Story of Mathematics* (Leicester: Brockhampton Press)

## Metoposcopy: Observation of the Forehead

The great exponent of metoposcopy was Jerome Cardan, a polymath whose many activities included medicine, astronomy, mathematics, astrology and the occult arts. He was born in 1500, the illegitimate son of a Milanese scholar who was both a doctor in medicine and law and also a competent mathematician. Jerome's education was rather irregular but, among other subjects, he studied medicine as well as mathematics and music and received an MD degree from Padua in 1526. He was an inveterate gambler, obstinate and cantankerous, and made many enemies. However, although his life was full of misfortune, he achieved great fame as an astrologer as well as a physician and travelled to Paris, London and Edinburgh to practice astrology and medicine. It was in London in 1552 that he cast the horoscope of King Edward VI, the son and successor of King Henry VIII. Cardan's predictions, though, were abysmally wrong and he foretold that Edward would live at least until his 55th birthday and then suffer from various diseases. In fact, he died in 1553 before reaching his 17th birthday. The excuse by Cardan, given in King (1973), is:

> It was unsafe to pronounce upon the term of life in weak nativities, unless all processes, and ingresses, and external movement that from month to month and year to year affect the ruling planets had been carefully inquired into. But to make such a calculation would have cost him, he said, not less than a hundred hours.

and then apparently he added a few more reasons!

Cardan, when he presented the subject of metoposcopy, reminds us of some present day writers on palmistry. First, he provided some blanket statements that could serve as excuses-in-advance. Metoposcopy, he said is an art, but like navigation, agriculture or medicine, it is by no means infallible. By regarding the lines on the forehead and the configuration of the head, we cannot really have complete confidence in the future, but we can predict what will happen *for the most part*. Knowledge is increased as we take into account other factors, including parents, education, age, palmistry and the constitution of the stars. Using all this information, we can make fairly safe judgements.

To indicate the validity of metoposcopy, Cardan cited a case from Suetonius, from the life of Titus. A person skilled in metoposcopy correctly predicted that Britannicus, the son of the Emperor Claudius, would not ascend the throne, but that Titus, then a child, would do so. This type of citation is of course a familiar method of *proving* the validity of prophecy by pointing to a particular case where a prediction came true. Cardan bolstered this argument by claiming that we never find a wicked man who entirely lacks the signs of wickedness on his forehead, but no upright person is contaminated by these signs.

He presented various rules that seemed to govern the art of metoposcopy and distinguished seven horizontal lines in the forehead each correlated with a different planet. This was necessary since the planets and the zodiac are central in all divinatory arts.

In addition to the major horizontal lines, there are others that intersect, forming small figures such as crosses, circles, squares, triangles and stellate figures, each having its own significance. Lines are to be judged by various qualities: if continuous, they indicate good fortune; if interrupted, adversity of some type; if prominent, they

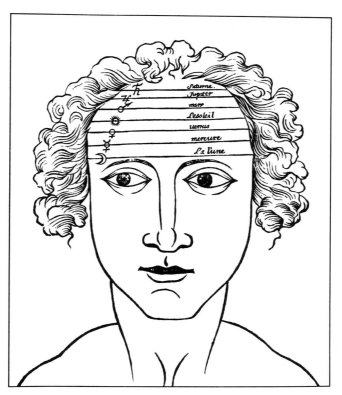

*Cardans's map of the forehead, indicating the spheres of influence of the planets.*

signify great and noteworthy events or a manifest effect of the planet represented; but if small or delicate, then obscure events or feeble effects. Curving lines reveal an exaltation that subsides. Branching lines mean trouble. When crossed, they portend danger and harm from the planet represented. The major lines are divisible into three portions: the left third have to do with early life, up to age 30; the middle third to age 60; and the right side to old age, approximately 90. Lines that ascend to the right are favourable, those that descend are bad.

Furthermore, various other marks on the forehead or face, such as naevi, have a significance of their own and may change the interpretation. Just as the lines on the forehead are associated with the planets, so are the various naevi on the face.

Cardan presented a large number of individual case reports with diagrammatic reproductions of the facial ines and marks, together with a brief forecast. But he did not link up his individual cases with general principles and hence there is no way to see how he derived his prophecies. Nor was there any follow-up to say if the predictions were valid!

Examples of case reports are:

*The man who is marked with these lines will die of plague.*

*This man will die of bladder stone.*

*The three semicircular lines with two straight ones, and a naevus on the right cheek, promises him a long life and great honour outside his country.*

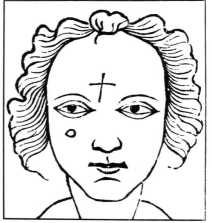

*This is the forehead of an unfortunate woman. The lines in the form of a cross denote a woman corrupted before and after her marriage, who nevertheless will have two husbands. The naevus here against the right zygoma means that she will have intercourse* (commercium) *with churchmen* (ecclesiasticus) *and will die of poison.*

297

Unlike palmistry, metoposcopy and the more obscure forms of prediction have fallen on hard times and are largely forgotten, except for one notable reference in the *Journal of the American Medical Association* in August 1968 (**205** 470). In this article, the authors noted a correlation between certain lines in the forehead (the vertical furrows between the eyebrows) and the coexistence of chronic duodenal ulcers. In a single-blind study they showed in one series that 87% of their patients with ulcer had three or more vertical furrows, but only 5% of the control patients with no ulcer were in this metoposcopical group. Their study presented the graphical distribution of vertical forehead lines in 400 patients and those with ulcers seemed significantly separated from the controls. However, it must be emphasised that these authors gave their explanations in terms of modern medical science—the planets were ignored!

REFERENCES
Text adapted from King L S 1973 'Metoposcopy and kindred arts' *J. Am. Med. Assoc.* **224** 42–6
Illustrations from 1658 *Metoposcopia* (Paris: Thomas Jolly)

## Charms against Teething

Shown below is a charm against teething consisting of a necklet of small shells. It was used in Jersey.

One of the most common charms for teething was the orris root, used in Norfolk and east London among other places, but there were

Left: *Necklet of shells from Seville*. Right: *Orris root, which was used not only in England but was also rubbed on the gums of Jewish children in the Balkan States and in Poland. (Courtesy Wellcome Institute Library, London.)*

also pigs' feet (south London), a necklet of nightshade (Gloucestershire and Warwickshire), calves' teeth (south London), a pimento necklace (Yorkshire), a bag of split ash twigs (Devon) and a wolf's tooth worn on a cap not only to guard against teething troubles but also to avert evil spirits (Palestine).

## Charm against Nightmares

*Holed stones to be hung at the head of the bed as a charm against nightmares. (Courtesy Wellcome Institute Library, London.)*

## Rabbit Fights Back

Natural justice, but not the police, has caught up with two men in South Australia who killed a rabbit with high explosive. According to a free magazine distributed to Australian expatriates in London, police are looking for the two who apparently tied a stick of gelignite to the bunny and lit the fuse before setting the terrified animal free. 'The trick backfired,' the magazine reports, 'when the rabbit took cover

under their four-wheel drive vehicle. The men escaped but their
$20 000 utility was wrecked.'

REFERENCE
*New Scientist* 16 August 1984 p. 41

## Charm against Rheumatism

*This amulet against rheumatism was two mole's feet in a satchel. (From
Sussex, England.) Mole's feet were also used as a charm for* toothache *in
Norfolk, an alternative in Wiltshire was joined hazel-nuts and in Devon,
stones or teeth. (Courtesy Wellcome Institute Library, London.)*

# Suicide

A monologue written by Roy Clegg and Harold Clegg-Walker in 1931.

Now I am a fellow who's lived—
    I've lived and I've loved and I've lost,
I loved a fair damsel who didn't love me,
    So my fate to the winds must be tossed.
She laughed when I tried to be serious,
    Her pretty heels trampled my heart,
The future means nothing, so now I must die,
    But the question is—how do I start?
To jump in a river and drown
    Is a good plan, or so I've been told.
But if some poor simpleton does fish you out,
    You get such a terrible cold.
To turn on the gas and be smothered
    Is one way, but you never can tell
If someone will biff in and spoil the whole thing,
    And just curse you for making a smell.
To hang by the neck on one's braces
    Is sometimes considered the thing,
But I fear mine would hardly stand up to the strain,
    For already they're tied up with string.
I like the idea of revolvers,
    I'd prefer to be shot than to hang,
But ever since childhood I've always been scared
    Of things that go off with a bang.
I once had a vague sort of notion
    Of leaping beneath an express,
But one has to consider one's fellows, you know,
    The chappies who clear up the mess.
Still the Underground might offer scope,
    One could sit down upon a live rail,
But that involves trespassing—what a disgrace
    If the corpse had to serve time in jail.
Now Keatings kills bugs, moths and beetles,
    In that case it ought to kill me.
The butler must make me some sandwiches, James!
    No. Dash it! he's gone out to tea.
Never mind though, there must be some way,
    Such as hurling yourself from a cliff,
Even then you might find that you'd only been
        stunned,
    And you'd wake up most frightfully stiff.

I've given up thinking of razors,
   I've tried and I'm wondering yet
How those fellows who do it can cut their own throats,
   It's a thing beyond my Gillette.
No, really, I'm finding that this sort of thing's
   Not as easily done as it's said.
So I think I'll pop off for the week-end or so—
   And perhaps shoot some rabbits instead.

REFERENCE
Marshall M (ed) 1981 *The Book of Comic and Dramatic Monologues* (London: Elm Tree Books/EMI Publishing). Reproduced by permission of EMI Music Publishing Ltd.

---

# Anointing

It may come as a surprise to many that the anointing is no inconsiderable part of the coronation service; but so it is, and it would even be correct to describe it as more important than the crowning itself—more important, that is, from the inner and significant aspect of the ceremony. As part of the coronation ritual anointing was introduced into England in 871, when King Alfred came to the throne, and into Scotland in 1097, when King Edgar was crowned; it now forms a culminating point in the solemn spectacle enacted in Westminster Abbey, when a new monarch comes to rule as King of the United Kingdom of Great Britain and Ireland and of the British Dominions beyond the seas. It is interesting, further, to recollect that anointing has a long history behind it, and that it probably took origin, as A E Crawley asserts [*Encyclopaedia of Religion and Ethics* 1908 vol. 1 pp 549–57], in pre-theistic and even pre-fetish times. At first unguents, varying in nature from a crude animal fat to a perfumed vegetable oil, were used for cosmetic purposes; they gave to the body, after bathing, a feeling of comfort and they were useful in hot climates to protect the skin against the sun's rays and in cold countries to prevent the escape of the body heat. But soon a new meaning—the aesthetic—was added to the cosmetic; the oil gave a gloss to the skin, and the effect was increased by the admixture of a colouring substance. 'Of the majority of early peoples,' says Crawley, 'it may be said that grease and ochre constitute their wardrobe.' In this way personal attractiveness was sought for, and, in the opinion of those interested, was obtained. What, then, could be more natural than to anoint the body for such special occasions as festivals and holy days? The honoured guest also was welcomed with bathing and anointing, and inanimate objects were polished and to some extent preserved from decay by being rubbed with unguents. It is easy to link on these uses to the magical and religious employment of ointments, but in the process a new notion is found to have come in. Organic matter is believed to be instinct with some Divine force or vital essence and to have a magical or supernatural power; further, this power can be transmitted either by eating and drinking the specially sacred matter, or, more easily and doubtless more safely, by external application of it to the body of the worshipper. Fat has come to occupy a curiously important place among the substances which are held to transmit the desirable gifts of life and strength; and it is reported, for instance; that the Australian savage will kill a man to get his kidney fat for anointing himself with. A further stage in development sees fat consecrated in various ways, so that it may serve as the vehicle for the transmission of a Divine Essence. It can now be seen how anointing has entered into medical practice, not by one, but by several routes. There has been the belief in the magical effect of the fat as well as in the sacred influence which it carries; there has been in an interesting way (too intricate to enter into here) the principle of the removal of tabu; there has been the discovery of the invigorating effect of rubbing the skin with oil; and there has been the knowledge of the action of the unguent upon the cutaneous secretions and upon transpiration. If to these be added the ascertained therapeutic effects of various medicinal substances, we are possessed of a series of contributing causes which sufficed to introduce anointing into medicine and to keep it among the accredited means of treating disease.'

REFERENCE
*Br. Med. J.* 1911 **1** 1479–80 (published at the time of the coronation of King George V)

---

## Leonardo's Last Supper

A most bizarre representation of the Leonardo masterpiece in the Sistine Chapel in the Vatican—on the back of Mrs Emma de Burgh of the USA. Nineteenth century photograph in the publication *La*

*Nature.* The tatoos were executed by the Riley brothers on the husband and wife Frank and Emma de Burgh.

*(Courtesy Mary Evans Picture Library.)*

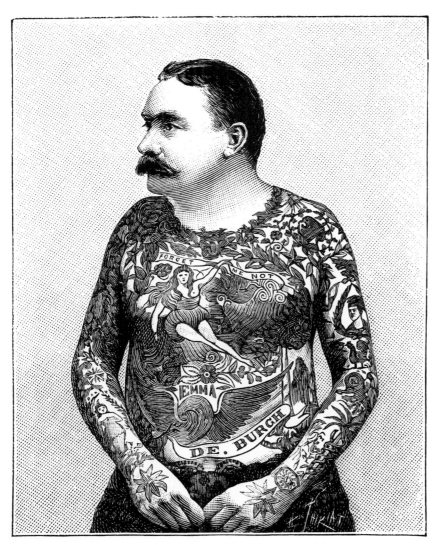

(*Courtesy Mary Evans Picture Library.*)

## The Royal Wedding

The following is an extract from a young schoolgirl's essay on the wedding of Prince Charles and Lady Diana Spencer. It was first overheard in 1982 as part of an after-dinner speech by an eminent

Archdeacon—*and is true.*

. . . Prince Charles rode in an open cart, but Dinah and Father Earl Spencer went by First Class Carriage. As the bride walked the red mat in St. Paul's we saw her veil was held to her head by the Spencer Terror . . .

When the Prince and his Lady had exchanged vowels and the ring was put on to make man and wife, the marriage was consummated in front of the high altar by the Archbishop of Canterbury followed by several of London's top clergy . . .

After the ride back, the Prince and She waves from Buck and Ham Palace from the garage roof . . .

---

## Treatment of Political Prisoners Under a Liberal Government

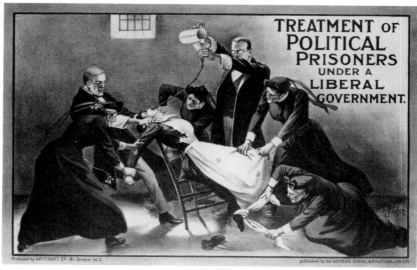

*A suffragette poster, circa 1910, published by the Women's Social and Political Union. (Courtesy Mary Evans Picture Library.)*

---

## How Diet Changes Your Personality

The changes in personality which can be caused by strict diets have been given considerable publicity in the last few days following the sentencing of champion racing jockey Lester Piggott for tax evasion on Friday. Piggott virtually starved himself for 30 years to stay at 8st 3lbs—two stones below his normal body weight—and it is thought that his diet might have led to his obsession with hoarding money.

A different effect is reported in the current issue of the *British Medical Journal* which tells the tale of a 68-year-old man who was admitted to hospital after his family had complained that his behaviour had changed dramatically.

This man's normal manner was quiet and reserved, but in his new guise he became over-excited and sleepless. When a doctor visited his home he was racing round non-stop, stark naked, and cracking jokes.

A month before this change, the man had consulted another doctor about heart problems, and had been put on a strict diet of two pints of milk a day with nothing else, to control a rise in his blood pressure.

Once in hospital, tests revealed that the diet had triggered his thyroid gland into over-drive, which in turn had led to his manic behaviour.

Within a month of being taken off the diet, his behaviour had returned to normal.

REFERENCE
The *Independent* 27 October 1987

## *Medical English*

The journal *World Medicine*, which sadly ceased publication in 1984, was well known as a source of anecdotal material. This included regular snippets from the papers and from communications between doctors, patients and hospitals. This is a selection from the years 1977–84.

> The condomned murderer and his girlfriend took overdoses today in an apparent suicide pact.
>
> **Glasgow Herald**

> The operation was carried out by a Scot, as was the anaesthetist.
>
> **Sunday Post**

> From our records, we know that 17 patients are using a mechanical method, though this may be the tip of the iceberg.
>
> **Health Bulletin**

My doctor wants me to have the baby in hospital because I'm an elderly primate.

**Patient in Sidmouth**

I don't wish to be put on the duty rooster at the chicken factory.

**Lady patient in Sussex**

Lost: green silk umbrella belonging to lady with one broken rib and curiously shaped ivory head.

*Herts Advertiser*

We shall be at the Curragh for the annual orgasm of the Hospitals' Trust Derby. Of course it is going to be a most gorgeous afternoon . . .

*Dublin Evening Press*

Thank you for sending this woman along. I have started the bull rolling by requesting a semen analysis.

**Letter from infertility clinic, Hertfordshire**

Although suicide by a shot in the neck was atypical, the experts said, there were a number of references in scientific literature which described it as a safe method.

*The Guardian*

Hand painted pub signs could be on their way back in the North West. More than 350 artists submitted entries when a Warrington based brewery invited people to be hung outside their pubs.

*Glazebury Journal*

Reminiscing about Empire Days past, she told the Policy & Resources Committee: 'In the morning we saluted the flag, and in the afternoon we had it off'.

*Worthing Gazette*

Impotence as a presenting symptom in this condition is said to be very uncommon, but there is very little hard data.

*British Journal of Sexual Medicine*

British babies like children casserole best.

*The Times*

Some men are so disturbed and depressed on receiving a life sentence that they end up getting physiotherapy.

*Medical News*

Event Nine: The Norden Trophy for the cow with the best udder. Presented by Mr Norden in memory of Mrs Norden.

**Guernsey Agricultural Show brochure**

Dear Doctor, Could you please leave a note to cover me, as I went into hospital for a scrap on Friday, and to be quarterized.

**Note from an Essex patient**

Using the embassy diplomatic bog to smuggle weapons and ammunition into Britain.

*Daily Telegraph*

Dear Doctor, Please come quickly—diarrhoea very bad and going downhill rapidly.

**Request for visit, Edinburgh**

The hospital's new diagnostic services include three x-ray rooms and nine extra places in the mortuary.

*Ormskirk Advertiser*

Two hundred and twenty three patients have been waiting for gynaecological surgery on their ears, nose or throat.

*Manchester Evening News*

The unusual series of sex change operations involves a bilateral mastectomy, hysterectomy and penal surgery.

*Aberdeen Evening Express*

Wanted: Assistant or associate professor in plastic, reconstructive and bum surgery.

**University of Kuwait advertisment**

An American surgeon lecturing medical students in Bristol was asked if he considered the operation he was describing was a valuable one. 'Valuable' demanded the surgeon, a little taken aback. 'I raised five kids on it'.

*Financial Times*

I saw this patient of yours in the clinic. She does appear to be pregnant, and there are social difficulties, with the possibility of the child not being hers.

**Note from Manchester obstetrician**

Quack Chinese doctors are paying £40 000 for bodies snatched from graves. They use parts of the bodies to make medicines that guarantee immortality.

*Yorkshire Evening Post*

For Sale: Rolls Royce hearse, with 1965 body.

*Droitwich Gazette*

Members of the family request mourners to omit floral tributes. The deceased was allergic to flowers.

**New York obituary column**

Throwing caution to the wind, I ordered a tournedos and half a giraffe of wine.

*Manchester Evening News*

The patient changed to Guinness under the mistaken belief that it would not affect his lever.

**Surrey consultant's letter**

Come to enjoy us in our spectacular LA HACIENDA with mariachis, music, beautiful open terrace and air conditioned entrails. Nice. Clean.

*Acapulco Hotel News*

*Nuclear Leak*. The underground control centre to be used at Kettering, Northants, in the event of a nuclear attack, has only an outside lavatory, according to a report to be considered by county councillors.

*Daily Telegraph*

*Gallstones go with a bang*. Doctors at a hospital in the ancient Chinese capital of Xian have perfected a new way of disposing of gall bladder stones. They blow them up with micro-explosives. Physicians insert an apparatus into the bladder, bore a hole in the stone, insert tiny explosives and reduce the stone to fragments which are discharged in the urine.

**Xinhua: official Chinese news agency**

Even the post mortem failed to establish a precise diagnosis and no useful treatment was subsequently initiated.

*Hospital Doctor*

---

## When the Bull Beats the Matador

The *British Medical Journal* of 12 July 1986 carried the following item previously reported in the *Journal of the Royal Society of Medicine* concerning a conference in Spain.

A Spanish surgeon who contributed to the meeting included several bullfight injuries in his series of patients with traumatic lesions of the anal sphincter. Apparently only cowardly matadors come to the proctologist; the brave full frontal ones are said to provide business for the urologists.

310

# The Many Pains of Snoring

Heavy snoring may sometimes be a sign of serious heart disease, but it can also lead to other unexpected painful conditions, as described in a letter in the *New England Journal of Medicine*.

A 66-year-old man from Toronto consulted his GP about a dull pain deep in the muscle of his calf. The GP could not find any major abnormalities, prescribed pain-killers and suggested the man return if the pain intensified.

Two nights later, the patient had just fallen asleep when he was abruptly awoken by a sharp pain in the same calf, the result of a kick by his wife. His remonstration: 'Don't kick me there, that's just where my leg hurts,' was greeted with the reply: 'You were snoring again, and that's where I always kick you to stop it.'

After some discussion his wife agreed to stop kicking. Unfortunately, he could not keep his side of the bargain and continued snoring, but the pain in his calf disappeared within three days.

REFERENCE
The *Independent* 6 October 1987

---

# A Student's Criminals

The statistical significance test, often used in medicine, is a test of the difference between two mean values. The *t*-distribution which under-lies the test was discovered in 1908 by W S Gosset, who was a brewer working for the Guinness Company. The data which he studied for the *t*-distribution came from records of the heights of 3000 criminals. Gosset, who published under the pseudonym 'student', took 750 samples, each of the heights of four criminals, and used these data to calculate 750 means.

REFERENCE
Mould R F 1989 *Introductory Medical Statistics* 2nd edn (Bristol: Hilger)

---

## An Old-Time Apothecary

This figure of the apothecary with his pestle and mortar forms one of the details in the choir of Amiens Cathedral, and since these date from 1508–19 this representation of a pharmacist is from about the time of the Field of the Cloth of Gold.

REFERENCE
*The Chemist & Druggist* 4 September 1920 (*Courtesy Wellcome Institute Library, London.*)

---

# *The Aetiology and Treatment of Childhood*[1,2]

This paper originally appeared in the *Journal of Polymorphous Perversity* 1985 **2** 3–7. © Wry-Bred Press Inc. 1985

Childhood is a syndrome which has only recently begun to receive serious attention from clinicians. The syndrome itself, however, is not at all recent. As early as the eighth century, the Persian historian Kidnom made reference to 'short, noisy creatures,' who may well have been what we now call 'children.' The treatment of children, however, was unknown until this century, when so-called 'child psychologists' and 'child psychiatrists' became common. Despite this history of clinical neglect, it has been estimated that well over half of all Americans alive today have experienced childhood directly (Suess 1983). In fact, the actual numbers are probably much higher, since these data are based on self-reports which may be subject to social desirability biases and retrospective distortion.

The growing acceptance of childhood as a distinct phenomenon is reflected in the proposed inclusion of the syndrome in the upcoming *Diagnostic and Statistical Manual of Mental Disorders, 4th Edition,* or *DSM-IV*, of the American Psychiatric Association (1985). Clinicians are still in disagreement about the significant clinical features of childhood, but the proposed *DSM-IV* will almost certainly include the following core features:

1. Congenital onset
2. Dwarfism
3. Emotional lability and immaturity
4. Knowledge deficits
5. Legume anorexia

## CLINICAL FEATURES OF CHILDHOOD

Although the focus of this paper is on the efficacy of conventional treatment of childhood, the five clinical markers mentioned above merit further discussion for those unfamiliar with this patient population.

[1] The author would like to thank all the little people.
[2] This research was funded in part by a grant from Bazooka Gum.

*Congenital onset.* In one of the few existing literature reviews on childhood, Temple-Black (1982) has noted that childhood is almost always present at birth, although it may go undetected for years or even remain subclinical indefinitely. This observation has led some investigators to speculate on a biological contribution to childhood. As one psychologist has put it, 'we may soon be in a position to distinguish organic childhood from functional childhood' (Rogers 1979).

*Dwarfism.* This is certainly the most familiar clinical marker of childhood. It is widely known that children are physically short relative to the population at large. Indeed, common clinical wisdom suggests that the treatment of the so-called 'small child' (or 'tot') is particularly difficult. These children are known to exhibit infantile behaviour and display a startling lack of insight (Tom and Jerry 1967).

*Emotional lability and immaturity.* This aspect of childhood is often the only basis for a clinician's diagnosis. As a result, many otherwise normal adults are misdiagnosed as children and must suffer the unnecessary social stigma of being labelled a 'child' by professionals and friends alike.

*Knowledge deficits.* While many children have IQs within or even above the norm, almost all will manifest knowledge deficits. Anyone who has known a real child has experienced the frustration of trying to discuss any topic that requires some general knowledge. Children seem to have little knowledge about the world they live in. Politics, art and science—children are largely ignorant of these. Perhaps it is because of this ignorance, but the sad fact is that most children have few friends who are not, themselves, children.

*Legume anorexia.* This last identifying feature is perhaps the most unexpected. Folk wisdom is supported by empirical observation—children will rarely eat their vegetables (see Popeye 1957, for review).

## CAUSES OF CHILDHOOD

Now that we know what it is, what can we say about the causes of childhood? Recent years have seen a flurry of theory and speculation from a number of perspectives. Some of the most prominent are reviewed below.

*Sociological model.* Emile Durkind was perhaps the first to speculate about sociological causes of childhood. He points out two key observations about children: 1) the vast majority of children are unemployed, and 2) children represent one of the least educated segments of our society. In fact, it has been estimated that less than 20% of children have had more than a fourth grade education.

314

Clearly, children are an 'out-group.' Because of their intellectual handicap, children are even denied the right to vote. From the sociologist's perspective, treatment should be aimed at helping assimilate children into mainstream society. Unfortunately, some victims are so incapacitated by their childhood that they are simply not competent to work. One promising rehabilitation program (Spanky and Alfalfa 1978) has trained victims of severe childhood to sell lemonade.

*Biological model.* The observation that childhood is usually present from birth has led some to speculate on a biological contribution. An early investigation by Flintstone and Jetson (1939) indicated that childhood runs in families. Their survey of over 8000 American families revealed that over half contained more than one child. Further investigation revealed that even most non-child family members had experienced childhood at some point. Cross-cultural studies (e.g. Mowgli and Din 1950) indicate that familial childhood is even more prevalent in the Far East. For example, in Indian and Chinese families, as many as three out of four family members may have childhood.

Impressive evidence of a genetic component of childhood comes from a large scale twin study by Brady and Partridge (1972). These authors studied over 106 pairs of twins, looking at concordance rates for childhood. Among identical or monozygotic twins, concordance was unusually high (0.92), i.e. when one twin was diagnosed with childhood, the other twin was almost always a child as well.

*Psychological models.* A considerable number of psychologically based theories of the development of childhood exist. They are too numerous to review here. Among the more familiar models are Seligman's 'learned childishness' model. According to this model, individuals who are treated like children eventually give up and become children. As a counterpoint to such theories, some experts have claimed that childhood does not really exist. Szasz (1980) has called 'childhood' an expedient label. In seeking conformity, we handicap those whom we find unruly or too short to deal with by labelling them 'children'.

## TREATMENT OF CHILDHOOD

Efforts to treat childhood are as old as the syndrome itself. Only in modern times, however, have humane and systematic treatment protocols been applied. In part, this increased attention to the problem may be due to the sheer number of individuals suffering from childhood. Government statistics (DHSS) reveal that there are more children alive today than at any time in our history. To paraphrase P T Barnum: 'There's a child born every minute.'

The overwhelming number of children has made government intervention inevitable. The nineteenth century saw the institution of what remains the largest single programme for the treatment of childhood —so-called 'public schools.' Under this colossal programme, individuals are placed into treatment groups based on the severity of their condition. For example, those most severely afflicted may be placed in a 'kindergarten' programme. Patients at this level are typically short, unruly, emotionally immature, and intellectually deficient. Given this type of individual, therapy is of necessity very basic. The strategy is essentially one of patient management and of helping the child master basic skills (e.g. finger painting).

Unfortunately, the 'school' system has been largely ineffective. Not only is the programme a massive tax burden, but it has failed even to slow down the rising incidence of childhood.

Faced with this failure and the growing epidemic of childhood, mental health professionals are devoting increasing attention to the treatment of childhood. Given a theoretical framework by Freud's landmark treatises on childhood, child psychiatrists and psychologists claimed great successes in their clinical interventions.

By the 1950s, however, the clinicians' optimism had waned. Even after years of costly analysis, many victims remained children. The following case (taken from Gumbie and Pokey 1957) is typical.

> Billy J., age 8, was brought to treatment by his parents. Billy's affliction was painfully obvious. He stood only 4' 3" high and weighed a scant 70 pounds, despite the fact that he ate voraciously. Billy presented a variety of troubling symptoms. His voice was noticeably high for a man. He displayed legume anorexia and, according to his parents, often refused to bathe. His intellectual functioning was also below normal —he had little general knowledge and could barely write a structured sentence. Social skills were also deficient. He often spoke inappropriately and exhibited 'whining behaviour.' His sexual experience was non-existent. Indeed, Billy considered women 'icky.'
>
> His parents reported that his condition had been present from birth, improving gradually after he was placed in a school at age 5. The diagnosis was 'primary childhood'. After years of painstaking treatment, Billy improved gradually. At age 11, his height and weight have increased, his social skills are broader, and he is now functional enough to hold down a 'paper route.'

After years of this kind of frustration, startling new evidence has come to light which suggests that the prognosis in cases of childhood may not be all gloom. A critical review by Fudd (1972) noted that studies of the childhood syndrome tend to lack careful follow-up. Acting on this observation, Moe, Larrie and Kirly (1974) began a large scale longitudinal study. These investigators studied two groups.

The first group comprised 34 children currently engaged in a long-term conventional treatment programme. The second was a group of 42 children receiving no treatment. All subjects had been diagnosed as children at least 4 years previously, with a mean duration of childhood of 6.4 years.

At the end of one year, the results confirmed the clinical wisdom that childhood is a refractory disorder—virtually all symptoms persisted and the treatment group was only slightly better off than the controls.

The results, however, of a careful 10-year follow-up were startling. The investigators (Moe, Larrie, Kirly and Shemp 1984) assessed the original cohort on a variety of measures. General knowledge and emotional maturity were assessed with standard measures. Height was assessed by the 'metric system' (see Ruler 1923), and legume appetite by the Vegetable Appetite Test (VAT) designed by Popeye (1968). Moe et al found that subjects improved uniformly on all measures. Indeed, in most cases, the subjects appeared to be symptom-free. Moe et al report a spontaneous remission rate of 95%, a finding which is certain to revolutionize the clinical approach to childhood.

These recent results suggest that the prognosis for victims of childhood may not be so bad as we have feared. We must not, however, become too complacent. Despite its apparently high spontaneous remission rate, childhood remains one of the most serious and rapidly growing disorders facing mental health professionals today. And, beyond the psychological pain it brings, childhood has recently been linked to a number of physical disorders. Twenty years ago, Howdi, Doody and Beauzeau (1965) demonstrated a six-fold increased risk of chicken pox, measles and mumps among children compared with normal controls. Later, Barby and Kenn (1971) linked childhood to an elevated risk of accidents—compared with normal adults, victims of childhood were much more likely to scrape their knees, lose teeth and fall off their bikes.

Clearly, much more research is needed before we can give any real hope to the millions of victims wracked by this insidious disorder.

## REFERENCES

American Psychiatric Association 1985 *The diagnostic and statistical manual of mental disorders, 4th edition: A preliminary report.* (Washington, DC: APA).
Barby B and Kenn K 1971 The plasticity of behavior. In B Barby and K Kenn (eds) *Psycho-therapies R Us* (Detroit: Ronco Press).
Brady C and Partridge S 1972 My dad's bigger than your dad *Acta Eur. Age* **9** 123–6
Flintstone F and Jetson G 1939 Cognitive mediation of labor disputes *Industrial Psychology Today* **2** 23–35

Fudd E J 1972 Locus of control and shoe-size *Journal of Footwear Psychology* **78** 345–56

Gumbie G and Pokey P A cognitive theory of iron smelting *Journal of Abnormal Metallurgy* **45** 235–9

Howdi C, Doodi C and Beauzeau C 1965 Western civilization: A review of the literature *Reader's Digest* **60** 23–5

Moe R, Larrie T and Kirly Q 1974 State childhood vs. trait childhood *TV Guide* May 12–19 1–3

Moe R, Larrie T, Kirly Q and Shemp C 1984 Spontaneous remission of childhood. In W C Fields (ed) *New Hope for Children and Animals* (Hollywood: Acme Press)

Popeye T S M 1957 The use of spinach in extreme circumstances *Journal of Vegetable Science* **58** 530–8

Popeye T S M 1968 Spinach: A phenomenological perspective *Existential Botany* **35** 908–13

Rogers F 1979 *Becoming My Neighbour* (New York: Soft Press)

Ruler Y 1923 Assessing measurement protocols by the multi-method multiple regression index for the psychometric analysis of factorial interaction. *Annals of Boredom* **67** 1190–260

Spanky D and Alfalfa Q 1978 Coping with puberty *Sears Catalogue* 45–6

Suess D R 1983 A psychometric analysis of green eggs with and without ham *Journal of Clinical Cuisine* **245** 567–78

Temple-Black S 1982 Childhood: An ever-so sad disorder *Journal of Precocity* **3** 129–34

Tom C and Jerry M 1967 Human behavior as a model for understanding the rat. In M de Sade (ed) *The Rewards of Punishment* (Paris: Bench Press, 1967)

## FURTHER READINGS

Joe G I 1965 Aggressive fantasy as wish fulfillment *Archives of General MacArthur* **5** 23–45

Leary T 1969 Pharmacotherapy for childhood *Annals of Astrological Science* **67** 456–9

Kissoff K G B 1975 Extinction of learned behavior. Paper presented to the Siberian Psychological Association, 38th Annual Meeting, Kamchatka

Smythe C and Barnes T 1979 Behavior therapy prevents tooth decay *Journal of Behavioral Orthodontics* **5** 79–89

Potash S and Hoser B 1980 A failure to replicate the results of Smythe and Barnes *Journal of Dental Psychiatry* **34** 678–80

Smythe C and Barnes T 1980 Your study was poorly done: A reply to Potash and Hoser *Annual Review of Aquatic Psychiatry* **10** 123–56

Potash S and Hoser B 1981 Your mother wears army boots; A further reply to Smythe and Barnes *Achives of Invective Research* **56** 570–8

Smythe C and Barnes T 1982 Embarrassing moments in the sex lives of Potash and Hoser: A further reply *National Enquirer* May 16

# An Italian Quack

The few lines of old Italian poetry at the bottom of the picture translate as:

The man who from acuteness of intellect,
Wishes to make some anatomical investigation,
is a sharp snake,
He shows over and above his obvious absolute power the privilege
of deceiving the world.

*(Courtesy Wellcome Institute Library, London.)*

The snake is a reference to the Greek god of medicine and healing, Asklepios, who as Aesculapius, was introduced into Rome in 293 BC, in the form of a snake, during a pestilence. Mythology has it that Aesculapius was the son of Apollo and was trained as a physician of miraculous powers by Chiron, the most learned of the Centaurs. Once, it was said, to the annoyance of his grandfather Zeus, he restored a dead man to life. Zeus became jealous and at his request, Pluto hurled a thunderbolt at Aesculapius, which ended his life on earth. Mythology also states that Aesculapius was the ship's doctor on the *Argo* when Jason and the Argonauts searched for the Golden Fleece.

REFERENCE
Mitelli G M 1660 *L'arti per Via* (Bologna: Longhi)

## Anecdote of Field Marshall Lord Montgomery

(Contributed by Harold Winter.)

Before the end of the 1939–45 war, a staff officer with high political connections was caught with his 'fingers in the till' in a petrol racket in Tripoli, Syria. At the subsequent Court Martial he was reduced to the lowest officer rank, that of Second Lieutenant, and in a note to Military Intelligence, Monty wrote: 'Fit to fart but not to fight'.

## A Scene of Mesmeric Therapy

Mesmer was an Austrian, born in 1734 near Lake Constance, who first trained for the priesthood, later changed his mind and studied law at Vienna University, and then finally transferred to medicine, obtaining his degree in 1766 at the age of 32. Early in 1768 he married a rich widow and it was his wife's cousin Francisca Oesterlin on whom he first tried his 'magnetic treatment'. Mesmer described her medical problem in the following terms: 'She was attacked several years before by a convulsive disease, the most troublesome symptoms of which were that blood rushed with impetuosity to the head, and excited in that part agonising pain in the teeth and ears which was followed by delirium, maniacal elatement, vomiting and syncope'.

Mesmer's belief in animal magnetism arose from his MD thesis in which he studied the suupposed influence of the Sun and the Moon on

320

the human body. He believed that they caused an ebb and flow in the body analogous to the tides. His treatment was intended to restore harmony between the individual and the universe. He treated his patients by fitting magnets to various parts of the body and was able to affect many dramatic 'cures'—but this was in effect an early example of successful psychotherapy, rather than proof of animal magnetism, and more orthodox physicians in Austria were always sceptical.

Eventually, his fate as a physician and magnetiser in Austria was sealed by the outcome of his treatment of Marie-Therese Paradis, a pianist who had been blind since the age of four. Mesmer had restored her eyesight (the problem would now be recognised as hysterical blindness). Her father was a secretary to the Emperor of Austria and was afraid that his daughter's pension and other advantages might now be forfeit. His attitude caused his daughter to relapse into her previous state and in the aftermath Mesmer was disgraced and left Vienna for Paris in 1778.

*This oil was painted c. 1800 and shows Franz Anton Mesmer holding a wand, right background of the painting. (Courtesy Wellcome Institute Library, London.)*

He bcame financially successful in Paris with many patients, and typically he would induce a state of trance by touching them with a wand, and after a short while they would break into a sweat and be siezed by convulsions. Many then threw themselves at his feet in gratitude.

321

However, his popularity made enemies of French physicians and in 1784 King Louis XVI set up a Royal Commission to investigate animal magnetism. Members of the Commission included the then American Ambassador to France, Benjamin Franklin; Joseph Guillotin the inventor of the guillotine; and Antoine Lavoisier the famous chemist. They failed to prove the existence of animal magnetism, and Dr Charles d'Elson (physician to the future King Charles X), Mesmer's most prominent supporter, was expelled from the Paris Medical Faculty. Mesmer was forced to retire, fled through Europe and eventually settled at Meersburg, Switzerland, on Lake Constance, where he died in 1815.

---

## Samuel Butler's Description of the Human Body

A pair of pincers set over a bellow and a stewpan and the whole fixed on stilts.

---

## Anecdote of Sir Clement Price Thomas

Sir Clement Price Thomas was an eminent thoracic surgeon at the Westminster and Brompton Hospitals who was in charge of the lung cancer operation on King George VI. This anecdote concerns a less eminent occasion.

A new firm of medical students had just been appointed to Sir Clement Price Thomas and were standing round the bed of a patient waiting for the great man to arrive. He duly did so, and asked whether certain matters had yet occurred, such as the taking of blood pressure, pulse rate, etc preparatory to making a clinical decision.
    'Yes', said the students, meekly.
    'Well, what is your decision?' asked Sir Clement.
There was a pause, but no answer.
    'Well, would you operate?'
The students were over-awed, but one managed to summon enough courage to reply, decided to be conservative, and said 'No'. The other five, having even less courage, followed suit and also said 'No'.
    'But *I* would' said the eminent surgeon, at which point a voice from the bed said in a loud Cockney London accent,

'No you won't mate, what, with 6 to 1 against, do you think I'm daft?'.—The patient was a bookmaker!

---

## Lettsom's Moral and Physical Thermometer

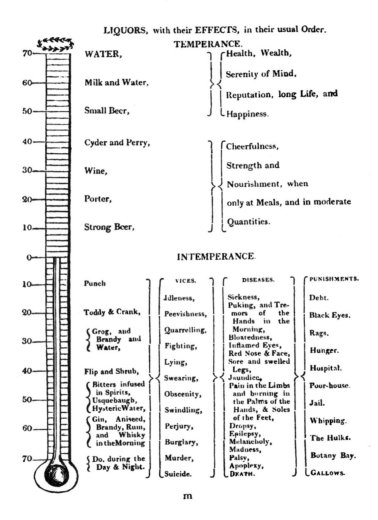

LETTSOM'S MORAL AND PHYSICAL THERMOMETER, OR THE SCALE OF THE PROGRESS OF TEMPERANCE AND INTEMPERANCE.

*(Courtesy Wellcome Institute Library, London.)*

REFERENCE
Abraham J J 1933 *Lettsom* (London: Heinemann) p. 484

## Impromptu Speech

How wonderful is the human brain. It starts working before we are born, it continues, waking or sleeping, until the day we die and the only time it stops is when someone asks us to stand up and say a few words.

REFERENCE
Kenny P 1982 *A Handbook of Public Speaking for Scientists and Engineers* (Bristol: Adam Hilger) p. 50

## Anecdote of Virchow and Bismarck

The eminent physiologist Rudolf Virchow criticised the German Chancellor Bismarck who challenged him to a duel. Since Virchow was the challenged, he chose the weapons. He held up two large sausages and said 'One contains the germ of trichinosis and the other is harmless. Let his Excellency choose and eat and I will eat the other'. Bismarck decided to laugh off the affront—and the duel.

## The Dinner-plate Twins

On Nelson's flagship there were twins, one with dinner-plate ears and one with dinner-plate eyes. The twin with the dinner-plate eyes was always posted to the crow's nest at the top of the tallest mast to look for enemy ships. He spotted one and called down to his Admiral: 'Admiral, there is a French ship three miles away on the horizon'.

Nelson was most impressed but asked how he knew it was a French ship. The reply was: 'My brother overheard them talking'.

# Advertisement for a Surgeon–Dentist

*(Courtesy Wellcome Institute Library, London.)*

## Persian Wisdom

The best thing you can bring back from your travel is yourself unscathed.

REFERENCE
*Manila Bulletin* 12 February 1985

## The Pestäule in Vienna

This monument, the Pestäule, is in the Graben in Vienna, near to St Stephen's Cathedral. It commemorates Leopold I's thanksgiving after the plague of April 1679 which took thousands of lives. The column is made of white Salzburg marble surrounded by cloud spirals and crowned by the Trinity in gilded copper. Nine angels in groups of three carry the emblems of Leopold's power, the lance and the Duke's hat, coat of arms and sword, and crown and sceptre. On the bottom part of the monument are the coats of arms of the House of Austria and of the Kingdoms of Hungary and of Bohemia.

## *Miraculous Medical Properties of Russian Vodka*

Shortly after the Chernobyl accident on 26 April 1986 various newspapers reported that the local population of Kiev, a city of 2.5 million inhabitants some 120 kilometres distant from the Chernobyl nuclear power station, were taking a radiation exposure antidote. This was apparently a mixture of Kiev vodka and strong red wine. The only logical reason for such a theory was presumably that after an adequate dose the patient would be incapable of worrying about nuclear fallout, civil or military!

I ignored these accounts as old wives tales until I actually visited the Chernobyl site and its evacuated dormitory town of Pripyat on 2 December 1987. The stories were true. People actually did believe that vodka 'helped'. Indeed, a scientist who should have known better even had a biological theory to justify the 'medicine'. The vodka goes straight to the cells of the body and the radiation always attacks the vodka first rather than any other cellular material and therefore vodka gives a certain amount of protection.

The prescription was: *'Take a bottle of vodka and a bottle of strong red wine. Pour a glass of vodka and add six drops of red wine. Drink. Repeat the prescription until the bottle of red wine is empty.'.*

A medical experiment with mice was also supposed to have been made with one group being given vodka and the control group no vodka. Eventually all were sacrificed for post-mortem studies. No evidence of leukaemia was found in either group, but the vodka-mice had cirrhosis of the liver.

The particular brand of vodka was not specified, but perhaps it was that described to me by Alan Frank of Radio London when my first book of anecdotes was being reviewed. The label is shown below and was originally located in the Bahamas!

100 % NEUTRAL GRAIN SPIRITS
CHARCOAL FILTERED

5238

To Russia with Love!

BOGOROV

БОГОРОВ

100% Pure Vodka

80 PROOF

FIVE TIMES DISTILLED AND FILTERED
THROUGH ACTIVATED CHARCOAL
100% NEUTRAL GRAIN SPIRITS.

327

# Abortions in Greece, January 1987

As soon as the Chernobyl accident on 26 April 1986 was reported in newspapers and on TV, the actual events (initial deaths at the power plant turned out to be 31: firefighters and plant workers) rapidly became magnified in several countries and guesses by reporters were wildly inaccurate. Thus in the United Kingdom the *Daily Mirror* on 30 April headlined 'Please get me out Mummy', 'Terror of Trapped Britons' and '2,000 are feared dead in Nuclear Horror'. The *Detroit Medical News* of 12 May stated 'So the Russians have started to self-destruct. That is the good news, the bad news is that they are exporting the fallout across the globe.'. The *New York Post* was most outlandish with '15,000 Dead in Mass Grave'. TV was not always much better with an American TV network buying for 20 000 US dollars a video purporting to be Chernobyl burning. Unfortunately for them, the views were recognised by Italians as being those of a fire at a Trieste cement factory. An ABC TV newscaster was reported to have said 'It is one mistake we will try not to make again.'.

The more horrific an accident, the 'better news' it is supposed to make and this will not alter. I even found this attitude when trying to find a publisher for my book *Chernobyl—The Real Story*. I believe that I quite rightly claimed the contents included 'dramatic pictures'. How-ever, one publisher rejected this claim because, as he said, 'there were no pictures of dead bodies with legs and arms missing'.

In the aftermath of this accident at Chernobyl, the panic caused by the incorrect and inadequate available information was particularly poignant in Greece. Pregnant mothers, because of the 'scare', had otherwise wanted pregnancies terminated. The expected number of live births in Greece in January 1987 was some 9000 but in practice, there were only some 7000 live births, a 23% reduction in those expected. This is considered by Greek physicians to be virtually entirely due to the worry concerning the accident and this is corroborated by the number of live births in February and March 1987 being as expected, since for the births expected in these months, relevant information on radiation effects from Chernobyl was available to the Greek physicians and obstetricians.

REFERENCES

R F Mould 1988 *Chernobyl—The Real Story* (Oxford: Pergamon Press Ltd)

D Trichopoulos 1988 Victims of Chernobyl in Greece *British Medical Journal* **295** 1100

# Eyewitness Account of the First Physician at the Scene of the Chernobyl Accident, 1986

The following report is, as far as I know, only to be found in the Russian language publication *Youth*, June 1987 issue, and refers to the eyewitness account of the first doctor to arrive at the scene of the accident, Dr Valentin Petrovich Belokon. He was 28 years old at the time of the accident, with two young daughters, Tania aged 5 years and Katya aged 1½ months. He was also a sportsman who specialised in weight lifting and was employed in Pripyat (the nearest town to the power station) as an accident and emergency physician. This interview took place probably some 2 months after the accident and Dr Belokon was again interviewed in the autumn of 1986. He was by then working as a paediatrician in Donetsk and suffered breathing problems as an after-effect of his experiences. The interviewer was a journalist, Yuri Scherbak.

On 25 April 1986 at 2000 hours I started my work in Pripyat, where there is an accident and emergency medical brigade consisting of one physician (myself) and a doctor's assistant, Sasha (Alexander) Ckachok, and six ambulances. On 25 April Sasha and I worked separately and my driver was Anatoly Guymarov. At 0135 on 26 April on my return to the medical centre I was told that there had been a call from the nuclear power plant and that two or three minutes earlier Sasha had left for the power plant. At 0140 he telephoned to say that there was a fire, with several people burnt and that they needed a doctor. I left with my driver and arrived in seven to ten minutes. When we arrived, the guard asked:

'Why don't you have
special clothes?'

I did not know they would be needed and was only wearing my doctor's uniform, and since it was an April evening and the night was warm I did not even wear a doctor's cap. I met Kibenok (a fireman lieutenant who later died) and asked:

'Are there patients
with burns?'

Kibenok's reply was that:

'There are no patients with burns but the situation
is not clear and my boys feel like vomiting.'

My talk with Kibenok was near the energy block (Unit Number 4) where the firemen stood. Pravik (also a fireman lieutenant who later died) and Kibenok had arrived in two cars and Pravik quickly jumped out of the car but did not come to see me. Kzbenok was excited a little and alarmed.

Dr Belokon then described the first of his patients.

Sasha Ckachok had already taken Shashenok (the second power plant worker to die, the first was buried in the rubble of Unit Number 4 whereas Shashenok died of extensive burns) from the nuclear power plant from which he had been pulled by workers after being burnt and crushed by a falling beam. He died on the morning of 26 April in a medical recovery room. The second patient was a young boy about 18 years old. He had vomiting and severe headache, and as I did not yet know about the high level of radiation I asked him:
    'What have you eaten and how
    did you spend the previous evening?'
His blood pressure at 140 or 150 over 90 was slightly higher than the normal 120 over 80 for an 18 year old. However, the boy was very nervous. At this time, workers who came out of the nuclear power plant were very disturbed and only exclaimed:
    'It is horrible.'
and that
    'The instruments went off scale.'
Three or four men from the technical staff all had the same symptoms of headache, swollen glands in the neck, dry throat, vomiting and nausea. They all received medication and were then put into a car and sent to Pripyat with my driver Gumarov. After that, several firemen were brought to me and they could not stand on their feet. They were sent to hospital.

The damaged nuclear power plant after the accident (Courtesy: TASS).

Dr Belokon now begins to feel unwell and records that at 1800 on 26 April, many hours after his arrival:

I felt something wrong in my throat and had a headache. Did I understand that it was dangerous? Was I afraid? Yes, I understood. Yes, I was afraid, but when people see a man in a white uniform is near, it makes them quieter. I stood as all of us stood, without any breathing apparatus, without any other means of protection. When it became lighter on 27 April, there was no fire to be seen in the block (Unit Number 4), but there was black smoke and black soot. The reactor was splitting, but not all the time, only as follows: smoke, smoke, then belch. Gumarov arrived back from Pripyat after taking the injured to hospital and I felt weakness in my feet. I did not notice it when I walked, but now it has happened. Gumarov and I waited another five minutes to see if anyone else asked for assistance, but nobody did. That is why I said to the firemen:

'I am going to hospital,
if there is a need, call us again.'

I went home, but before I washed and changed my clothes, I passed iodine to those in the hostel, asked them to close all windows and to keep the children inside. Then I was taken to the treatment department of a hospital by our Dr Diyakanov and given an intravenous infusion. I felt very bad and started to lose my memory, at first partially and then totally. Later, in Moscow, in Clinic No 6, I was in one ward with a dosimetrist. He told me that just after the explosion, all instruments were off-scale, they called to the safety engineer and that engineer answered:

'What is the panic? Where is the shift chief?
When he is available tell him to call me, you
yourself don't panic, such a report (about off-scale
measurements) is not correct.'

REFERENCES
R F Mould 1988 *Chernobyl—The Real Story* (Oxford: Pergamon Press Ltd)
Y Scherbak 1987 *Unost* (June)

# *Medical Papyri from Egypt*

This selection of five papyri show:
1  Care of hands and feet: chiropody.
2  Poliomyelitis: note the thin leg on the left.
3  This was dedicated by Amenhotep to Amon Ra: presumably to cure earache.
4  Effects of famine.
5  A case history of a fracture of the clavicle, *c.* 3000 BC

331

*Examination.* If thou examinest a man having a break in his collar bone, thou shouldst find his collar bone short and separated from its fellow.
*Diagnosis.* Thou shouldst say concerning him 'One having a break in his collar bone. An ailment which I will treat.'
*Treatment.* Thou shouldst place him prostrate on his back with something folded between his two shoulder blades. Thou shouldst spread out with his two shoulders in order to stretch apart his collar bone until that break falls into its place. Thou shouldst make for him two splints of linen and thou shouldst apply one of them both on the inside of his upper arm and the other the underside of his upper arm. Thou shouldst bind it with YMRW. Treat it afterwards with honey everyday until he recovers.

Other case histories *c.* 3000 BC include fracture of the skull, a wound in the head, a slit in the outer ear, a dislocation of the mandible, dislocation of the cervical vertebra and breast abscess.

1

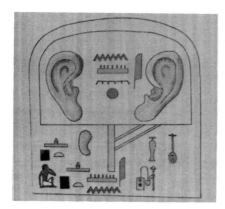

2

3

4

5

## Most-sought-after Plant

The malunggay plant has become a most-sought-after vegetable. Its pods, flowers, leaves and twigs contain 75 calories of food energy, 6 grams protein, 13 grams carbohydrates and 353 milligrams of calcium per 100 grams edible portion. It has more niacin, thiamine, phosphorous and ascorbic acid than other vegetables, and contains an iron compound that helps prevent anaemia.

REFERENCE
The *Manila Bulletin* 11 February 1985

## Manioc is a Root of Good and Evil for the Third World

'A day without manioc is like a day without sunshine,' says Richard Brown, an American who works for Zaire's health ministry.

Maybe so, but manioc has its cloudy side, too.

Also known as cassava and a slew of other aliases, manioc is a scraggly looking plant with elephantine starchy roots. It grows in tropical lands and has many uses.

In Europe, manioc is fed to cattle. In the US, it is turned into tapioca pudding. In Brazil, it is put into cars' gasoline tanks in the form of methanol. In parts of Africa, it is used to starch shirts, roof houses and stiffen leather sheaths.

But above all, manioc gets eaten. Five hundred million people in Africa, Latin America and parts of Asia eat it every day, sometimes two or three times a day. It is one of the world's staple foods, the potato of the Third World.

With drought afflicting more than 20 countries in Africa, manioc's importance in Africans' diet is growing. The plant can live on 50 centimetres of rain a year and survive prolonged dry spells after other crops have perished. For Africa's hungry, its roots are an important source of calories. Many Africans grow manioc, and those who don't can buy it for very little.

As with most good things in the Third World, there are catches. The Food and Agriculture Organization notes that manioc 'isn't a balanced food, consisting, as it does, largely of starch.' It 'compares very poorly in vitamins' with other staples, the FAO says.

Manioc rates especially poorly in protein. Nutritionists say that a child can't consume enough manioc alone to avoid severe protein deficiency.

Also, hardy though it may be, manioc can't survive the kind of drought that much of Africa now is experiencing. Another thing: raw manioc contains lethal amounts of cyanide. But it isn't hard to remove most of the poison.

On the dinner plate, manioc looks thick and mushy, and it tastes the same way. Even some of its fans concede that its taste isn't sublime.

'It has no real taste of its own,' says William O Jones, a professor at Stanford University's Food Research Institute and the author of the 1959 book *Manioc in Africa*.

In North America, manioc, despite its use in tapioca pudding, is unlikely to gain fad-food status akin to that of tofu or kiwi. Goya Foods Inc., a New York food wholesaler, sells processed, sliced and frozen yuca—another name for manioc—imported from Costa Rica. It also sells a dry cassava bread imported from the Dominician Republic.

But Conrad Colon, the purchasing agent for Goya, says manioc has a limited market in the US outside Brazilian and Cuban restaurants. 'It's nothing that people go crazy over,' he says.

Most Westerners prefer their manioc fried, if at all. Many Africans like it ground up as a kind of porridge or in cakes or as a moist substitute for rice: you just pound the peeled manioc into a lumpy flour, boil it in water, pour off the excess water and allow the manioc to sit and absorb moisture. Then, if you like, you can pour a meat or a vegetable dish over the manioc.

'If you are organized, you have a pot of manioc flour soaking in reserve all the time and you just change the water once in a while,' says Musa Joh, an accounting student from Gambia who helps manage the Calabash Restaurant in London's African Centre.

Stanford's Mr Jones says it is the ideal dish for the busy bachelor. Just toss the mix in boiling water for a while and eat.

As for cyanide, it takes only about 50 to 60 milligrams of it to kill an adult. 'Bitter' manioc plants contain an average of 200 to 300 milligrams of cyanide in the form of prussic acid in every kilogram of manioc. So-called sweet strains of the plant contain 70 milligrams of cyanide per kilogram.

Traditional methods of preparing manioc by soaking it, peeling it and drying it in the sun get most of the cyanide out. Medical researchers have nonetheless linked chronic cyanide poisoning from residual prussic acid in manioc to a variety of diseases, including cretinism, optic neuritis, liver degeneration and skin problems.

Some scientists believe that manioc may have been the root of the trouble on Henry Morton Stanley's 1887–9 expedition to Africa, during which nearly half of his 600 porters died. Many of the men, who came from parts of Africa where manioc wasn't well-known, unsuspectingly munched raw manioc and keeled over.

But years before that, Stanley was already griping about the food. He said the 'unvarying slices' of manioc 'deserve three-quarters of the blame now lavished on "murderous Africa".'

But the plant remains a popular crop in the Third World. Originally found in Latin America, it was carried around the world by Portuguese colonists. It can be cultivated to heights ranging from 1.5 metres to 3.5 metres and requires almost no care. It is edible from head to toe: roots, stems and leaves.

'It is an extremely cheap source of calories, it is drought resistant and relatively easy to take care of,' says Stanford's Mr Jones.

Stick a 15-centimetre piece of stem in the ground, and a new plant will sprout up. Manioc grows in bad soil, resists pests, and thrives at altitudes ranging from sea level to nearly 2000 metres.

It can be harvested at any time of the year, but it spoils quickly once harvested and develops brown fibrous streaks and freckles. For this reason importers of manioc in Europe or the US run high risks if they can't sell the stuff fast. Britain's main manioc importer failed last year.

What does the future hold for manioc?

The US Agency for International Development is experimenting in Zaire with a new, fast-growing, low-cyanide hybrid of manioc called Pronam F100. The new breed produces 50% more manioc and doesn't require soaking to remove the cyanide.

But Zairian health officials still advise soaking it just in case: 'Some of the old stuff just might slip in.'

It may be safer, but unfortunately Pronam F100 has pretty much the same taste as the manioc eaten by Stanley when he went to find Livingstone in 1871. On that trip he said manioc 'is insipid food for breakfast and dinner throughout a term of three years. A few months and this diet makes the European sigh for his petit verre, caviar, mock-turtle, salmon—with sauce Hollandaise.'

He added that 'British paupers and Sing Sing convicts might fare better.'

REFERENCE
The *Asian Wall Street Journal* 12 February 1985

## Charlatan Extracting Teeth

*An engraving by Theodor Rombouts. (Courtesy Wellcome Institute Library, London.)*

## His First Operation: A Short Story by Sir Arthur Conan Doyle

Sir Arthur Conan Doyle was a medical student in Edinburgh from

1876–81 and this story is about a young student's initiation in the rites of surgery.

It was the first day of a winter session, and the third year's man was walking with the first year's man. Twelve o'clock just booming out from the Tron Church.

'Let me see,' said the third year's man, 'you have never seen an operation?'

'Never.'

'Then this way, please. This is Rutherford's historic bar. A glass of sherry, please, for this gentleman. You are rather sensitive, are you not?'

'My nerves are not very strong, I am afraid.'

'Hum! Another glasss of sherry for this gentleman. We are going to an operation now, you know.'

The novice squared his shoulders and made a gallant attempt to look unconcerned.

'Nothing very bad—eh?'

'Well yes—pretty bad.'

'An—an amputation?'

'No, it's a bigger affair than that.'

'I think—I think they must be expecting me at home.'

'There's no sense funking. If you don't go today you must tomorrow. Better get it over at once. Feel pretty fit?'

'Oh, yes, all right.'

The smile was not a success.

'One more glass of sherry, then. Now come on or we shall be late. I want you to be well in front.'

'Surely that is not necessary.'

'Oh, it is far better. What a drove of students! There are plenty of new men among them. You can tell them easily enough, can't you? If they were going down to be operated upon themselves they could not look whiter.'

'I don't think I should look as white.'

'Well, I was just the same myself. But the feeling soon wears off. You see a fellow with a face like plaster, and before the week is out he is eating his lunch in the dissecting rooms. I'll tell you all about the case when we get to the theatre.'

The students were pouring down the sloping street which led to the infirmary—each with his little sheaf of notebooks in his hand. There were pale, frightened lads, fresh from the High Schools, and callous old chronics, whose generation had passed on and left them. They swept in an unbroken, tumultuous stream from the University gate to the hospital. The figures and gait of the men were young, but there

337

was little youth in most of their faces. Some looked as if they ate too little—a few as if they drank too much. Tall and short, tweed coated and black, round-shouldered, bespectacled and slim, they crowded with clatter of feet and rattle of sticks through the hospital gate. Now and again they thickened into two lines as the carriage of a surgeon of the staff rolled over the cobblestones between.

'There's going to be a crowd at Archer's,' whispered the senior man with suppressed excitement. 'It is grand to see him at work. I've seen him jab all round the aorta until it made me jumpy to watch him. This way, and mind the whitewash.'

They passed under an archway and down a long, stone-flagged corridor with drab coloured doors on either side, each marked with a number. Some of them were ajar, and the novice glanced into them with tingling nerves. He was reassured to catch a glimpse of cheery fires, lines of white-counterpaned beds and a profusion of coloured texts upon the wall. The corridor opened upon a small hall with a fringe of poorly-clad people seated all round upon benches. A young man with a pair of scissors stuck, like a flower, in his button-hole, and a notebook in his hand, was passing from one to the other, whispering and writing.

'Anything good?' asked the third year's man.

'You should have been here yesterday,' said the outpatient clerk, glancing up. 'We had a regular field day. A popliteal aneurysm, a Colles' fracture, a spina bifida, a tropical abscess, and an elephantiasis. How's that for a single haul?'

'I'm sorry I missed it. But they'll come again, I suppose. What's up with the old gentleman?'

A broken workman was sitting in the shadow, rocking himself to and fro and groaning. A woman beside him was trying to console him, patting his shoulder with a hand which was spotted over with curious little white blisters.

'It's a fine carbuncle,' said the clerk, with the air of a connoisseur who describes his orchids to one who can appreciate them. 'It's on his back, and the passage is draughty, so we must not look at it, must we, daddy? Pemphigus,' he added carelessly, pointing to the woman's disfigured hands. 'Would you care to stop and take out a metacarpal?'

'No thank you, we are due at Archer's. Come on;' and they rejoined the throng which was hurrying to the theatre of the famous surgeon.

The tiers of horseshoe benches, rising from the floor to the ceiling, were already packed, and the novice as he entered saw vague, curving lines of faces in front of him, and heard the deep buzz of a hundred voices and sounds of laughter from somewhere up above him. His companion spied an opening on the second bench, and they both squeezed into it.

'This is grand, the senior man whispered; you'll have a rare view of it all.'

Only a single row of heads intervened between them and the operating table. It was of unpainted deal, plain, strong and scrupulously clean. A sheet of brown waterproofing covered half of it, and beneath stood a large tin tray full of sawdust. On the further side, in front of the window, there was a board which was strewed with glittering instruments, forceps, tenacula, saws, canulas and trocars. A line of knives, with long, thin, delicate blades, lay at one side. Two young men lounged in front of this; one threading needles, the other doing something to a brass coffee-pot-like thing which hissed out puffs of steam.

'That's Peterson,' whispered the senior. 'The big, bald man in the front row. He's the skin-grafting man, you know. And that's Anthony Browne, who took a larynx out successfully last winter. And there's Murphy the pathologist, and Stoddart the eye man. You'll come to know them all soon.'

'Who are the two men at the table?'

'Nobody—dressers. One has charge of the instruments and the other of the puffing billy. It's Lister's antiseptic spray, you know, and Archer's one of the carbolic acid men. Hayes is the leader of the cleanliness-and-cold-water school, and they all hate each other like poison.'

A flutter of interest passed through the closely-packed benches as a woman in petticoat and bodice was led in by two nurses. A red wollen shawl was draped over her head and round her neck. The face which looked out from it was that of a woman in the prime of her years, but drawn with suffering and of peculiar bees-wax tint. Her head drooped as she walked, and one of the nurses, with her arm round her waist, was whispering consolation in her ear. She gave a quick side glance at the instrument table as she passed, but the nurses turned her away from it.

'What ails her? asked the novice.'

'Cancer of the parotid. It's the devil of a case, extends right away back behind the carotids. There's hardly a man but Archer would dare follow it. Ah, here he is himself.'

As he spoke, a small brisk, iron-grey man came striding into the room, rubbing his hands together as he walked. He had a clean-shaven face of the naval officer type, with large, bright eyes, and a firm, straight mouth. Behind him came his big house surgeon with his gleaming pince-nez and a trail of dressers, who grouped themselves into the corners of the room.

'Gentleman,' cried the surgeon in a voice as hard and brisk as his manner. 'We have here an interesting case of tumour of the parotid, originally cartilaginous but now assuming malignant characteristics, and therefore requiring excision. On to the table, nurse! Thank you! Chloroform, clerk! Thank you! You can take the shawl off, nurse.'

The woman lay back upon the waterproofed pillow and her murderous tumour lay revealed. In itself it was a pretty thing, ivory white with a mesh of blue veins, and curving gently from jaw to chest. But the lean, yellow face, and the stringy throat were in horrible contrast with the plumpness and sleekness of this monstrous growth. The surgeon placed a hand on each side of it and pressed it slowly backwards and forwards.

'Adherent at one place, gentlemen,' he cried. 'The growth involves the carotids and jugulars, and passes behind the ramus of the jaw, whither we must be prepared to follow it. It is impossible to say how deep our dissection may carry us. Carbolic tray, thank you! Dressings of carbolic gauze, if you please! Push the chloroform, Mr Johnson. Have the small saw ready in case it is necessary to remove the jaw.'

The patient was moaning gently under the towel which had been placed over her face. She tried to raise her arms and to draw up her knees but two dressers restrained her. The heavy air was full of the penetrating smells of carbolic acid and of chloroform. A muffled cry came from under the towel and then a snatch of a song, sung in a high, quavering, monotonous voice.

He say, says he,
If you fly with me
    You'll be the mistress of the ice-cream van;
    You'll be the mistress of the—

It mumbled off into a drone and stopped. The surgeon came across, still rubbing his hands, and spoke to an elderly man in front of the novice.

'Narrow squeak for the Government,' he said.

'Oh, ten is enough.'

'They won't have ten long. They'd do better to resign before they are driven to it.'

'Oh, I should fight it out.'

'What's the use. They can't get past the committee, even if they get a vote in the House. I was talking to—'

'Patient's ready, sir, said the dresser.'

'Talking to M'Donald—but I'll tell you about it presently.'

He walked back to the patient, who was breathing in long, heavy gasps. 'I propose,' said he, passing his hand over the tumour in an almost caressing fashion, 'to make a free incision over the posterior border and to take another forward at right angles to the lower end of it. Might I trouble you for a medium knife, Mr. Johnson.'

The novice, with eyes which were dilating with horror saw the surgeon pick up the long, gleaming knife, dip it into a tin basin and balance it in his fingers as an artist might his brush. Then he saw him pinch up the skin above the tumour with his left hand. At the sight, his nerves, which had already been tried once or twice that day, gave way utterly. His head swam round and he felt that in another instant he might faint. He dared not look at the patient. He dug his thumbs into his ears less some scream should come to haunt him, and he fixed his eyes rigidly upon the wooden ledge in front of him. One glance, one cry, would, he knew, break down the shred of self-possession which he still retained. He tried to think of cricket, of green fields and rippling water, of his sisters at home—of anything rather than of what was going on so near him.

And yet, somehow, even with his ears stopped up, sounds seemed to penetrate to him and to carry their own tale. He heard, or thought he heard, the long hissing of the carbolic engine. Then he was conscious of some movement among the dressers. Were there groans too breaking in upon him, and some other sound, some fluid sound which was more dreadfully suggestive still? His mind would keep building up every step of the operation, and fancy made it more ghastly than fact could have been. His nerves tingled and quivered. Minute by minute the giddiness grew more marked, the numb, sickly feeling at his heart more distressing. And then suddenly, with a groan, his head pitching forward and his brow cracking sharply upon the narrow, wooden shelf in front of him, he lay in a dead faint.

When he came to himself he was lying in the empty theatre with his collar and shirt undone. The third year's man was dabbing a wet sponge over his face, and a couple of grinning dressers were looking on.

'All right,' cried the novice, sitting up and rubbing his eyes; 'I'm sorry to have made an ass of myself.'

'Well, so I should think,' said his companion. 'What on earth did you faint about?'

'I couldn't help it. It was that operation.'

'What operation?'

'Why, that cancer.'

There was a pause, and then the three students burst out laughing.

'Why, you juggins,' cried the senior man, 'there never was an operation at all. They found the patient didn't stand the chloroform well, and so the whole thing was off. Archer has been giving us one of his racy lectures, and you fainted just in the middle of his favourite story.'

REFERENCES
Goldman M 1987 *Lister Ward* (Bristol: Adam Hilger) pp 70–7

Conan Doyle A 1934 *Round the Red Lamp* 2nd edn (London) pp 9–19
Conan Doyle A 1930 *Memories and Adventures* 2nd edn (London) p. 33

---

## Mr Chamberlen's Secret

It was not very ethical, for an English accoucheur as famous as Hugh I Chamberlen, momentarily in Paris, to try and sell to the French Government (for 10 000 *scudi*) a family secret jealously held for nearly a century. It was, he argued, a portentous secret: an instrument with which a woman in labour 'could be delivered in eight minutes'. Hugh's motivation for effecting the sale was anything but humanitarian. His whole family was heavily involved in the consequences of a financial disaster amounting to bankruptcy. At the time (1670), obstetrical and perinatal mortality were rampant in Paris; and the Government, well aware of the fact, was more concerned with actuarial figures than with Dr Chamberlen's questionable ethics. Thus a meeting was arranged between the owner of the secret instrument and Dr F Mauriceau, then regarded as 'the oracle of accoucheurs of his century'. And who can tell that it was not Mauriceau himself, aware of Chamberlen's practices, to attract him to France—partly to find out more about the business and partly, perhaps, to discredit him with malice aforethought?

Right then, the French accoucheur had been called upon to assist a woman in childbirth: a primiparous dwarf with a pelvis so narrow and misshapen as to make vaginal delivery practically impossible. Hence the agreement: should Dr Chamberlen succeed in delivering the dwarf with his secret instrument, Mauriceau would authorise payment of the sum requested; otherwise, the instrument would be forever condemned.

So on 19 August 1670 an apparently confident Dr Chamberlen shut himself in a room with the poor dwarf in labour. But after three hours of torturing as well as fruitless attempts at extracting the fetus (by then mercifully dead) with his forceps, he was forced to admit defeat; and clinical disaster was compounded soon afterwards, when a Caesarean section was made and the mother also died.

### CHAMBERLEN ENTERPRISE

Chamberlen was not a complete stranger to France. His great-grandparents were in fact French. Probably the whole blame for what follows must be fixed on a wooden splinter from a spear, which penetrated the visor of King Henry the Second of France during a tournament and gouged out one of his eyes—in consequence of which the sovereign died. The ill-fated monarch was succeeded by François II, a sickly teenager who promised a world of religious tolerance but soon fell to persecuting the Protestants (called Huguenots after the corrupted German *eidgnoss*, or 'confederate'), a party that was active in Geneva in the XVI century.

There were three wars against the Huguenots in France between 1562 and 1570. At last in 1572 Catherine de Medici (then regent for her second-born son Charles IX, after the sudden death of the first-born Francis II), organised an attempt on the life of Admiral Gaspard de Coligny, chief of the Huguenots; but the killer failed, merely wounding his victim. Obviously the Huguenots made a great ado about this, with loud demands that those responsible be punished. Things like that are bound to end badly. Catherine ordered a massacre. And sure enough, on the feast of St Bartholomew (24 August 1572) 3000 Huguenots were butchered in Paris and 15 000 more in the provinces. Many survivors, smelling trouble, left France in a hurry. Some migrated to North America (where they colonised North and South Carolina); others scattered throughout Europe. Some, more enlightened than others, had left France before the massacre—among them the surgeon and accoucheur William Chamberlen, who fled Paris with his wife Geneviève and son Peter I in 1569 to repair first to Southampton, later to London.

It seems that the original family name was Chambrelein—at least, it is so spelled in the actuary books of the French Church at Southampton; later documents show the name variously as Chambellan, Chamberlain, or Chamberleine. William's sons always signed their name 'Chamberlan'; the great-grandchildren switched to 'Chamberlen'. The family founder's name must have been Villame Chambrelein. After fleeing France, he had two more children, one again named Peter and so becoming Peter II.

It was probably Peter I who constructed with his own hands the instrument from which the obstetric forceps was derived: a 'noble tool capable of limiting human suffering and saving more lives than any other in the surgical armamentarium'. The instrument made by Peter I consisted simply of two curved iron blades shaped somewhat like two spoons. Simple as it was, the tool was guarded as a family secret for fully three generations, or almost one century. When Peter I or one of his obstetrical epigones went abroad to deliver a woman in labour, they did not carry the forceps in their bag. No sir: they had it carried by two strong men in an apparently heavy wooden chest ornamented with gold inlays, or transported on a special coach to increase mystery and make people talk. What Peter's original instrument looked like was discovered by mere chance in 1813, a century and a half after his death, when a hidden trap-door was found in his old house at Woodham Mortimer Hall, Essex County, leading to a hollow space between the floor and the ceiling below. And there was a large box full of odds and ends: old letters, trinkets, coins, fans—and as you have guessed, no fewer than four pairs of obstetrical forceps. One was just a rough-hewn model, possibly an experimental prototype. The other three consisted of two flat metal levers curved to fit the fetal head. The two levers could be placed separately into the birth canal and then hinged together for better purchase and more effective pulling. Perhaps the main intuition of Peter I was to exploit the fact that once the fetal head was extracted, the rest would follow without any difficulty.

That the secrets of forceps handling be divulged by indiscreet observers was practically impossible since no outsiders were ever admitted to the delivery room and the woman herself was carefully blindfolded. So Peter I became rich and famous and had many illustrious customers, including Queen Anne and Mary Henrietta, wife of Charles I of England.

Peter II was second to none. He was admitted with his elder brother to the Barber Surgeon's Company of London, where the two were noted not only because of their professional abilities but also because of their quarrelsome and greedy disposition. Their proposal of creating an association of London midwives was rejected by the College of Physicians on the grounds that it was motivated by personal interest, namely trying to monopolise the profession for the sake of gain. The College also mentioned a 'deprecable family secret' that was handed on from father to son ('. . . *It seems they use metal instruments . . .*'). Neither brother abode by the laws of regulatory agencies. In 1612 Peter I was prosecuted by the Barber Surgeon's Company for illicit practice of the medical profession and jailed at Newgate, later freed by intercession of the Queen and Archbishop of Canterbury; and Peter II was investigated for the same reason in 1620.

MIDWIVES INCORPORATED

Another example of Chamberlen enterprise was the son of Peter II, named Peter III (the Chamberlens seemed to lack imagination in naming their offspring—they called almost everybody Peter and gave each Peter a consecutive number). Peter III (soon nicknamed 'Doctor Peter') started his medical studies at age 14 at the Emmanuel College of Cambridge and continued them at Heidelberg and Padua, where he obtained his doctorate in 1619. In 1628 he was admitted to the Royal College of Physicians—upon condition that he change his 'eccentric way of dressing' and adopt the sober garb of his fellow-members. In 1630 he succeeded his uncle as Royal Accoucheur, later to manage the birth of King Charles II.

Doctor Peter, however, was also famous for another enterprise. Revamping the old idea of Daddy and Uncle, he approached London midwives with promises of benefits galore if they would make a Corporation where he himself would train them and give them a patent. In return, the ineffable Master would graciously accept a tax or levy on each delivery conducted by one of his certified pupils; and of course he would be summoned to attend all difficult deliveries. The idea was not entirely without merit. The need was felt throughout Europe for some sort of training and qualification of *commères*, for centuries a breed of practical 'old hands' completely innocent of medical knowledge. It seems that the first 'sworn' or certified midwife was Madame Louise Bourgeois, the wife of a French surgeon, who in the latter part of the XVI century asked permission to attend the *Hôtel Dieu* for five years; after which she was examined by a committee of physicians who made her a 'sworn and patented *accoucheuse*', or certified midwife. (At the time, ignorant 'collecting midwives' were heard in court as 'expert witnesses'.) Madame Bourgeois' training enabled her to assist Catherine de Medici when she gave birth to the future Louis XIII, write articles on obstetrics (she gave a good description of face presentation) and discuss head conversion—to which she preferred podalic version.

In Italy Scipione Mercuri devoted himself to obstetrics. Being a man of the cloth, he was strongly opposed by his brethren, who could hardly tolerate that a minister of God should busy himself with the female genital function. Scipione, however, worked steadily toward making the obstetrician's position official—lest, perhaps, he should suffer the fate of Hamburg's Dr Wertt, who had been burned alive in 1552 for assisting, disguised in feminine garb, a woman in labour—woe to the man who would dare so much in those times!

## HUGH I SELLS HIS SECRET

In England, also, things were slowly turning to the advantage of people like the Chamberlens. Peter II (who lived to be 82) had three sons: Hugh I, Paul and John. And Hugh I (the future father of Hugh II) began to meditate that he could live like a prince if he succeeded in selling the family secret: hence his trip to Paris and trial by ordeal on Mauriceau's poor dwarf woman. The fiasco was nothing short of grandiose. But Hugh I was not a man to be easily discouraged. He hooked up a Dutch obstetrician, Eric van Roonhuyze, and gave him a swindle most memorable, as we shall see in a moment.

345

His professional background, as you will have guessed, was far from adamantine. He had his own queer theories about treating diseases—at one point the Royal College of Physicians enjoined that he submit to board examinations for authority to practice medicine; but he ignored the summons. Later on, he was inquired over the death of a woman with pneumonia, whom he had 'treated' by bloodletting, diaphoretics, purgatives, and emetics. Then he was removed from the post of Royal Accoucheur on the grounds that when the Queen gave birth to James III, he was late in arriving at Court!

Now about the swindle. Some 50 years after the sale of his 'secret' to van Roonhuyze, the latter resold it to two well-meaning physicians, de Visscher and van de Poll, who intended to divulge it. And after disbursing the handsome sum requested, they found themselves holding nothing more than a plain metal lever. So it became apparent that, in order to preserve his superiority on all other obstetricians, Hugh I had not even sold his forceps to van Roonhuyze, but only a piece of iron!

Be that as it may, Hugh I should be credited with being the first to make public mention of his family secret:

> My father, my brothers and myself (but no one else in Europe) have by the grace of God and our own industry developed an instrument which we have been using for a long time to deliver women in childbirth, in cases of head presentation, without any risk to them or the infants; whereas others must place in jeopardy, if not destroy with hooks, one or the other life, or both.

So started the great adventure of the obstetrical forceps. Rogier van Roonhuyze (the son of the original buyer), himself an obstetrician, had a good flair for business. For a suitable consideration he would sell his secret to a few close friends, who would in turn resell it to others. The traffic became so important that it came to involve the Amsterdam College of Physicians and Pharmacists, in the form of a 1746 decree to the effect that no one could deliver children who could not demonstrate that he was in possession of 'the secret'—which the examiners themselves were currently selling for 2000–2500 guldens a throw!

REFERENCE

Sterpellone L 1986 *Instruments for Health* (Farmitalia Carlo Erba: Freiburg i. Br.) pp 28–36

## First Orbiting Space Mausoleum Okayed

The US government yesterday approved an American consortium's plan to rocket the cremated remains of 10330 people into space for a fee of $3900 each.

The Transportation Department announced tentative approval for a commercially built rocket launch of the world's first space mausoleum from Wallops Island, Virginia, by mid-1987.

'It doesn't interfere with national defence, with international treaty obligations or with public health and safety,' a Transportation Department spokeswoman said.

While plans for the unique venture have been approved, she said, ashes of those who want to circle the earth or rest among the stars will not be launched until final technical details are reviewed.

REFERENCE
The *Bangkok World* 13 February 1985

---

## Man Become Violent After Eating Crisps

An allergy to potatoes caused a 'pleasant and likeable' young Irishman to change into a 'surly and aggressive individual,' a court heard yesterday.

Tony Doherty, 24, was charged in March with assaulting his father whom he tried to strangle with a tie after eating two packets of crisps, Belfast Crown Court was told.

Doherty, of Ballymacombs Road, Portglenone, Co. Antrim, admitted the offence but sentence was deferred because of suspicion about an allergy.

Imposing a two-year conditional discharge on Doherty, Judge Peter Gibson called the case 'a highly exceptional one.'

REFERENCE
The *Daily Telegraph* 20 September 1986

# Anatomical Rhyming Slang

Rhyming slang is not only limited to London's Cockney population, but is also used in Australia and in the American underworld, and occasionally in Ireland. When spoken, the second word of the two-word phrase is omitted and therefore sounds like some sort of unintelligible code to those new to rhyming slang.

In some cases there are more than one or two rhyming phrases for the same part of the anatomy. Thus for *leg* there are *clothes* peg, *Scotch* peg, *Dutch* peg, *bacon* and egg, *fried* egg; and for the famous four-letter word there are *friar* tuck, *colonial* puck, *goose* and duck. Parts of London are sometimes used as in *Hounslow* Heath or *Hampstead* Heath for *teeth*; *Barnet* Fair for hair; and *Hampton* Wick for *penis* or alternatively *almond* rock or *giggle* stick. The origin is not always clear and for *goose* and duck, Franklyn in his *Dictionary of Rhyming Slang* gives the following logical reasoning! The term was current in 1944 on the Pacific Coast of America but it is assumed by Americans to be of English origin. It is not now used in London and there is never any trace of it having been. It is emphatically not known in Australia. Therefore, it must be Irish.

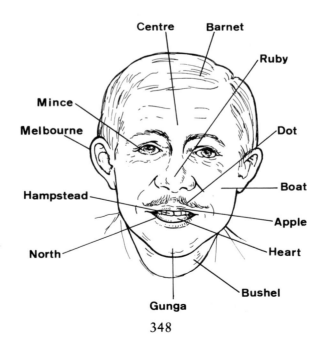

Hair = Barnet Fair
Head = alive or dead
Ear = Melbourne pier, bottle of beer, sighs and tears
Eyes = mince pies
Face = boat race
Nose = ruby rose, these and those
Tongue = heart and lung, brewers bung
Forehead = centre lead
Lips = apple pips
Mouth = north and south
Teeth = Hounslow Heath, Hampstead Heath
Chin = Gunga Din
Neck = bushel and peck
Throat = hairy goat
Moustache = dot and dash
Hand = German band
Breast = Bristol City
Feet = plates of meat
Knees = bugs and fleas
Testicles = cobblers awls, cobblers stalls
Penis = almond rock, giggle stick, Hampton Wick
Leg = clothes peg, Scotch peg, Dutch peg, bacon and egg, fried egg
Anus = Khyber Pass, Elephant and Castle

A few medical conditions are also the subject for rhyming slang such as:

Syphilis (pox) = coachman on the box
Gonorrhoea (clap) = hat and cap
Deaf = Mutt and Jeff
Gout = salmon and trout, in and out

REFERENCE
Franklyn J 1975 *A Dictionary of Rhyming Slang* (London: Routledge and Kegan Paul)

## The Death of John Partridge in 1708

A London man called John Partridge, who claimed to be an astrologer, was publishing predictions in the form of almanacs, which prognostications annoyed Dean Swift because they were, he felt, simply Whig political propaganda. Accordingly, in 1708, Swift produced a rival almanac in which he forecast the death of Partridge on 29 March. Then on 30 March he published an eyewitness account of Partridge's death together with a funeral elegy. Partridge indignantly protested and advertised in the papers that he was still alive. Swift retorted in a *vindication* proving that the poor man was really dead, and other writers took up the cry. In the end, no one believed the unhappy Partridge, and so, in effect, he ceased to exist.

## Anecdote of The Earl of Shrewsbury and a Mouse

John Talbot, Earl of Shrewsbury, born near Whitchurch in 1373, was a renowned soldier who, after many battles, fell at Castillon at the battle of Bordeaux in 1453. Shakespeare in *Henry VI*, Act 2, describes him as:

> The scourge of France,
> The Talbot so much feared abroad,
> That with his name, the mothers still their babes.

Even after his death, the old soldier has remained the focus of much interest and controversy. He was buried initially at Rouen and 50 years later his heart, which was embalmed separately, and his bones were brought to Whitchurch and buried in the church, with his heart lying under the porch.

In 1874 his tomb was restored and at a solemn ceremony the bones were reinterred. Great interest was aroused at the condition of the old warrior's remains, for he was discovered to be resting not in a coffin, but in a strong box three feet long, which soon crumbled to dust on exposure to air. It was found that every bone, from the skull to the smallest metacarpal, had carefully been wrapped in rare cloth, and in the skull itself was found the body of a mummified mouse with her three young ones.

An interesting controversy ensued. A prominent dental surgeon insisted that it must have been a French Catholic mouse which had found its way through the base of the skull before the bones had been prepared for burial. The then Rector of St Alkmund's Church disagreed, and claimed to have found torn leaves of an English Prayer Book in the mouse's nest and a gnawed hole in the cloth on which the skull was wrapped. The fatal gash in the Earl's skull provided the access point. It was an Anglican mouse! Whatever the religion, though, the mummified mouse was awarded an inscription:

A mighty mountain from its womb one day,
With groanings loud brought forth a mouse they say:
The skull which once upreared great Talbot's crest
Gashed by a foeman, gave this mouse a nest.


REFERENCE

Adapted from *A Short History and Guide of the Parish Church of St. Alkmund, Whitchurch, Shropshire* 1978

---

## *Florence Nightingale and the Crimean War Hospitals, 1854–6*

Miss Nightingale reported the arrival of her expedition at Constantinople in a short note to her parents:—

Constantinople, *November* 4, on board *Vectis*.—Dearest People—Anchored off the Seraglio point, waiting for our fate whether we can disembark direct into the Hospital, which, with our heterogeneous mass, we should prefer.

At six o'clock yesterday morn I staggered on deck to look at the plains of Troy, the tomb of Achilles, the mouths of the Scamander, the little harbour of Tenedos, between which and the mainshore our *Vectis*, with steward's cabins and galley torn away, blustering, creaking, shrieking, storming, rushed on her way. It was in a dense mist that the ghosts of the Trojans answered my cordial hail, through which the old Gods, nevertheless, peered down from the hill of Ida upon their old plain. My enthusiasm for the heroes though was undiminished by wind and wave.

We made the castles of Europe and Asia (Dardanelles) by eleven, but also reached Constantinople this morn in a thick and heavy rain, through which the Sophia, Sulieman, the Seven Towers, the walls, and the Golden Horn looked like a bad daguerrotype washed out.

We have not yet heard what the Embassy or Military Hospital have done for us, nor received our orders.

Bad news from Balaclava. You will hear the awful wreck of our poor cavalry, 400 wounded, arriving *at this moment* for us to nurse. We have just built another hospital at the Dardanelles.

You will want to know about our crew. One has turned out ill, others will do.

(*Later*) Just starting for Scutari. We are to be housed in the Hospital this very afternoon. Everybody is most kind. The fresh wounded are, I believe, to be placed under our care. They are landing them now.

The Hospital, to which Miss Nightingale refers, was to be the chief scene of her labours for the next six months, and a few particulars about it and other hospitals, in which the nursing was under her superintendence, must be given in order to make future proceedings intelligible. The principal hospitals of the British army during the Crimean War—four in number—were at Scutari (or in its immediate neighbourhood), the suburb of mournful beauty which looks across to Constantinople from the Asiatic side of the Bosphorus.

The first hospital to be established was in the Turkish Military Hospital. This was made over to the British in May 1854, and was called by them *The General Hospital*. Having been originally designed for a hospital, and being given up to the English partially fitted, it was, wrote Miss Nightingale, 'reduced to good order early, by the unwearied efforts of the first-class Staff Surgeon in introducing a good working system. It was then maintained in excellent condition till the close of the war.' It had accommodation for 1000 patients, but the Battle of the Alma showed that much larger accommodation would be wanted.

North of the General Hospital, and near to the famous Turkish cemetery of Scutari, are the Selimiyeh Barracks—a great yellow building with square towers at each angle. This building was made over to the British for use as a hospital after the Battle of the Alma, and by them was always called *The Barrack Hospital*. This is the hospital in which Miss Nightingale and her band of female nurses were first established, and in which she herself had her headquarters throughout her stay at Scutari. It is built on rising ground, in a beautiful situation, looking over the Sea of Marmora on one side, towards the Princes' Islands on another, and towards Constantinople and up the Bosphorus on a third. 'I have not been out of the Hospital Walls yet,' wrote Miss Nightingale ten days after her arrival, 'but the most beautiful view in all the world, I believe, lies outside.' Her quarters were in the north-west tower, on the left of the Main Guard (or principal entrance). There was a large kitchen or storeroom, of which we shall hear more presently, and out of it on either side various other rooms opened. Mr Bracebridge and the courier slept in one small room; Miss Nightingale and Mrs Bracebridge in another. The nurses slept in other rooms. The whole space occupied by Miss Nightingale and her nurses was about equal to that allotted to three medical officers and their servants, or to that occupied by the Commandant. 'This was done,' she explained, 'in order to make no pressure for room on an already overcrowded hospital. It could not have been done with justice to the women's health, had not Miss Nightingale later taken a house in Scutari at private expense, to which every nurse attacked with fever was removed.' The quarters were as uncomfortable as they were cramped. 'Occasionally,' wrote Miss

Nightingale, 'our roof is torn off, or the windows are blown in, and we are under water for the night.' The Hospital was infested also with rodents and vermin; and, among other new accomplishments acquired under the stress of new occasions, Miss Nightingale became an expert rat-killer. This skill was afterwards called into use at Balaclava. In the spring of 1856, one of the nuns whom she had taken with her to the Crimea—Sister Mary Martha—had a dangerous attack of fever. Miss Nightingale nursed the case; and one night, while watching by the sick-bed, she saw a large rat upon the rafters over the Sister's head; she succeeded in knocking it down and killing it, without disturbing the patient. The condition of physical discomfort in which, sur-rounded by terrible scenes of suffering, she had to do her work, should be remembered in taking the measure of her fortitude and devotion.

The maximum number of patients accommodated at any one time (Dec. 23, 1854) in the Barrack Hospital was 2434. It was half-an-hour's walk from the General Hospital, and an invalided soldier records that he used to accompany Miss Nightingale from one hospital to another in order to light her home on wet stormy nights, across the barren common which lay between them.

Farther south of the General Hospital, in the quarter of Haidar Pasha, was what was known as *The Palace Hospital*, consisting of various buildings belonging to the Sultan's Summer Palace. These were occupied as a hospital in January 1855. Miss Nightingale had no responsibility here; but in the summer of 1855, the female nursing of sick officers, quartered in one of these buildings, was placed under the superintendence of Mrs Willoughby Moore, the widow of an officer who had died a noble death in the war, and four female nurses, sent out specially from England.

Finally, there were hospitals at *Koulali*, four or five miles farther north, upon the same Asiatic shore of the Bosphorus. These hospitals were opened in December 1854. The nursing in them was originally under Miss Nightingale's supervision, but she was presently relieved of it. The hospitals were broken up in November 1855, when, of the female nursing establishment, a portion went home, and the rest passed under Miss Nightingale into the hospitals at Scutari.

There were also five hospitals in the Crimea, but particulars of these may be deferred till the time comes for following Miss Nightingale upon her expeditions to the front. For the nursing in the Civil Military Hospitals at Renkioi (on the Dardanelles) and at Smyrna, and for the Naval Hospital at Therapia, Miss Nightingale had no responsibility, though there is voluminous correspondence among her papers show-ing that she was constantly consulted upon the site and arrangements of these hospitals. The medical superintendent of the hospital at Renkioi was Dr E A Parkes, with whom Miss Nightingale formed a friendship which endured to the end of his life.

The state of the hospitals when Miss Nightingale arrived requires some description, which, however, need not be long. The treatment of the sick and wounded during the Crimean War was the subject of Departmental Inquiries, Select Committees, and Royal Commissions, which, when they had finished sitting upon the hospitals, began sitting upon each other. Enormous piles of Blue-books were accumulated, and in the course of my work I have disturbed much dust upon them. The conduct of every department and every individual concerned was the subject of charge, answer, and countercharge innumerable. Each generation deserves, no doubt, the records of mal-administration which it gets; but one generation need not be punished by having to examine in detail the records of another. Some of the details of the Crimean muddle will indeed necessarily be disinterred in the course of our story; but all that need here be collected from the heaps aforesaid are three general conclusions.

The reader must remember, in the first place, that, apart from controverted particulars, it was made abundantly manifest that there was gross neglect in the service of the sick and wounded. The conflict of testimony is readily intelligible. It was easy to give an account based upon the facts of one hospital or of one time which was not applicable to another. At Scutari, for instance, the General Hospital was from the first better ordered than the Barrack Hospital. Then, again, different witnesses had different standards of what was 'good' in War Hospitals; to some, anything was good if it was no worse than the standard of the Peninsular War. Of Sir George Brown, who commanded the Light Division in the Crimea, it was said: 'As he was thrown into a cart on some straw when shot through the legs in Spain, he thinks the same conveyances admirable now, and hates ambulances as the invention of the Evil One.' Miss Nightingale had much indignant sarcasm for those who seemed content that the soldier in hospital should be placed in the condition of 'former wars,' instead of perceiving that he 'should be treated with that degree of decency and humanity which the improved feeling of the nineteenth century demands.' But the principal reason for the conflict of testimony was that the very facts of protest and inquiry put all the officials concerned upon the defensive. Any suggestion of default or defect was resented as a personal imputation. There is a curious illustration in the letter which the Head of the Army Medical Department wrote to his Principal Medical Officer in view of the Roebuck Committee. 'I beg you to supply me, and that immediately'—with what? with the truth, the whole truth, and nothing but the truth? No—'with every kind of information which you may deem likely to enable me to establish a character for it [the Department], which the public appear desirous to prove that it does not possess.' But though there was much conflict of evidence, the final verdict was decisive. What Greville wrote in his Journal—'the accounts published in the *Times* turn out to be true'—was established by official inquiry and admitted by Ministers. In consequence of the indictment in the *Times*, a Commission of Inquiry was dispatched to the East by the Secretary of State. The Commission arrived at Constantinople simultaneously with Miss Nightingale, and four months later it reported to the Duke of Newcastle. I need not trouble here with many particulars of its Report; for they were adopted and confirmed by a Select Committee of the House of Commons a few months later (the famous 'Roebuck Committee'), which pronounced succinct sentence that 'the state of the hospitals was disgraceful.' The ships which brought the sick and wounded from the Crimea were painfully ill-equipped. The voyage from Balaclava to Scutari usually took eight days and a half. During the first four months of the war, there died on a voyage, no longer than from Tynemouth to London, 74 out of every 1000 embarked. The landing arrangements added to the men's sufferings. To an unpractised eye the buildings used as hospitals at Scutari were imposing and convenient; and this fact accounts for some of the rose-coloured descriptions by which persons in high places were for a time misled. Even the Principal Medical Officer on the spot was naively content with whitewash as a preparation to fit the Barrack for use as a hospital. In fact, however,

*One of the wards in the hospital at Scutari, from a coloured lithograph by Simpson. (Courtesy Mansell Collection.)*

the buildings were pest-houses. Underneath the great structures 'were sewers of the worst possible construction, loaded with filth, mere cesspools, in fact, through which the wind blew sewer air up the pipes of numerous open privies into the corridors and wards where the sick were lying.' There was also frightful overcrowding. For many months the space for each patient was one-fourth of what it ought to have been. And there was no proper ventilation. 'It is impossible,' Miss Nightingale told the Royal Commission of 1857, 'to describe the state of the atmosphere of the Barrack Hospital at night. I have been well acquainted with the dwellings of the worst parts of most of the great cities in Europe, but have never been in any atmosphere which I could compare with it.' Lastly, hospital comforts, and even many hospitals necessaries, were deficient. The supply of bedsteads was inadequate. The commonest utensils, for decency as well as for comfort, were lacking. The sheets, said Miss Nightingale, 'were of canvas, and so coarse that the wounded men begged to be left in their blankets. It was indeed impossible to put men in such a state of emaciation into those sheets. There was no bedroom furniture of any kind, and only empty beer or wine bottles for candle-sticks.' Necessary surgical and medical appliances were often either wanting or not forthcoming. There was no machinery, until Miss Nightingale came, for providing any hospital delicacies. The result of this state of things upon patients arriving after a painful voyage in an extreme state of weakness and emaciation, from wounds, from frost-bite, from dysentery, may be imagined, and it is no wonder that cholera and typhus were rife. In February 1855 the mortality per cent of the cases treated was forty-two. No words are necessary to emphasize so terrible a figure.

Mr Herbert had not waited for the reports of Commission and Committee to reach the conclusion that things were wrong:—

'I have for some time,' he wrote on December 14, 1854, to the Commandant at Scutari, 'been very anxious and very much dissatisfied as to the state of the hospital. I believe that every effort has been made by the medical men, and I hear that you have been indefatigable in the conduct of the immediate business of your department. But there has been evidently a want of co-operation between departments, and a fear of responsibility or timidity, arising from an entire misconception of the wishes of the Government. No expense has been spared at home, and immense stores are sent out, but they are not forthcoming. Some are at Varna, and for some inexplicable reason they are not brought down to Scutari. When stores are in the hospital, they are not issued without forms so cumbrous as to make the issue unavailing through delay. The Purveyor's staff is said to be insufficient. The Commissariat staff is said to be insufficient, your own staff is said to be insufficient,' etc.

By admission, then, and by official sentence, there were things amiss at Scutari which urgently called for amendment. This is the first general conclusion which has to be remembered in relation to Miss Nightingale's work.

To what individuals the disgrace of 'a disgraceful state of things' attached, it is happily no concern of ours here to inquire. But as I have called Mr Sidney Herbert as a witness to the fact of the disgrace, I must add my conviction that his own part in the business was wholly beneficent. Some research among the documents entitles me, perhaps, to express entire agreement with Mr Kinglake's remark upon 'what might have been if the Government, instead of appointing a Commission of *enquiry* on the 23rd of October, had then delegated Mr Sidney Herbert to go out for a month to the Bosphorus, and there *dictate* immediate action.' At home, Mr Herbert was a good man struggling in the toils. The fact is that, though there were some individuals palpably to blame, the real fault was everybody's or nobody's. It was the fault of a vicious system, or rather the vice was that there was no system at all, no co-ordination, but only division of responsibility. The remarks of Mr Herbert, just quoted, point to the evil, and on every page of the Blue-books it is written large. There were at least eight authorities, working independently of each other, whose co-operation was yet necessary to get anything well done. There was the Secretary of State; there was the War Office; there were the Horse Guards, the Ordnance, the Victualling Office, the Transport Office, the Army Medical Department, and the Treasury. The Director-General of the Medical Department in London told the Roebuck Committee that he was under five distinct masters—the Commander-in-Chief, the Secretary of State, the Secretary-at-War, the Master-General of Ordnance, and the Board of Ordnance. The Secretary of State said that he had issued no instructions as to the hospitals; he had left that to the Medical Board. But the Medical Director-General said that it would have been impertinent for him to take the first step. If I were writing the history of the Crimean War, or of the Government Offices, other fundamental reasons for the disgraceful state of things in the hospitals—notably the miscalculated plan of military campaign—would have to be taken into account; but I am writing only the life of Miss Nightingale, and all that under this head the reader need be asked to bear in mind is this: That the root of the evils which had to be dealt with was division of responsibility, and reluctance to assume it.

The third conclusion of the official inquiries, which I want to emphasize, is contained in a passage in the Roebuck Committee's Report, which prefaced a reference to Miss Nightingale's mission: 'Your Committee in conclusion cannot but remark that the first real improvements in the lamentable condition of the hospitals at Scutari are to be attributed to private suggestions, private exertions, and

357

private benevolence.'

So, then, we see that there were disgraceful evils at Scutari needing amendment, and that in order to amend them what was needed was bold initiative. This it was that Miss Nightingale supplied. The popular voice thought of her only or mainly as the gentle nurse. That, too, she was; and to her self-devotion in applying a woman's insight to a new sphere, a portion of her fame must ever be ascribed. But when men who knew all the facts spoke of her 'commanding genius,' it was rather of her work as an administrator that they were thinking. 'They could scarcely realize without personally seeing it,' Mr Stafford told the House of Commons, 'the heartfelt gratitude of the soldiers, or the amount of misery which had been relieved' by Miss Nightingale and her nurses; and, he added, 'it was impossible to do justice, not only to the kindness of heart, but to the clever judgment, the ready intelligence, and the experience displayed by the distinguished lady to whom this difficult mission had been entrusted.' These were the qualities which enabled her to reform, or to be the inspirer and instigator of reforms in, the British system of military hospitals. She began her work, where it lay immediately to her hand, in the Barrack Hospital at Scutari. She did the work in three ways. She applied an expert's touch and a woman's insight to a hospital hitherto managed exclusively by men. She boldly assumed responsibility, and did things herself which she could find no one else ready to do. And, thirdly, she was instant and persistent in suggestion, exhortation, reproaches, addressed to the authorities at home.

REFERENCE
Cook E 1913 *The Life of Florence Nightingale* vol. I (1820–1861) (London: Macmillan) pp 171–81

---

# A Dropped Catch—The Case of the Bouncing Baby

The second child I ever delivered slipped between my trembling fingers and fell head first into the bucket. To this day, those gleaming steel receptacles stationed three feet below the bulging perineum remind me of that awesome moment in 1959. The noise of impact of head against bucket has remained with me throughout my obstetrical career.

Of all the sensory stimuli, a noise is the most difficult to describe in writing. If you are prepared to drop a large turnip into a pail containing one inch (2.5 cm) of water, you will get some idea of the sound the baby's head produced. The metallic plop was followed by a momentary silence lasting a thousand years. There was an anguished cry from the widwife. No one, in living memory, had dropped a babe before. Neonatal baby battering by budding obstetricians had not, as yet, been recorded in the literature.

'Call the house surgeon! Fetch the paediatric house officer! Tell the midwifery sister I want her immediately!' The demands were rattled out with sten-gun precision. 'And you, Mr McGarry, pick up that infant and place it on the cot before the mother stops breathing the gas and air.'

This I did; the infant was by now crying lustily and, with hindsight, I admit that the cry had a hurtful note about it—rather like a four-year-old who has dropped his ice-cream down a roadside drain.

'Fetch the registrar,' said the obstetric house surgeon when he arrived. 'Fetch the senior registrar,' simpered the junior as the procession continued. 'We will have to tell the professor,' said the embryo consultant.

While this trio was glaring at me, the paediatric house physician, an above-average scrum half in his spare time, moved into action. He winkled the baby from the melee, placing it in a small cot on wheels. Running like a man possessed, he headed for the x-ray department.

Meanwhile, the recently delivered Mrs Smythe had abandoned the delights of nitrous oxide and was beginning to surface. The widwife, although still snarling, had recovered sufficient poise to declare that the baby had indeed arrived, but had been taken away for a rest as he had a slight headache!

In those days, delivered mums seemed to have no rights and were usually unable to stand up to people in uniform. The midwife's statement was accepted with a bleary 'What does he weigh?' and 'Will I need stitches?'

Meanwhile, I had been transformed from an apparently normal medical student into a quivering blob of jelly. Beckoned into the corridor by a professoral digit, I almost passed out.

The professor looked stern. 'McGarry, in many years of teaching midwifery, I have never known a medical student to drop a child.' He glanced at the senior registrar. 'I suppose an x-ray will be taken?'

'It has been arranged as a matter of great urgency, sir,' the registrar replied, though he knew full well the idea had never entered his head. In any case, the paediatric department, always an enemy, had snatched the child. It was no longer under obstetric jurisdiction.

The professor decreed that, after the results of the x-ray were known, I was to be the sole bearer of tidings to Mrs Smythe. I was instructed to explain that an accident had happened, that I was entirely to blame, and that I hoped that no subsequent harm was to befall the child.

It was with tears in my eyes that I faced the exhausted mother and did as I had been told. However, on hearing my story, the good lady burst into tears and apologised for being such a burden to me. Astonishingly, she said she hoped I would take up obstetrics permanently. She insisted I was not to blame.

For many years, on each anniversary of this occasion, I received a letter detailing the progress of her son, which has been excellent. Recently, I was told that he had gained a first-class honours degree at one of our better universities, and had been recommended to stay on and do a Ph.D. You know, in a way, I can't help feeling responsible!

REFERENCE
McGarry J 1984 *World Medicine* **19** (October) 39

---

## The Mystery of Byron's Corpse

Byron had, at various times before his death in 1824, expressed wishes to be buried with his dog Bo'sun under the high altar in the ruins of the old priory at Newstead; in his beloved Greece; and, in one of his early poems, among his ancestors.

His friends, particularly John Hobhouse, wanted an Abbey funeral, but this was refused by the Dean and Chapter. The obvious place seemed to be the family vault at the small village, as it then was, of Hucknall Torkard, near Nottingham.

In the little church a final service was held by the parson, the Rev. Charles Nixon, and the vault containing at least 27 members of the Byron family was opened. The remains of the poet were lowered and the vault closed. These remains consisted of the body in a coffin, and an urn containing the heart and brains placed in an oak chest. Thus Byron was brought home to rest with his ancestors. But it was not the end of the story.

Owing to the extravagances of the 6th Lord Byron and his one-man campaigns in Greece, the Byron coffers were sadly depleted. When Captain George Anson Byron succeeded as 7th baron, Newstead had long since been sold and the bulk of what remained of the Byron money had been left to members of the poet's immediate family. Thenceforward no other member of the Byron family was interred in the vault at Hucknall except the poet's daughter, the Countess of Lovelace, whose coffin was deposited there in 1852.

Then, in the 1880s, the little church became inadequate to accommodate the increasing population, so it was almost all taken down and rebuilt. During these alterations the vault was not disturbed structurally but it must have been opened and workmen probably had access to it.

The final chapter in the story of the Byron vault is told by Canon Thomas Gerrard Barber, who was Vicar of Hucknall from 1907 to 1946. The Canon had heard rumours that the body of the poet was not in the family vault at all. So to clear up the point and to determine whether there was, in fact, a crypt under the church (the witnesses to Lady Lovelace's burial said that the vault was arched), he obtained the consent of the Home Office to excavate.

It was on June 15, 1938, that a group of people saw for the first and only time in this century the interior of the Byron vault. It was all done in strict secrecy. Canon Barber rightly guessed that if word got about as to what was afoot, the church would be besieged by what he charitably called 'pilgrims'.

Canon Barber had expected a catacomb with Gothic arches soaring upwards and recesses all around containing the sarcophagi of long-dead Byrons. What he actually saw was a small cramped cell, roughly eight feet high by six feet square with coffins piled untidily one on top of the other, the lower ones having been crushed to dust.

Two of the top coffins, as would be expected, were the last two to be deposited, those of Lord Byron and his daughter, lying almost touching. Either the sightseers at Lady Lovelace's funeral or the workmen who enlarged the church over 30 years later had been busy. Byron's coffin had been opened and the pearls of his coronet resting on it had been stolen. The caps of maintenance inside both Byron's and Lady Lovelace's coronets had gone too, probably also taken as relics.

'Upon closer examination of the coffin' wrote Canon Barber, 'I made the discovery that the lid was loose . . . someone had deliberately opened the coffin.

'A horrible fear came over me that souvenirs might have been taken from within the coffin. The idea was revolting, but I could not dismiss it. Had the body itself been removed? Terrible thought . . .

'Reverently, very reverently, I raised the lid, and before my very eyes lay the embalmed body of Byron in as perfect a condition as when it was placed in the coffin 114 years ago. His features and hair easily recognisable from the portraits with which I was so familiar. The serene, almost happy expression on his face made a profound impression on me . . . I gently lowered the lid of the coffin—and as I did so, breathed a prayer for the peace of his soul.'

Barber's account appeared in 1939. How different from 1824!

Harold Nicholson's book *Byron: The Last Journey* (published to mark the centenary of the poet's death) describes the lying in state at Sir Edward Knatchbull's house in Great George Street, London: 'The room was draped with black and lighted by wax tapers. Mr Hanson [Byron's solicitor] informed Hobhouse that it would be his duty as executor to identify the body.

'Mastering his repulsion, Hobhouse entered the room and gazed upon the coffin. The decaying face that glimmered in the light of the candles was not the face of Byron: Hobhouse was able to identify the body by raising the red velvet pall and glancing at the foot.'

Having read several volumes on the life of Byron during a long convalescence, I went eagerly to Hucknall Torkard for the first time on May 5, 1966. I found a seedy sprawl as unlike the peaceful village of Byron's day as could be imagined. Had the poet returned he would not have recognised any part of it, though he would have viewed with surprise and amusement his statue above the Co-operative store and his name writ large upon the wall of the local cinema.

The church was exciting. Although altered, enlarged and rebuilt almost beyond recognition from its original form, it was dark and mysterious, with a magnificent feeling of space. Victoriana, including some pre-Raphaelite wall mosaics *à la* Burne Jones and fine glass by Kempe, were everywhere.

I learnt from the vicar that there were still one or two people alive who had been present on the day that Canon Barber opened the vault. Among them was Mr James Bettridge, the then caretaker, who lived nearby.

My wife and I eventually met Mr Bettridge, a charming man with great knowledge of the church, and love for it. His description of the vault was more detailed than Canon Barber's. I persuaded him to write it all down, and this is part of his account:

'In early June the Canon informed me that he had received permission from the Home Office to go ahead with the work, and he also invited me to "Come along and lend a hand".

'The date fixed for the opening was June 15, 1938. All were bound to secrecy, to forestall undesirable sightseers. Accordingly, on that lovely summer's afternoon, Mr J. C. Woodsend, a mason from Nottingham, and three of his workmen, arrived at the rear of the church with a lorry containing picks, shovels, planks etc. and work began.

'Old prints show that the entrance to the vault was covered by two large flagstones into which four iron rings were fastened. One of these stones is visible, the second one is now covered by the chancel steps. The masons—directed by Mr Woodsend—soon got to work with hammers and chisels, breaking the tiles around the stone, then the

362

lower chancel step was removed and both stones were lifted. We could then see the interior of the vault.

'To everyone's disappointment the vault was found to measure a mere seven-and-a-half feet by six feet. The coffins were piled one on top of each other, in three stacks; on the left the Lords, in the centre the Ladies and on the right a few children.

'After Mr Lane [the diocesan surveyor] had made measurements of the vault and staircase, closer examination of the coffins began.

'The uppermost coffins of the first and second stacks were of striking appearance, being of oak. That on the left bore a coronet from which the cap of maintenance and pearls were missing, as also were the name-plates and some of the coffin handles. The coffin below this was in rather poor condition, the wood having decayed, leaving part of the leaden coffin exposed. The third one was entirely lead and had already lost its shape, having been crushed by those above. Lower still all the coffins were flattened.

'The floor of the vault was covered to a depth of eight or nine inches in debris, consisting of dust, decayed wood, pieces of lead and numerous bones. All this debris must have been the remains of the early burials.

'Later, Canon Barber decided to make a closer examination of what we were pretty certain was the poet's coffin. The oak lid was found to be loose, and raising it we could see the lead shell which had been cut open. Inside this leaden case was a wooden coffin, the lid of which had never been fastened.

'After some deliberation, Canon Barber very reverently raised the lid, and suddenly I gazed upon the face of Lord Byron . . . The features, with the slightly protruding lower lip and the mass of curly hair, were easily recognisable from the many pictures we had seen. The body had been covered but the funerary clothing had decayed.

'The head was slightly raised as though it had rested on a pillow; parts of which could still be seen in the corners of the coffin. The colour of the body was dark stone. Both feet seemed to be normal. The right foot, however, had been severed just above the ankle, and lay in the corner of the coffin. I heard afterwards that in the doctor's opinion this was a skilful amputation made after death; and the cause of the poet's lameness was a deformed Achilles tendon. Neither foot was clubbed. The vault was finally sealed on the following day.'

There is a question to be answered as a result of all this. Why could Byron's closest friend apparently not recognise him soon after his death and yet 114 years later a number of rational and intelligent people gave evidence of recognising a face already familiar to them through portraits?

Was it a question of auto-suggestion? Did they recognise Byron because they *wanted* to recognise him? The real test would have been to confront a third person with the visage—someone who had not the faintest idea beforehand. We shall never know now.

*Lord Byron. (Courtesy Mansell Collection.)*

Canon Barber seemed sure and the late Mr Bettridge was in no doubt whatever. What a pity that Mr Bullock, the photographer, did not take a photograph of Byron. It would have been easy, but Canon Barber thought it a very irreverent thing to do, so photographs were confined to coffins and coffin plates. Mr Bettridge's family had lived in Hucknall since the time of Lord Byron, and he remembered his grandmother telling him how she had seen the funeral of Byron's daughter Lady Lovelace in 1852, and his great grandmother had seen the cortège of Byron himself.

To have gazed upon the face of Lord Byron must have been an uncanny but highly privileged occasion. Apart from an admittedly morbid appeal, such experiences bridge time effortlessly and bring us nearer to our forebears. The past suddenly becomes the present as the fog of history rises for a moment and we are dazzled by the light.

REFERENCE
Innes-Smith R *Sunday Telegraph* 7 February 1988

---

## *Bathtub Elixir puts Estonian behind Bars*

You can't be a profit-maker in your own official basement—that was the lesson learned by an Estonian biochemist when he concocted a cure-all 'health elixir'. The biochemist was arrested last December, but details of his bizarre case have only just been made public—and investigations are still continuing.

I Khint, a citizen of Tallin, capital of the Soviet republic, installed several large bathtubs in the depths of the Special Design and Technological Bureau, of which he was director, and brewed and sold large quantities of his quack drink, making a handsome profit—before he was arrested.

The ingredients of the concoction were rather likes those of Macbeth's witches' cauldron: pigs' feet, blood and entrails, mushrooms, beets, cabbage, dandelion, carrots, nettles, sugar and a 'secret fermentation' formula. After being strained, the liquid was bottled for sale to eager consumers.

Though the contents were neither analysed nor officially approved, Khint managed to talk the Soviet Ministry of Health into giving him a permit to sell his 'Health Elixir' at a 'middle ceiling price'. The elixir was widely recommended as a panacea for every malady—including cancer.

The enterprising Khint organised seminars, conferences, and lectures to promote his elixir. He even managed to attract the interest of the Soviet State Planning Commission by claiming that when the elixir was diluted in the drinking water of livestock, productivity increased from 20–40 per cent.

Khint's come-uppance came when the medical authorities finally checked his concoction and determined that it was toxic, addictive and devoid of medical value. He and his numerous accomplices were tried and convicted on a string of charges. Khint got a maximum 15-year sentence.

REFERENCE
*The Sunday Times* May 13 1984

## Capgras Syndrome

People who believe that a relative or close friend has been replaced by their double are said to suffer from Capgras Syndrome. While these delusions can be devastating and emotionally hurtful for the relative, they are not usually in any danger.

However, a report in the *British Journal of Psychiatry* contains a chilling account from a secure psychiatric unit in Missouri. A man there accused his step-father of being a robot and subsequently decapitated him to look for batteries and microfilm inside his head.

REFERENCE
The *Independent* 23 February 1988

---

## French X-Ray Finds a Screw Loose

It was reported from Cannes that doctors could scarcely believe their eyes when a routine x-ray of a man complaining of headaches showed a seven inch screwdriver embedded in his skull.

On further investigation they found the tool was not in the man's head—but in the x-ray machine where it had been left by a careless technician.

REFERENCE
*London Evening Standard* 22 February 1985

---

## Self-Inflicted Frontal Lobotomy

Brain surgery which can change people's personalities or behaviour has had a controversial history. Until high-powered scanners made it easier to make tiny and accurate incisions in the tissue, the success of operations was hit and miss.

A bizarre tale in the *British Journal of Psychiatry* describes a 19-year-old man who had suffered for years from severe compulsive and obsessive behaviour. He became increasingly depressed and attempted to commit suicide by shooting himself in the mouth. He survived the injuries but coincidentally performed a precise operation on the frontal lobes of his brain: when he eventually recovered from his ordeal his ritualistic behaviour was much less pronounced.

REFERENCE
The *Independent* 20 January 1988

---

## *Almost-failed EEC Contraception*

(Contributed by Mr J M McGarry, FRCOG, Barnstaple.)

Prior to their marriage contraception had been by the sheath (now more contemporaneously called the condom). Although their pre-marital sexual activity had been reasonably satisfying to both partners, a decision was taken during the pre-nuptual preparation that she would better be fitted with a diaphragm. It being 1961, when the Pill had only just come on to the market, the choice of oral contraception was not made because in that immediate post-thalidomide era this young couple wanted to wait until the new 'pill' had been tested by time. In any case the low risk of pregnancy when using the cap and spermicides was one that the espoused were prepared to take.

After the wedding breakfast they hurriedly left for a honeymoon that was to be taken exploring the Mosel valley in Germany as they both had oenophilic interests and this seemed to be a good opportunity to combine several pleasures at the same time. The Weymouth–Cherbourg ferry crossing was rough so the first night in France did not involve any love-making. The next day's drive across France to the Luxembourg border was similarly exhausting: so marital bliss was again postponed. In a cafe in Longwy in eastern France a late coffee was followed by a trip to the Ladies room. Here she truly 'powdered her nose', her powder compact being similar to the plastic case containing the contraceptive cap. Owing to tiredness and hurry something was left in front of the mirror in the 'Ladies'.

Their open-topped Morris soon crossed into Luxembourg. Passports were duly stamped with Gallic indifference by both sets of Customs and in that late evening one of the smallest countries of the world was quickly traversed, a distance of 45 km. The Luxembourg/ German border was crossed with Prussian efficiency and they were soon ensconced in a Mosel valley hotel inside Germany. A suitable honeymoon repast was washed down with Bernkastel and they retired early to bed.

Just as his undressed unrequited loins were beginning to tingle she gasped that she had lost her cap. Memories of the shelf in front of the mirror in the Ladies in the cafe in Longwy quickly dawned and with a sigh he set out to traverse Luxembourg for the second time that evening. Each set of Customs examined his passport with greater diligence this time but he was eventually let through.

On arrival at the French cafe it was of course shut. An irate Patron was roused from his slumbers, and a request to go into the ladies 'loo' was reluctantly acceded to. Fortunately there was the plastic case and its contents intact and undamaged. It was hastily pocketed and with profuse apologies to the cafe owner the luckless groom prepared to cross Luxembourg for the third time in one day.

The Customs officers by now had become increasingly suspicious and during the next four hours his person and car were rigorously searched by four different sets of Customs officers. All they found was his passport and a Dutch Cap. Each time the cap was discovered he tried to explain in halting schoolboy French and German the reason for it being on his person.

The Mosel valley hotel was eventually reached at 4 am. Our frustrated suitor had to wake up the owner of the small hostelry who was not amused; this time with German bad grace instead of French.

He crawled into bed clutching the cased cap. Unfortunately the lady's unfamiliar sampling of the Bernkastel had induced a deep soporiferous state. Also his phenomenal achievement of becoming the first person to completely cross the same country three times in one day (albeit a small country) had so exhausted him that amorous cavorting did not seem to be worth the effort.

The marriage was consummated on the fourth day and both considered it was worth waiting for. At conventions of French, German and Luxembourg Customs officers, the antics of this strange English doctor in the open car are still talked about with unbelieving awe. No wonder it took us so long to get into the EEC.

---

## Premature Burial

In some instances simulation of death has been so exact that it has led to premature internment. There are many such cases on record[1] and it is a popular superstition that all the gruesome tales of persons buried alive and returning to life, only to find themselves hopelessly lost in a narrow coffin many feet below the surface of the earth, are true.

Indeed, because of this dread by members of the public, doctors for generations have been denounced for their inability to discover an infallible sign of death. There was even a suggestion in the *Scientific American* of 1896[3] that x-rays could be used for this purpose:

> Dead flesh offers more resistance to the penetration of the rays than living, and a glance at the radiograph of the person would determine whether it was that of a corpse or not.

Most of the instances of premature burial on record, and particularly those from lay journals, are vivid exaggerations, drawn possibly from such trivial signs as a corpse found with a fist tightly clenched or a face distorted—which are the inspiration of the horrible details of the dying struggles of the person in the coffin. However, in all modern methods of burial, even if life were not extinct, there could be no possibility of consciousness or of struggling. Absolute asphyxiation would soon follow the closing of the coffin lid. Nevertheless, mistakes have been made, particularly in the instances of catalepsy or trance, and during epidemics of fevers and plagues, in which there was an absolute necessity of hasty burial for the prevention of contagion. It is also said that Vesalius, sometimes called the Father of Anatomy, having been sent for to perform an autopsy on a woman subject to hysterical convulsions, and who was supposed to be dead, on making the first incision perceived by her movements and cries that she was still alive. This circumstance becoming known, it rendered him so odious that he had to leave the community in which he practiced, and it is believed that he never entirely recovered from the shock it gave him.

It is probably far more likely that any stories of a return to life which are true, are those when the event occurred after certification of death but before burial. One such event[2] has been related to me by an impeccable source, a health services correspondent of a well known paper who was the daughter of the Irish GP who was an eyewitness:

> The GP had pronounced one of his patients dead and the usual arrangements were made for the wake. The GP was invited and, there they all were—the family of the dead man, the priest, the district nurse, *et al*, all taking their place around the scrubbed clean kitchen table on which Patrick lay in his open coffin. The whisky and the porter flowed freely: the mood grew merrier and merrier as they celebrated the arrival of the dear departed soul at the pearly gates. Then suddenly, the 'corpse' sat up, assessed the situation and asked 'Is this a private party, or can anyone join in?'. On being given the facts, the poor man had a coronary and dropped dead. The wake continued from where it had briefly left off and in due course Patrick was buried and sent on his heavenly way.

REFERENCES
[1] Gould G M and Pyle W L 1900 *Anomalies and Curiosities of Medicine* (Philadelphia: Saunders)
[2] Harding M 1984 Personal communication
[3] Mould R F 1980 'X-ray mystery' *BIR Bulletin* **6** No 4

---

# Yellow Fever Experiment in Cuba

In 1900 there was an epidemic of yellow fever among the US troops stationed in Havana after the Spanish–American war. Fifteen years earlier it had been suggested by Charles Finlay[1] that mosquitoes carried the disease from man to man, but this had been largely ignored and the US Secretary of War appointed a four-man medical board of Walter Reed and three assistants, Carroll, Agramonte and Lazear, 'to pursue scientific investigations with reference to the infectious disease prevalent on the island of Cuba.[2] Following the theory of Finlay, they hatched mosquitoes from eggs, induced the insects to bite patients with known yellow fever and then maintained colonies of presumably infected mosquitoes for experiments. Some of these bit Carroll, Lazear and a volunteer from the US Army, Private Dean, and all contracted yellow fever. Carroll and Dean recovered but unfortunately Lazear died. However, it had been shown that mosquitoes carried the disease from man to man.

Final proof that yellow fever was transmitted only by mosquito bite came when three other volunteers never exposed to Cuban mosquitoes slept for 20 nights in a room free of mosquitoes, but full of mattresses, pillows, towels, sheets, underwear, pyjamas and blankets soiled with blood and a variety of discharges from known yellow fever patients. None of the volunteers became sick.

REFERENCES
[1] Finlay C 1886 'Yellow fever: its transmission by means of the culex mosquito' *Am. J. Med. Sci.* **92** 395–409
[2] Comroe J H 1977 *Retrospectroscope* (Von Gehr Press: Menlo Park, CA)

---

# Vaccination

Dr Edward Jenner's 1796 discovery of vaccination, the immunisation for smallpox, is well known. What is not so well known are the following anecdotes:

The first child to be vaccinated in Russia was named Vaccinov and educated at the expense of the nation as a tribute to the discovery.

During the 1870–1 Franco-Prussian War, vaccination was compulsory in the Prussian army where only 297 died of smallpox. It was not compulsory in the French army and 23 400 died from smallpox. The French also lost the war!

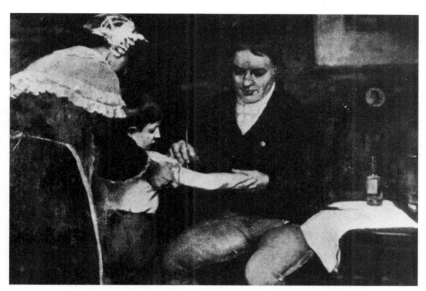

*The first vaccination. Edward Jenner inoculating an eight-year-old boy, James Phipps, with cow pox matter on 14 May 1796. (Courtesy Mansell Collection.)*

# A Deaf Australian Magistrate

A slightly deaf Magistrate had just heard a case of larceny, to which the defendant had pleaded guilty. The Prosecutor had read out a long list of prior convictions for dishonesty, assaults and resisting arrests. The Magistrate then asked the defendant, who was unrepresented by Counsel, if he had anything to say in extenuation. The defendant said, 'Bugger all'.

The Magistrate did not hear what the defendant said and he leaned over the Bench and whispered to the Clerk of Sessions, sitting immediately below and in front of him and said, 'What did he say?'

The Clerk said, 'Bugger all'.

The Magistrate said, 'That's funny, I could have sworn I saw his lips move'.

REFERENCE
Gillespie-Jones A S 1978 *The Lawyer Who Laughed* (Hutchinson of Australia: Richmond, Australia) p. 141

---

## Sir Anthony Oyster

In the late 1830s *The Lancet* was looking for any evidence of abuse of power or inefficiency in the hospitals of the day. The blow fell on the Westminster Hospital surgeon Sir Anthony Carlisle, President of the Royal College of Surgeons in 1828 and 1837, and a pupil of John Hunter. He had delivered the 1820 Hunter Oration at the Royal College, taking as his title *The Anatomy of the Oyster*.

*Sir Anthony Carlisle FRS, Surgeon to Westminster Hospital, 1793–1840.*

372

Thomas Walkley, the editor of *The Lancet* attacked Carlisle, calling him Sir Anthony Oyster, on the subject of a man called Thomas Holmes, admitted to Westminster Hospital with an injured arm, and who subsequently developed delirium tremens. Sir Anthony had not seen him on admission to hospital, but passing through the wards some days later he noticed that Holmes was in delirium and was causing discomfort to the other patients. Thereupon he signed a certificate of insanity which he left with the House Surgeon so that if necessary, and if the patient needed it, the certificate could be signed by a second medical man and the patient could then be removed as insane.

A subsequent enquiry by the Board of Governors established that the patient was a heavy drinker, had had previous attacks and that the certificate was supplied by Sir Anthony at the request of the Chairman of the Board's investigating committee, who had previously ordered Holmes to be put in a straitjacket.

Further evidence showed where *The Lancet* had got its story: from the hospital Apothecary and the House Surgeon. The former rather ingenuously said 'Our only object in publishing was to preserve an instance of the efficiency of large doses of opium in the treatment of delirium tremens' but also added that the Board must now consider Carlisle's fitness for hospital duties.

Sir Anthony was exonerated, but one year later *The Lancet* published an account of a surgical round by Carlisle which rather ridiculed the old gentleman, written by one Clarke. Carlisle was annoyed by this since he had not realised that Clarke had attended the round for such journalistic purposes and told his medical students that if they ever saw Clarke in the hospital again he would be obliged if they would 'rough handle him a little'. When Clarke was told of this, he paid a fee and became a Governor of Westminster Hospital.

However, at his first Board meeting he was recognised and had to jab the point of his umbrella into the eye of one student in order to escape to the Board Room. Then, after the meeting the students pursued Clarke with so much fury that he was forced to take refuge in a local public house.

Sir Anthony Carlisle died in 1840 but Clarke remained a member of the Board of Governors at least until 1845 with Board meetings continuing to be stormy and Clarke (by then an Assistant Editor of *The Lancet*) suggesting that the hospital must have been built over the old Westminster Cockpit!

REFERENCE
Humble J G and Hansell P 1974 *Westminster Hospital, 1716–1974* 2nd edn (London: Pitman Medical)

# The Birth of Anaesthesia

(By permission of Professor Harold Ellis.)

If I were asked to name the most dramatic moment in surgery, it would not be when the clamps were taken off the first heart transplant, nor the first esophagectomy, or the ablation of any other major organ but a very simple, almost trivial affair—the removal of a benign tumour of the neck. The date was October 16, 1846, the place was Massachusetts General Hospital in Boston, and the event was the watershed between the past agonies of surgery and the modern era, now only a little over a century old, when patients enjoy the blissful oblivion of anesthesia.

William Morton, the dentist who was to become the father of modern anesthesia, was twenty-seven years of age. He had been experimenting with ether for dental extractions, and his first patient, Eben Frost, had a tooth pulled out under the influence of ether saturated into a handkerchief on September 30, 1846. The patient testifies as follows:

> This is to certify that I applied to Dr Morton at 9 o'clock this evening suffering under the most violent toothache; that Dr Morton took out his pocket handkerchief, saturated with a preparation of his, from which I breathed for about half a minute, and then was lost in sleep. In an instant more I awoke and saw my tooth lying upon the floor. I did not experience the slightest pain whatever. I remained 20 minutes in his office afterward, and felt no unpleasant effects from the operation.

Dr H J Bigelow, son of the great Jacob Bigelow, was attracted by the newspaper reports of Morton's work and went to see him. Impressed by what he saw, he introduced Morton to Professor John Collins Warren, then sixty-eight years of age and one of the country's leading surgeons. A few days later Morton received the following letter:

> I write at the request of Dr John Collins Warren to invite you to be present Friday morning, October 16, at 10 o'clock at the hospital to administer to a patient who is then to be operated upon, the preparation you have invented to diminish the sensibility to pain.

The letter was signed by Dr C F Heywood, House Surgeon.

It was now only two weeks since ether had been administered to Frost, but already Morton had progressed from a soaked handkerchief to a simple anesthetic machine. This consisted of a two-necked glass globe, one neck allowing the inflow of air, the other fitted with a wooden mouthpiece through which the patient inhaled air across the

374

surface of the ether in the bottom of the jar.

All the anxieties of a great clinical trial are summed up by Morton's young wife, who wrote:

> The night before the operation my husband worked until 1 or 2 o'clock in the morning upon his inhaler. I assisted him, nearly beside myself with anxiety, for the strongest influences had been brought to bear upon me to dissuade him from making this attempt. I had been told that one of two things was sure to happen; either the test would fail and my husband would be ruined by the world's ridicule, or he would kill the patient and be tried for manslaughter. Thus I was drawn in two ways; for while I had bounded confidence in my husband, it did not seem possible that so young a man could be wiser than the learned and scientific men before whom he proposed to make his demonstration.

The operating theatre at the Massachusetts General Hospital was situated just below the central dome of the old building. It is preserved to this day. On the morning of October 16, it was crowded with surgeons and medical students. The audience included both Jacob Bigelow and his son, Henry. The patient was Gilbert Abbott, twenty years of age, who had a benign vascular tumor of the neck. Petrified at the thought of the pain of his operation, he had readily agreed to the experiment. Professor Warren explained to the audience how much he had always wished to free his patients from the pain of operation and for that reason had agreed to the experiment. The time of the operation arrived and passed. By ten minutes past ten Professor Warren picked up his knife and said 'As Dr Morton has not arrived, I presume he is otherwise engaged.' (I personally believe that this was simply Morton creating a tradition, which has been handed down from one generation of anesthetists to another, of keeping the surgeon waiting.)

Just as Abbott was being strapped down on the operating chair, a breathless and flustered Morton arrived; he had been modifying his apparatus up to the very last moment. Warren said, 'Well, sir, your patient is ready.'

'Are you afraid?' Morton asked the patient. 'No, I feel confident that I will do precisely what you tell me.' Morton applied his ether, the smell heavily disguised with some aromatic agent to prevent people from recognizing it and discovering its secret. Turning to Warren, Morton was now able to say, '*Your* patient is ready, doctor.' Many years later Mrs Morton described the scene:

> Then in all parts of the ampitheatre there came a quick catching of breath, followed by silence almost deathlike, as Dr Warren stepped forward and prepared to operate. . . . The patient lay silent, with eyes closed as if in sleep; but everyone present fully expected to hear a shriek of agony ring out as the knife struck down into the sensitive nerves,

375

but the stroke came with no accompanying cry. Then another and another, and still the patient lay silent, sleeping while the blood from the severed artery spurted forth. The surgeon was doing his work, and the patient was free from pain.

The operation took thirty minutes, and at the end Abbott agreed that the whole affair had been free from pain. Warren turned to the audience and said, 'Gentlemen, this is no humbug.' It took a few moments before the sensational importance of what they had seen struck the audience, who then rushed forward to congratulate Morton, to examine the patient, and to ask him over and over again if the operation really had been painless. Everyone in that room must have realized that they had witnessed an historic occasion.

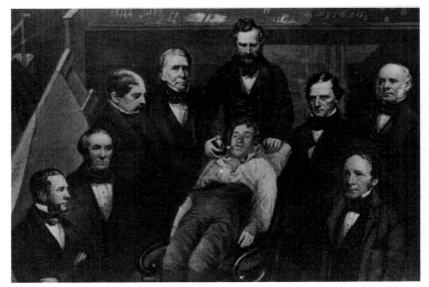

*William Morton making the first demonstration of etherisation at the Massachusetts General Hospital, surrounded by the medical staff of that institution 16 October 1846 (Courtesy Mansell Collection.)*

It was now necessary to proceed to the crucial experiment. The new agent might be effective in the removal of a subcutaneous lump from the neck, but would it work in a capital operation, an amputation? A case was duly scheduled, therefore, for November 7. Before this could be put to the test, a burning ethical issue arose. Should Morton be allowed to administer a secret agent, beneficial though it might be, or should its use be prohibited until its nature was revealed to the medical profession? Warren was prepared to go ahead, his only concern being relief of pain, but the Massachusetts Medical Society

376

resolved unanimously—no formula, no patients. Even though Morton offered to supply the preparation free for use in the Boston hospitals, the doctors remained adamant.

On the very day of the operation the argument continued, with the patient waiting in the anteroom and the theatre packed to the ceiling with expectant doctors and students. Unable to bear the thought of the patient's suffering, Morton quietly announced that his liquid was indeed sulfuric ether.

The patient was a twenty-one-year-old servant girl, Alice Mohan, who had been in the hospital since the previous March with tuberculosis of the knee joint. Dr George Hayward was to perform the amputation, with Warren and Bigelow in attendance. Morton administered the ether, and after some coughing, the patient fell into a deep sleep. Hayward stuck a pin into her arm and, when there was no reaction, rapidly amputated her leg. As he finished, Alice began to groan and move. Hayward bent over her and said, 'I guess you've been asleep, Alice.' 'I think I have, sir,' she replied. 'Well, you know why we brought you here; are you ready?' 'Yes sir, I am ready.' Hayward then reached down, picked up the amputated limb from the sawdust, showed it to her and said, 'It's all done, Alice.' (What Alice said when she saw her leg has not been recorded.)

Scenes of intense excitement were then seen, with the medical audience clapping and shouting with amazement. Morton described the affair modestly: 'I administered the ether with perfect success. This was the first case of amputation.' The patient did well and was discharged from the hospital in time for Christmas.

News of 'the most glorious, nay, the most god-like discovery of this or any other age' spread with amazing speed. Bigelow wrote a letter on November 28 to Dr Francis Boott, an American friend living in Gower Street, London, near the University College Hospital. He enclosed a copy of his son's account of the details of the operations and Morton's apparatus. Boott in turn conveyed the news to Robert Liston, Professor of Surgery at University College Hospital, London, and Liston got his colleague, Dr William Squire, to obtain ether from an uncle of his, who had a chemist's shop on Oxford Street. The apparatus was assembled, and on December 19, Squire anesthetized Boott's niece for a dental extraction carried out by James Robinson. The operation was uneventful, and plans were made for major surgery two days later.

The patient was Frederick Churchill, a butler who had been admitted a month previously with chronic osteomyelitis of the tibia. At two o'clock the operating theatre at the Univeristy College Hospital was packed to capacity. Squire called for a volunteer among the doctors and medical students present, saying that he had only tried the apparatus once before and would like one more rehearsal before submitting a patient to its influence for a capital operation. No one moved. The theatre porter, Shelldrake, was therefore asked to submit to the test. He was not a good choice to try out an anesthetic as he was fat, plethoric, and with a liver no doubt very used to strong liquor. After a few deep breaths of ether, Shelldrake leaped off the table and ran out of the room, cursing Squire and everybody else at the top of his voice.

Fifteen minutes later, Liston arrived, and Churchill was brought into the theatre by the now sober and recovered Shelldrake. Squire now took the precaution of choosing two hefty students to stand by in case the patient repeated the porter's performance. What happened next has been brilliantly described by Dr F W Cock:

> A firm step is heard, and Robert Liston enters—that magnificent figure of a man, six foot two inches in height, with a most commanding expression of countenance. He nods quietly to Squire and, turning round to the packed crowd of onlookers, students, colleagues, old students and many of the neighbouring practitioners, says somewhat dryly, 'We are going to try a Yankee dodge to-day, gentlemen, for making men insensible.' He then takes from a long narrow case one of the straight amputating knives of his own invention. It is evidently a favourite instrument, for on the handle are little notches showing the number of times he had used it before. . . . The patient is carried in on the stretcher and laid on the table. The tube is put into his mouth, William Squire holds it at the patient's nostrils. A couple of dressers stand by, to hold the patient if necessary, but he never moves and blows and gurgles away quite quietly. William Squire looks at Liston and says, 'I think he will do, sir.' 'Take the artery, Mr Cadge,' cries Liston. Ransome, the House Surgeon, holds the limb. 'Now gentlemen, time me,' he says to the students. A score of watches are pulled out in reply. A huge left hand grasps the thigh, a thrust of the long, straight knife, two or three rapid sawing movements, and the upper flap is made; under go his fingers, and the flap is held back; another thrust, and the point of the knife comes out in the angle of the upper flap; two or three more lightning-like movements and the lower flap is cut, under goes the great thumb and holds it back also; the dresser, holding the saw by its end, yields it to the surgeon and takes the knife in return—half a dozen strokes, and Ransome places the limb in the sawdust, 'Twenty-eight seconds,' says William Squire. The femoral artery is taken upon a tenaculum and tied with two stout ligatures, and five or six more vessels with the bow forceps and single thread, a strip of wet lint put between the flaps, and the stump dressed. The patient, trying to raise himself, says, 'When are you going to begin? Take me back, I can't have it done.' He is shown the elevated stump, drops back and weeps a little; then the porters come in and he is taken back to bed. Five minutes have elapsed since he left it. As he goes out, Liston turns

again to his audience, so excited that he almost stammers and hesitates, and exclaims, 'This Yankee dodge, gentlemen, beats mesmerism hollow.'

Liston could hardly have realized at that moment that the need for rapid surgery, which his skill had brought to such a pitch of perfection, was now to be replaced by the new era, when anesthesia would allow calm and unhurried operations.

Immediately after Christmas, ether anaesthesia reached Westminster Hospital, two or three miles down the road. The operation, however, was far less grand—the scraping off of some venereal warts from a lady of doubtful reputation.

As for Morton, the rest of his short life was not a happy one, although he had the compensation of using ether with great success during the American Civil War. He died in 1868 at only forty-eight years of age. The citizens of Boston erected a monument over his grave, the inscription on which was composed by Dr Jacob Bigelow:

Inventor and revealer of anaesthetic inhalation
By whom pain in surgery was averted and annulled
Before whom in all time surgery was agony
Since whom science has controlled the pain.

REFERENCES
Ellis H 1984 *Famous Operations* (Harwal Publishing Co.: Media, PA) pp 51–61
Cock F W 1915 'The first major operation under ether in England' *Am. J. Surg.* **29** (Suppl) 98
Coltart D J 1972 'Surgery between Hunter and Lister as exemplified by the life and works of Robert Liston (1794–1847)' *Proc. R. Soc. Med.* **65** 556
MacQuitty B 1969 *The Battle for Oblivion. Discovery of Anaesthesia* (Harrap: London)
Warren J C 1897 'The influence of anaesthesia on the surgery of the nineteenth century' *Trans. Am. Surg. Assoc.* **15** 1

---

# Florence Nightingale as Medical Statistician, 1859–61

Few books made a greater impression on Miss Nightingale than those of Adolphe Quetelet, the Belgian astronomer, meteorologist, and statistician; and she had few friends whom she valued more highly than Dr William Farr, the leading statistician of her day in this country. From his meteorological studies, Quetelet deduced a law of the flowering of plants. One of his cases was the lilac. The common

379

lilac flowers, according to Quetelet's law, when the sum of the squares of the mean daily temperatures, counted from the end of the frosts, equals 4264° centigrade. Miss Nightingale was greatly interested in such calculations, and the lilac had a special place in her year. Lady Verney's birthday was April 19, and a branch of flowering lilac was Florence's regular birthday present to her sister. Miss Nightingale used to talk of Quetelet's law with great delight, and commended it to gardening friends for verification in their Naturalist's Diaries. But this is a lighter example of Quetelet's researches. What fascinated Miss Nightingale most was his *Essai de physique sociale*, in which he showed the possibility of applying the statistical method to social dynamics, and deduced from such method various conclusions with regard to the physical and intellectual qualities of man. In regard to sanitation, we have heard already of the reforms which Miss Nightingale was instrumental in carrying out in Army Medical Statistics. She turned next to the question of Hospital Statistics, where improvement seemed desirable both for the surer advance of medical knowledge and in the interests of good administration.

THE CRIMEAN WAR

Miss Nightingale had been painfully impressed during the Crimean War with the statistical carelessness which prevailed in the military hospitals. Even the number of deaths was not accurately recorded. 'At Scutari,' she said, 'three separate registers were kept. First, the Adjutant's daily Head-roll of soldiers' burials, on which it may be presumed no one was entered who was not buried, although it is possible that some may have been buried who were not entered. Second, the Medical Officers' Return, in regard to which it is quite certain that hundreds of men were buried who never appeared upon it. Third, the return made in the Orderly Room, which is only remarkable as giving a totally different account of the deaths from either of the others.' When Miss Nightingale came home, and began examining Hospital Statistics in London, she found, not indeed such glaring carelessness as this, but a complete lack of scientific co-ordination. The statistics of hospitals were kept on no uniform plan. Each hospital followed its own nomenclature and classification of diseases. There had been no reduction on any uniform model of the vast amount of observations which had been made. 'So far as relates,' she said, 'either to medical or to sanitary science, these observations in their present state bear exactly the same relation as an indefinite number of astronomical observations made without concert, and reduced to no common standard, would bear to the progress of astronomy.'

# MODEL FORMS

Miss Nightingale set herself to remedy this defect. With assistance from friendly doctors on the medical side, and of Dr Farr, of the Registrar-General's Office, on the statistical, she prepared (1) a standard list, under various Classes and Orders, of diseases, and (2) model Hospital Statistical Forms. The general adoption of her Forms would, as she wrote, 'enable us to ascertain the relative mortality in different hospitals, as well as of different diseases and injuries at the same and at different ages, the relative frequency of different diseases and injuries among the classes which enter hospitals in different countries, and in different districts of the same countries.' Then, again, the relation of the duration of cases to the general utility of a hospital had never been shown. Miss Nightingale's proposed forms 'would enable the mortality in hospitals, and also the mortality from particular diseases, injuries, and operations, to be ascertained with accuracy; and these facts, together with the duration of cases, would enable the value of particular methods of treatment and of special operations to be brought to statistical proof. The sanitary state of the hospital itself could likewise be ascertained.' Having formed her plan, Miss Nightingale proceeded with her usual resourcefulness to action. She had her Model Forms printed (1859), and she persuaded some of the London hospitals to adopt them experimentally. Sir James Paget at St Bartholomew's was particularly helpful; St Mary's, St Thomas's, and University College also agreed to use the Forms. She and Dr Farr studied the results, which were sufficient to show how large a field for statistical analysis and inquiry would be opened by the general adoption of her Forms.

The case was now ready for a further move. Dr Farr was one of the General Secretaries of the International Statistical Congress which was to meet in London in the summer of 1860. He and Miss Nightingale drew up the programme for the Second Section of the Congress (Sanitary Statistics), and her scheme for Uniform Hospital Statistics was the principal subject of discussion. Her Model Forms were printed, with an explanatory memorandum; the Section discussed and approved them, and a resolution was passed that her proposals should be communicated to all the Governments represented at the Congress. She took a keen interest in all the proceedings, and gave a series of breakfast-parties, presided over by her cousin Hilary, to the delegates, some of whom were afterwards admitted to the presence of their hostess upstairs. The foreign delegates much appreciated this courtesy, as their spokesman said at the closing meeting of the Congress; 'all the world knows the name of Miss Nightingale,' and it was an honour to be received by 'the illustrious invalid, the

Providence of the English Army.' The written instructions sent by 'the Providence' to her cousin for the entertainment of the guests show her care for little things and her knowledge of the weaknesses of great men: 'Take care that the cream for breakfast is not turned.' 'Put back Dr. X's big book where he can see it when drinking his tea.' Miss Nightingale also induced her friend Mrs. Herbert to invite the statisticians to an evening party. The feast of statistics acted upon her as a tonic. 'She has been more than usually ill for the last four or five weeks,' wrote her cousin Hilary; 'now I cannot help thinking that her strength is rallying a little; she is much interested in the Statistical Congress.' Congresses, like wars, are sometimes 'muddled through' by our country, and Miss Nightingale was able here and there to smooth ruffled plumes. A distinguished friend of hers, though his name had been printed as one of the secretaries of a Section, had not received so much as an intimation of the place of meeting; he was disgusted at so unbusiness-like an omission, and was half inclined to sulk in his tents. Miss Nightingale's letter on the subject is characteristic:—

(*Miss Nightingale to Dr T Graham Balfour*) 30 OLD BURLINGTON ST., *July* 12 [1860]. You are quite right in what you say. We are all of us in the same boat. And, if it were not that England *would not be* the mercantile nation she *is*, if she had not business habits somewhere, I should wonder from my experience where they are. Certain of us, who were asked to do business for the Statistical Congress, had it all ready since December last—and were not able to get it out of the Registrar-General's Office till this week. Certain of us were asked to do business this morning, and to have it ready by to-night, which if *not* done, would arrest the proceedings of the Congress, and *if* done, must be the fruit of only five hours' consideration, when five months might just as well have been granted for it. I don't say this is so bad as the treatment of you who are Secretary. But still it is provoking to see a great International business worked in this way.

What I want now is to put a good face upon it before the foreigners. Let *them* not see our short-comings and disunions. Many countries, far behind us in political business, are far before us in organization-power. If any one has ever been behind the scenes, living in the interior, of the Maison Mère of the 'Sisters of Charity' at Paris, as I have—and seen their Counting House and Office, all worked by women—an Office which has twelve thousand Officials (all women) scattered all over the known world—an office to compare with which, in business habits, I have never seen any, either Government or private, in England—they will think, like me, that it is this mere business-power which keeps these enormous religious 'orders' going.

I hope that you will try to impress these foreign Delegates, then, with a sense of *our* 'enormous business-power' (in which I don't believe one bit), and to keep the Congress going. Many thanks for all your papers. I trust you will settle some sectional business with the Delegates here to-morrow morning. And I trust I shall be able to see you, if not to-morrow morning, soon.

Mind, I don't mean anything against *your* Office by this tirade. On the contrary, I believe it is one of the few efficient ones now in existence.

*Florence Nightingale, 1858.*

Having received the *imprimatur* of an International Congress, Miss Nightingale circulated her paper on Hospital Statistics widely among medical men and hospital officials. Thereby she produced immediate effect. She printed large quantities of her Model Forms, and supplied them, on request, to hospitals in various parts of the country. Through the good offices of M. Mohl, she also worked upon public opinion in France. 'Some months ago,' she wrote to Dr Farr (Oct. 20, 1860), 'I got inserted into the leading medical journals of Paris an article on the proposed Hospital Registers; and you see they are at work.' the London hospitals took the matter up. Guy's printed a statistical analysis of its cases from 1854 to 1861; St Thomas's, of its from 1857 to 1860; St Bartholomews, a table of its cases for 1860. With regard to the future, a meeting was held at Guy's Hospital on June 21, 1861, and it was unanimously agreed—by delegates from Guy's, St Bartholomew's, St Thomas's, the London, St George's, King's College, the Middlesex, and St Mary's—that the Metropolitan Hospitals should adopt one uniform system of Registration of Patients; that each hospital should publish its statistics annually, and that Miss Nightingale's Model Forms should as far as possible be adopted. She called further attention to her scheme in a paper sent to the Social Science Congress at Dublin in August 1861, and incorporated it in a later edition of her *Notes on Hospitals*. The statistics of the various hospitals which had accepted her Forms were published in the *Journal of the Statistical Society* for September 1862, but I do not find that the experiment has been continued. So far from there being any uniform hospital statistics, of the kind contemplated by Miss Nightingale, even in London some of the hospitals do not keep, or at any rate do not publish, any at all. The laboriousness, and therefore the costliness, of the work of compilation, the difficulty of securing actual, as well as apparent, uniformity, and a consequent doubt as to the value of conclusions deduced from the figures are presumably among the causes which have defeated Miss Nightingale's scheme. Some limited portion of her object is perhaps attained by the statistical data which the administration of King's Hospital Fund demands, but even here there are possibilities of misleading comparison. There is probably no department of human inquiry in which the art of cooking statistics is unknown, and there are sceptics who have substituted 'statistics' for 'expert witnesses' in the well-known saying about classes of false statements. Miss Nightingale's scheme for Uniform Hospital Statistics seems to require for its realization a more diffused passion for statistics and a greater delicacy of statistical conscience than a voluntary and competitive system of hospitals is likely to create.

At the time she was full of hope, and, having obtained a start with medical statistics, she next pursued the subject in relation to surgical operations. Sir James Paget had been in communication with her on this point. 'We want,' he had written (Feb. 18, 1861), 'a much more exact account and a more particular record of each case. Thus in some returns we have about 40 per cent of the deaths ascribed to "exhaustion" in others, referring to the same [kind of] operations, about 3 per cent or less; the truth being that in nearly all cases of "exhaustion" there was some cause of death which more accurate inquiry would have ascertained.' Miss Nightingale (May 1, 1861) congratulated him on 'St Bartholomew's having the credit of the first Statistical Report worth having,' but the table of operations was still, she thought, most unsatisfactory. 'It would be most desirable that an uniform Table should be adopted in all Hospitals, including all the elements of age, sex, accident, habit of body, nature of operation, after-accidents, etc., etc. Could you come in to-morrow between 2 and 4, and bring your list of the causes of death after operations? It would be invaluable, coming from such an authority, for constructing a Form.' She consulted other surgeons, civil and military, and wrote a paper, with Model Forms, for the International Statistical Congress held at Berlin in September 1863. These also were included in a revised edition of *Notes on Hospitals*. The Royal College of Surgeons referred the subject to a Committee, which, however, reported adversely upon Miss Nightingale's Forms.

Before the International Congress at London in 1860 separated, Miss Nightingale addressed a letter to Lord Shaftesbury (President of the Second Section), which was read to the whole Congress, and adopted by it as a resolution. The point of it was to impress upon Governments the importance of publishing more numerous abstracts of the large amount of statistical information in their possession. She gave various instances in which useful lessons might thus be enforced upon the public mind, and cited Guizot's words: 'Valuable reports, replete with facts and suggestions drawn up by committees, inspectors, directors, and prefects, remain unknown to the public. Government ought to take care to make itself acquainted with, and promote the diffusion of all good methods, to watch our habits and institutions, there is but one instrument endowed with energy and power sufficient to secure this salutary influence—that instrument is the press.' With Miss Nightingale statistics were a passion and not merely a hobby. They did, indeed, please her, as congenial to the nature of her mind. Her correspondence with Dr Balfour and Dr Farr shows how she revelled in them. 'I have a New Year's Gift for you,' wrote Dr Farr (Jan. 1860); 'it is in the shape of Tables, as you will conjecture.' 'I am exceedingly anxious,' she replied, 'as you may suppose, to see your charming Gift, especially those Returns showing the deaths, admissions, diseases,' etc., etc. But she loved statistics, not for their own sake, but for their practical uses. It was by the statistical method that she had driven home the lessons of the Crimean hospitals. It was the study of statistics that had opened her eyes to the preventable mortality among the Army at home, and that had thus enabled her to work for the health of the British soldier. She was already engaged on similar studies in relation to India. She was in very serious, and even in bitter, earnest a 'passionate statistician.'

## CENSUS RETURNS

Miss Nightingale made a valiant attempt to extend the scope of the

Census of 1861 in the interest of collecting statistical data for sanitary improvements. There were two directions in which she desired to extend the questions. One was to enumerate the numbers of sick and infirm on the Census day. For sanitary purposes it would be extremely useful to determine the proportion of sick in the different parts of the country. To those who said that it could not be done, because the people would not give the information, the answer was that it had been done in Ireland. The other point was to obtain full information about house accommodation; facts which, as would now be considered obvious, have a vital bearing on the sanitary and social conditions of the people. This point also had been covered in the Irish Census. Dr Farr entirely agreed with Miss Nightingale, but he could not persuade Sir George Lewis, the Home Secretary, to include these provisions in the Census Bill (1860). Miss Nightingale thereupon drew up a memorandum on the subject, and, through Mr Lowe (Vice-President of the Council), submitted it to the Home Secretary. Mr Lowe may have agreed with her, but he failed to persuade his colleague. 'Whenever I have power,' wrote Mr Lowe (May 9), 'you can always command me, but official omnipotence is circumscribed in the narrow limits of its own department.' Sir George Lewis replied that 'both of Miss Nightingale's points had been duly considered before the Census Bill was introduced. It was though that the question of health or sickness was too indeterminate.' 'With regard to an enumeration of houses, it was thought that this is not a proper subject to be included in a Census of population.' A very official answer! But Sir George added that he did not see how the result of such enumeration could be 'peculiarly instructive'—an avowal which he also made in the House of Commons. The cleverest of men are sometimes dense; and this remark of Sir George Lewis, added to his subsequent conduct of the War Office, earned for him, in Miss Nightingale's familiar correspondence, the sobriquet of 'The Muff.' In communicating the result of her first attempt to Dr Farr, she said, 'If you think that anything more can be done, pray say so. I'm your man.' But she had not waited to be spurred on. She had already bethought herself of a second string in the House of Lords. Lord Shaftesbury, to whom she had appealed, promised to do all he could. Lord Grey did the same, and asked her to send Dr Farr to coach him. She began to 'thank God we have a House of Lords':—

(*Miss Nightingale to Robert Lowe.*) OLD BURLINGTON ST., *May* 10 [1860]. I cannot forbear thanking you for your letter and for your exertions in our favour. Sir George Lewis's letter, *being interpreted*, means: 'Mr. Waddington does not choose to take the trouble.' It is a letter such as I have scores of in my possession, from Airey, Filder, and alas! from Lord Raglan, from Sir John Hall (the doctor) and from

Andrew Smith. It is a true 'Horse Guards' letter.

They are the very same arguments that Lord John used against the feasibility of registering the 'cause of death' in '37—which has now been the law of the land for 23 years. He was beaten in the Lords. And we are now going to fight Sir George Lewis in the Lords. And we hope to beat him too. It is mere child's play to tell us that what every man of the millions who belong to Friendly Societies does every day of his life, as to registering himself sick or well, cannot be done in the Census. It is mere childishness to tell us that it is not important to know what houses the people live in. The French Census does it. The Irish Census tell us of the great diminution of mud cabins between '41 and '51. The connection between the *health* and the *dwellings* of the population is one of the most important that exists. The 'diseases' can be obtained approximately also. In all the more important—such as small-pox, fevers, measles, heart-disease, etc.—all those which affect the *national* health, there will be very little error. (About ladies' nervous diseases there will be a great deal.) Where there *is* error in these things, the error is uniform, as is proved by the Friendly Societies; and corrects itself. . . .

The passionate statisticians were, however, hopelessly out-voted in the House of Commons. Mr Caird moved in her sense on the subject of fuller detail about house-acocommodation, and in sending her the printed notice of his amendment, said that 'his position would be greatly strengthened with the House if he could obtain Miss Nightingale's permission to quote her name in favour of the usefulness of such an inquiry.' I do not know whether she gave permission; the debate is reported very briefly in Hansard. But in any case Mr Caird's amendment was promptly negatived. As for the House of Lords, Miss Nightingale's reliance upon a better love of statistics in that assembly was cruelly falsified. The Census Bill came up late in the session, and I do not find that either Lord Grey or Lord Shaftesbury said a word upon the subject. The only critical contribution made to the debate proceeded from Lord Ellenborough, who, so far from wanting the Census Bill to include provision for more statistical data, proposed to exclude most of those that were already in. He could not for the life of him see what was the use of asking people so many questions. Here, then, Miss Nightingale was in advance of the time; in one case, by a generation, in the other, by two generations. Recent Censuses have included more particulars of the housing of the people, though still not so many as she wanted. Official statistics of the local distribution of sickness will presently be obtained, I suppose, in a different way, through the machinery of the National Health Insurance Act.

Deprived by the recalcitrance of the Home Secretary and Parliament of a fuller feast of statistics at home, Miss Nightingale turned to the Colonies and Dependencies. The Secretary for the Colonies gave her facilities for collecting much curious and instructive information;

and the Secretary of India accepted her aid in collecting and tabulating facts and figures which were the foundation of some of the most notable and beneficient of her labours. But, though she was already (1860–1) engaged in these inquiries, they belong in the main to a later period.

REFERENCE
Cook E 1913 *The Life of Florence Nightingale* vol I (1820–1861) (London: Macmillan) pp 428–38

## *Plato on Hiccoughing*

This was brought to my attention in a letter by S Zak to the Editor of the *Journal of the American Medical Association* (1972) **219** 88.

Aristodemus said that it came to the turn of Aristophanes to speak; but it happened that, from repletion or some other cause, he had an hiccough which prevented him; so he turned to Eryximachus, the physician, who was reclining close beside him, and said, 'Eryximachus, it is but fair that you should cure my hiccough, or speak instead of me until it is over.'—'I will do both,' said Eryximachus, 'I will speak in your turn, and you, when your hiccough has ceased, shall speak in mine. Meanwhile, if you hold your breath some time, it will subside. If not, gargle your throat with water, and if it still continue, take something to stimulate your nostrils, and sneeze; do this once or twice, and even though it should be very violent, it will cease'—'I will follow your directions.'

[Eryximachus] '. . . for I observe your hiccough is over.' 'Yes,' said Aristophanes, 'but not before I applied the sneezing. I wonder why the harmonious construction of our body should require such noisy operations as sneezing, for it ceased the moment I sneezed.'

## *The Cannon Effect*

Did Dmitri Shostakovich have music on the brain? An extraordinary article by a Chinese neurosurgeon in next month's *Musical Times* suggests that he had a piece of shell shrapnel lodged deep inside his brain, and that as a result each time he leaned his head to the side he heard musical melodies—different each time—which he could use when composing. Moving his head back level immediately stopped the music. Dr Dajue Wang claims to have had the story from the Soviet neurosurgeon whom Shostakovich consulted, and whose x-rays allegedly located the musical fragment in the temporal horn of the left ventricle. Shostakovich was in Leningrad during the siege, but there has previously been no mention of any injury. Dr Ronald Henson, a British neurologist consulted about Wang's story, says cautiously: 'I would hesitate to affirm that it could not happen.'

REFERENCE
*The Times* Diary 31 May 1983

---

## *Medicine in Ancient Egypt*

In Homer's *Odyssey* the Egyptian is mentioned as being unparalleled in his skill with medicine while Herodotus remarks that Egyptian schools of medicine were unrivaled in their fame and reputation. From all the known world students converged on Egypt to learn medicine, and foreign kings and princes sent to Egypt for physicians to heal them.

The origin of Egyptian superiority in this field lay in another highly developed art of those times—witchcraft. The first medicine men were priests, and as their skill was taken over by institutions of advanced learning, gradually a body of specialists developed to care for the sick in the community.

Fortunately, numerous papyri recording the medical knowledge of the ancient Egyptians have survived intact. Among these is the well-preserved Ibers papyrus, over sixty feet in length, and the most famous of such medical papyri. It is a virtual textbook of medicine and it is likely that the prescriptions recorded therein were handed down from a much earlier period, probably the Old Kingdom. Divided into 887 chapters, it embraces prescriptions for diseases of the eyes, skin, stomach, heart, arteries and bladder, as well as for gynaecological disturbances. 'Herewith begins the book of medical pharmacology for diseases of all parts of the body,' it states, and proceeds to diagnose a list of diseases and to prescribe proper treatment.

Another papyrus, that of Edwin Smith, meticulously describes surgical operations as well as the treatment of wounds in various parts of the body. It is composed of 48 chapters, each dealing with one particular case of illness, beginning with a description of the disease, followed by its symptoms, diagnosis, and treatment. The physician then gives his opinion of the case: either it is a disease he will 'treat', implying that the prospects for a cure are good; or it is a disease he will 'combat', implying that a cure is uncertain; or it is a disease he will 'not treat', implying that a cure is hopeless. This scrupulous work, obviously the product of experience and meticulous observation, is a far cry from witchcraft. In its mention of surgery, the American Egyptologist, Breasted, considers it the oldest document of its kind on record. Written during the New Kingdom the practices recorded were most likely followed in previous eras. The Hearst papyrus, too, unearthed at Dair-al-Ballas in 1899 and in poor condition, is composed of columns in which mention is made of 250 prescriptions, dating from the New Kingdom. There are many other papyri besides, preserved in museums throughout the world; they are named after their purchasers or the place of their discovery, and reveal the medical skill of the ancient Egyptian.

The earliest beginnings of medicine were probably the outcome of trial and error: a cure for a certain sickness was remembered and used at the next occurrence; and what did not work was eliminated. At this time the cause of sickness was seen to be the work of evil spirits, linking medicine to witchcraft, and most physicians were priests, who, besides prescribing medicine, used talismans and prayers to exorcize evil spirits.

It is clear from some of the medical papyri that different gods were believed to influence certain organs of the body. An appeal to the god Ra as the best cure for a scorpion sting was probably inspired by the myth of Ra extracting poison from the body of his daughter. In treating such cases the priest would call upon Ra, saying, 'O God Ra, hurry to your daughter, whom the scorpion has stung along a remote road. Her screams pierce the skies. The poison runs beneath her skin, spreading throughout her body. She licks the wounded spot like a cat.' Ra would reply, 'Don't be scared. Don't be scared, my daughter, for here I stand behind you.'

The medical profession gradually developed a vocational creed which was passed down from one generation to another, and eventually formed the basis of the Hippocratic oath. Centres for the teaching of medicine were also established. The author of the Ibers papyrus writes, 'I graduated from the University of Heliopolis, together with the princes of the great household, the masters of protection and the rulers of eternity.' That there were such schools of medicine in Heliopolis, Sais, and Memphis is certain.

One of the most famous of ancient physicians was Imhotep, who practiced during the reign of King Zoser in the 28th century BC. After his death he was elevated to godhood by the ancient Egyptians, and the Greeks knew him as Asclepius, the god of healing.

Herodotus pointed out that Egyptian medicine was divided into various fields, each of which had its specialists, but there were also special physicians for the populace, the army, and the royal court. One group of doctors specialized in the use of herbs and the recital of prayers, while another specialized in the use of bandages.

Treatment was divided accordingly into that of internal disorders and that of external disorders, the latter also included surgery, and required experience, skill, and accurate observation. Treatment of internal disorders called for a knowledge of medicine and witchcraft.

One medical papyrus describes catarrh and its treatment. A talisman, which was often read after administering the medicine, gives a precise description of the symptoms of the catarrh:

> Get away catarrh, son of catarrh, which breaks the bones, crushes the skull, pierces the brain, and affects the seven apertures of the head. I have brought a medicine to combat you.

The Ibers papyrus reveals a knowledge of the function of veins and arteries.

It also touched upon gynaecology, prescribed a method to judge a woman's fertility, as well as the quality of mother's milk by its odour.

Surgery is described in the Edwin Smith papyrus, which says specialists in this field were priests devoted to the god Sikhoset, who conducted minor operations such as circumcision and the opening of abscesses. They could differentiate between fractures and dislocations of the bones and they knew how to make splints. Wounds were bandaged after an application of fresh meat, honey, or fat.

After the diagnosis the appropriate medicine was selected from an apparently large number of available preparations, some of which were fast-working and others not, some prescribed for summer and others for winter. If the patient was an adolescent or an adult, medicine was given in the form of pills, but if the patient was still a child, it was mixed with the mother's milk.

I might mention here that in the margins of every medical papyrus, beside the prescriptions, there are such notations as 'good', 'useful', or 'good, I saw it administered and prepared it myself.'

Some of the medicines used, such as the eye medicine, were imported from the city of Jabail on the Phoenician coast. Other medicines were known as cure-alls, called repellents of death, and considered the concoctions of Ra, who had invented them for use against his own senility. Such medicine, made of honey, wax, and fourteen kinds of herbs, was applied in the form of a plaster to the painful area.

391

Medicine to be taken orally came in the form of pills or syrups, and that to be applied externally was made into ointments or plasters. The prescription indicated whether the medicine was to be mixed, ground, or pounded.

Perhaps the skill of the ancient Egyptians in medicine explains their expertise at mummification, the art of preserving the body from decay. It is uncertain exactly when Egyptians began to practice mummification, which ended with the spread of Christianity, but we know that in prehistoric times the dead were buried in relatively shallow pits after being first wrapped in mats, skins, flax, or the like. The dryness of the climate and the heat of the sun dehydrated the body and preserved it as long as it remained dry. Because of the difference in weather conditions, mummifying through dehydration could be successfully carried out in Upper Egypt but not in the delta.

The first Egyptians to note the effects of dehydration must have been greatly astonished to discover, as they went into the desert to bury the dead, that the body of a grandfather or great-grandfather was still so well preserved. It was perhaps this kind of experience that led the early Egyptians to a belief in an afterlife and resurrection. To the ancients the body was like a box containing various substances that could exist eternally, but only on condition that the body was preserved. It was also for this reason that tombs were provided with food and water at the time of burial.

The first reliable evidence of mummification, according to the famous chemist Lucas, dates from the beginning of the Fourth Dynasty. Lucas found, through examining the intestines of Queen Hetub Hars, mother of King Cheops, builder of the Great Pyramid of Giza, that the embalming fluid was composed of natron, sodium sulphate, and sodium chloride.

The process of dehydration of the body was done in two ways: by the sun or by fire; or by the use of absorbent materials. Dehydration of the body through the action of the sun is slow, since 75 per cent of body weight is water. If the body is wrapped in skin, mat, or flax and buried in a dry place exposed to the sun, it loses moisture through the absorbing action of the sand. This process was found accidentally in prehistoric times but later became one of two ways of mummification, though it was not suited for general use.

Herodotus left us a description of embalmment in which he says that the specialists of this trade made available to the family of the deceased three wooden models of embalmment at three different prices. The embalming process, which greatly impressed Herodotus, called for the extraction of most of the brain through the nostrils by means of an iron hook and medical dissolvents. The abdomen was cut to remove the intestines, and body cavity washed out with date wine mixed with spices, later to be filled with incense. The incision was then sewn up and the body placed in natron for seventy days, after which it was washed once again and wrapped with flax saturated in glue. This was then handed over to the family, who would encase it in a coffin to be placed in a tomb.

Herodotus also describes a second, cheaper method, in which the body was injected with rice oil through the anus. It was then placed in natron for seventy days, during which time the rice oil leaked out with the dissolved intestines.

The third method mentioned by Herodotus, one used by the poor, called for cleaning out the body with a laxative solution, then placing it seventy days in natron.

Diodorus Siculus, on the other hand, who presumably knew the words of Herodotus, mentions only one method of embalmment, similar to the first one, where the abdomen was cut to remove the intestines, heart, and kidneys, then washed with date wine mixed with spices and perfumed with incense.

In considering these accounts we must keep in mind that both Herodotus and Diodorus were describing embalming as practiced at a late stage in Egyptian history. Herodotus visited Egypt between 448 and 445 BC, Diodorus in 59 BC. In Lucas's opinion, Herodotus is generally more accurate in his description, although the details of his account would probably vary from those of earlier ages. The period during which the body was placed in natron, in particular, has been an object of debate among scholars.

REFERENCE

Abdel-Kader H 1982 *Life in Ancient Egypt* 2nd edn (Cairo: Al Ahram Commercial Press) 98–105

---

## Anecdote of Aesclepiades of Prusa

Aesclepiades was a Greek physician who practised in Rome during the first century BC. So confident was he of his own medical competence that he maintained he would no longer claim the title of physician should he ever become ill. Subsequently he had the good taste to die, while still in the bloom of health, by plunging down a staircase. His reputation survived his fatal fall!

# The First Whole-Body X-Ray

This radiograph was taken in 1896 in New York and published in the *Archives of the Roentgen Ray*, London. The exposure lasted 30 minutes. Clothing and jewellery which can be seen include: a hatpin, necklace, rings, bracelet, high buttoned boots with a buckle, nail heel shoes and a whalebone corset.

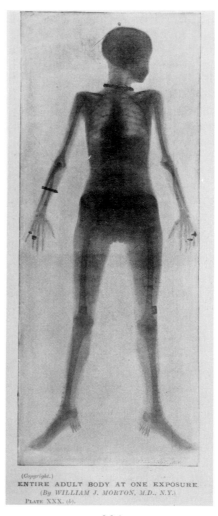

*(Copyright.)*
ENTIRE ADULT BODY AT ONE EXPOSURE.
(By *WILLIAM J. MORTON, M.D., N.Y.*)
PLATE XXX. (*b*).

# Meo Hill Tribe Herb Doctor in Northern Thailand

On a visit to Thailand in 1985 I saw the villag of the Meo Tribe, near the Golden Triangle, in the far north of Thailand beyond Chiangmai and beyond the temple of Doi Suthep at 1056 metres.

A village market was present with all the usual wood carvings, carpets and knick-knacks, but in addition there was the stall of the village 'herb doctor'. Most remedies cost some 20 Bahts (£0.75), such as:

> *Tiger strength*
>> for backache
>
> *The bandit has been killed from the sky*
>> for intestinal cancer and for haemorroids
>
> *God given*
>> for arthritis and for muscle pain
>
> *Cardamom*
>> for heart disease, for dizziness and for vertigo
>
> *Bitter root*
>> for diabetes, for nephritis and for cystitis

*The bandit has been killed from the sky ( for intestinal cancer).*

*God given ( for arthritis and muscle pain).*

*Cardamom ( for heart disease, dizziness and vertigo).*

395

*Erecting (for impotence).*

However, the most interesting hard sell, via an interpreter, came when the herb doctor tried to persuade me to buy a tiger's penis for 1500 Baht (£50). This medicinal item was baked in honey and was used by dissolving it in a jar of locally brewed alcohol, and then 15 ml was prescribed orally each day. One tiger's penis was said to last a year and was supposed to be the best aphrodisiac anywhere in the world.

*The herb doctor.*

I then had to explain that I was not a rich American, to which the herb doctor replied with a statement of why 1500 Baht was a reasonable price. 'First, you have to catch your tiger, and then there is only one penis per tiger, so it has to be 1500 Baht.'

Finally, the herb doctor must have felt sorry for me, or else the doctor friend interpreter gave him instructions, but whatever the reason, I was presented with the 20 Baht version of an aphrodisiac, which has the name Erecting.

---

## Anatomical Comparisons by Rabelais

Lungs like a fur-lined hood
Heart like a chasuble
Mediastinum like forceps
Arteries like a watchman's cloak
Diaphragm like a cockaded cap
Liver like a double axe
Veins like canvas window blinds

---

## Jail Fever in Newgate

In the early part of the eighteenth century, for every person hanged at Tyburn, four would die of jail fever in Newgate prison before they could be hanged. Jail fever was the result of bad sanitation and over-crowding and it was not limited to London; for example in 1577 at an assize in Oxford, some 300 persons died.

At Newgate, the epidemic also extended to the adjoining Sessions House and in 1750 the fever struck down more than 60 people, including jurymen, barristers, the Lord Mayor of London, Under Sheriff and also a former Lord Mayor. As a result, the jail had to be virtually sterilised and was to this end washed down with vinegar. Similarly, prisoners were washed down with vinegar before they were brought to trial, and from this time judges carried with them, and had on the bench, a nosegay of flowers to ward off the prison smells.

Thus from a 1772 document:

To keep out the foul air while in court, candied orange peel or lemon peel, preserved gingers, and garlic if not disagreeable, cardamon, carraway, or other confits, may be very useful, and should the mouth be clammed, dry raisins, currants, or lemon drops, will cool, and quench thirst, which, should it increase, may be assuaged by small draughts of old hock and water, or small punch. Smelling to good wine vinegar during the trials, will not only refresh, but revive, more agreeably and cool, than the use of spirituous waters distilled from lavender or rosemary, and more than any other scents.

and

To guard the seat of justice from the approaches of infection, it will certainly be most prudent to fumigate and steam the place, by means of large braziers, pans or coppers, put in the day before the sessions are to commence, and during that day to burn in them charcoal, with tobacco stalks, and dried aromatic herbs in winter, as mint, rosemary, southern-wood etc., bruised juniper berries may also be burnt; and on a hot iron shovel may be put wet gunpowder, and frankincense, but particularly the steams of boiling hot vinegar should be conveyed to all parts of the building. The next morning about an hour or two before the court meet, the braziers should be filled with coke cinders, as used by maltsters instead of charcoal, and after they shall have burned awhile, the ventilator should let in fresh air, and the floor should be sprinkled with cold vinegar of the sharpest sort. At the time of opening the court, the air-holes, made close to the ground, about a foot square, should be set open and the wooden flaps hooked up.

*The old Sessions House in the Old Bailey, 1750. (Courtesy Mansell Collection.)*

One suggestion for improving Newgate conditions was to construct a windmill on the roof of the gatehouse which it was hoped would draw off the poisonous air. The medical theory was that stale air putrifies and that this in turn dissolves the blood and humours of the human body to set up further putrefaction which would ultimately dissolve the human body itself! Several workmen engaged in the windmill exercise became ill from the smell and this mirrored what happened two years earlier for a similar windmill at London's Bridewell prison. Then, workmen had refused to carry out the work unless they were given tobacco and drink as precautionary measures. Surprisingly, they stayed immune, but some carpenters who worked with them and did not insist on the same conditions got jail fever, and two of them died.

REFERENCES

Anon 1772 *Directions to Prevent the Contagion of the Jail—Distemper Commonly Called Jail-Fever* (London)

Rumblelow D 1982 *The Triple Tree: Newgate, Tyburn and Old Bailey* (London: Harrap)

## Diderot on Doctors

The best doctor is the one you run for and cannot find.

## John Hunter on Operations

Never perform an operation on another person which, under similar circumstances, you would not have performed on yourself.

## On Alcoholism

An alcoholic is someone you do not like who drinks as much as you.

# Witch from Brittany

*Carved wood figure of a witch. Breton workmanship from Treanton Morlaix. The witch is using a pestle and mortar and has a cat on her shoulder. (Courtesy Wellcome Institute Library, London.)*

---

# Moscow Weddings

When young Muscovites get married, many of them follow a pattern which leaves parents, aunts and uncles at home getting an enormous party ready, whilst the happy couple tour the major sites of Moscow. These include the grave of the unknown warrior in the Alexandrovsky Gardens by the Kremlin Wall, the Lenin Hills below the University of Moscow (where photographs are taken with a panoramic view of Moscow in the background), and the large black wooden crosses a few kilometres from the airport, which mark the nearest point to Moscow reached by Hitler's army in what is termed, in the USSR, The Great Patriotic War (World War II). One quaint custom often observed is for the wedding car to have a Russian doll or a Russian teddy bear attached to the radiator grille. I was told on good authority that this would ensure either a girl baby (for those with a doll) or a boy baby (for those with a bear). Whenever I have seen a group of wedding cars, dolls always seem to outnumber bears in a ratio of 10:1. However, since this ratio bears no relation to the actual Russian population statistics, this predictive technique is a failure!

---

## On Filling in Life Insurance Forms

When the death penalty was still in force in the United Kingdom, an Englishman and an American were exchanging experiences on the problem of completing life insurance forms when, as in both their cases, their fathers were found guilty of murder and the ultimate sentence was passed on them. The American's solution was to write that his father met with an accident when he was in the Chair of Applied Electricity, whereas the Englishman's solution was to write that his father died when the platform gave way at a public event.

## Saint Pantaleon

*Patron saint of physicians in the Eastern Church and a patron saint of consumption. From a fourteenth-century Greek manuscript in the Bibliothèque Nationale, Paris. (Courtesy Wellcome Institute Library, London.)*

---

## On How to Start a Speech in the Apothecaries Hall

In a dinner in 1988 in the Apothecaries Hall in the City of London, the Master of one of the City Livery Companies began his speech with the following information. 'A recent survey of city businessmen gathered together for a meeting in a room at any time yielded the following statistic. 66% are thinking about sex. I therefore hope that two-thirds of you are enjoying yourselves. . . .'

# On Getting Matters into Perspective

Two Canadian backpackers encountered a grizzly bear on their trek through the woods. Immediately one of them removed his pack, and as quickly as possible put on his running shoes. The other stood in amazement and said 'That is no good, the grizzly can run a lot faster than you.' 'Ah' said his friend, 'but I don't have to run faster than the bear, I only have to run faster than you.'

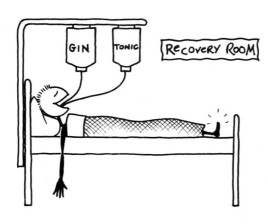

# A Moralistic Story About Two Peasants, a Bird and a Cow

In the bitter cold of winter, in a small village during a howling gale and with darkness falling, a peasant is wandering home to his meagre village. Suddenly he stops as he sees a small game bird on the ground, nearly dead from cold and privation. The peasant picks up the bird and warms it. The bird soon recovers and the peasant wonders what to do next. At that moment a herd of cattle come by and one of them drops a large dollop right in front of him. Realising that if he puts the bird in the steaming cow's dollop, the bird will stay warm until morning and then be able to fly away, he does this and then goes home. But a second peasant comes along after the first one has gone and hears the bird chirping happily to itself in the steaming mess. This

403

peasant seizes the bird, breaks its neck and takes it home for supper.
This story has three morals:
1 Do not believe that everybody who drops you in the shit is your enemy.
2 Do not believe that everybody who gets you out of the shit is your friend.
3 Whenever you *are* in the shit keep quiet about it.

## International Definitions of Heaven and Hell

*Heaven*: Italian cooks, English police, a French mistress, German car mechanics and the Swiss undertaking any necessary organising
*Hell*: English cooks, German police, a Swiss mistress, French car mechanics and the Italians doing the organising

These definitions were told by a Russian national working at the World Health Organisation in Geneva, as the WHO in-joke of 1987. It was noticeable that no Russian attributes figured in the anecdote!

## The Frog

The humble frog has found its way into medical applications, such as in 1896, one year after the discovery of X-rays, when it was used as a test object, one of the earliest in diagnostic radiology, to demonstrate the 'new light'.

Earlier than Röntgen's discovery of X-rays, its name was taken to describe a thermometer, invented around 1650 in Florence. The Frog Thermometer was a strange contraption shaped like a large hollow glass frog. It was filled half-way with water containing glass beads of different colours according to their specific gravity. Fixation to someone's body was by means of ribbons so that the belly of the frog would be in contact with the skin. The water would warm up and become lighter in proportion, and the glass beads would rearrange themselves according to their own specific gravity or colour. From their position and colour, the temperature was estimated.

The frog is still used in this context today. In a *New Scientist* editorial dwelling on the world's slow perception of the extent and importance of AIDS, deforestation, desertification, the greenhouse effect and the hole in the ozone layer, the following was written. 'If you take a frog from its pond and put it in a pan of hot water, it will jump out. If you put it in a pan of cold water and heat it slowly on the stove, the frog will sit there until it boils to death. The frog's senses are equipped to measure only large differences in temperature, not gradual ones. It has evolved that way because it normally has no need to measure gradual changes in temperature. Today, the human race has a lot in common with the frog in the pot'.

REFERENCES
*Mould R F 1985 A History of X-rays and Radium* (Sutton: IPC Business Press)
*New Scientist* 17 June 1989
Sterpellone L 1988 *Instruments for Health* (Milan: Farmitalia Carlo Erba)

## Sales Opportunities!

The following is a newspaper cutting of May 1982 which appeared with the caption 'Nuns Arrested'. It was contributed by Mr Tony O'Connor.

When a group of nuns spent over £2 million on a new convent in northern Belgium, officials launched an investigation into how the funds were raised. Now the Mother Superior and a senior nun are under arrest, charged with defrauding the Health Service of £7 million. It is alleged they encouraged doctors to sign blank treatment forms, then got other sisters to claim payment for health work never carried out.

## Bulgarian Proverb

Nature, time and patience are the three great physicians.

## Irish Proverb

Death is the poor man's best physician.

# A Medical Demon From Angkor Wat

*Ravana, King of the Rakshas, a demon to whom were attributed several medical works. Bas-relief from Angkor Wat. (Courtesy Wellcome Institute Library, London.)*

---

## Confusion

Contributed by a sales manager who wishes to remain nameless!

'I fully realise that I have not succeeded in answering all of your questions. . . . Indeed, I feel I have not answered any of them completely. The answers I have found only serve to raise a whole new set of questions, which only lead to more problems, some of which we weren't even aware were problems. To sum it all up . . . in some ways I feel we are as confused as ever, but I believe we are confused on a higher level, and about more important things.'

# Moscow Case History: 12-Year Follow-Up

**Moscow Case History, 1983** (pages 134–145) relates my experiences with the KGB in Moscow that ended with them questioning me about the *leukaemia radiation pill*. I now believe this referred only to the supplies of potassium iodide pills stored in British police stations for the purpose of blocking the thyroid in the event of any widespread necessity. Readers may be interested in what later transpired.

### The Newspapers

A number of newspaper reviews of *Mould's Medical Anecdotes* concentrated on **Moscow Case History** with headlines such as 'Heavy hand of the KGB' (*Guardian*), 'Sinister encounter with Moscow KGB' (*Daily Telegraph*), 'Blocking move' (*Times*), 'Cold shoulder for the KGB' (*Yorkshire Post*) and 'Doctor comes in from the cold ... just' (*Manchester Evening News*).

*New Scientist* (under the heading 'Take nothing for granted') published the cartoon below showing the reception desk at the Hotel Minsk in Gorky Street, just a short walk from Red Square, the Bolshoi Theatre and the Kremlin.

The *Guardian* review was written by Anthony 'Bill' Tucker whom I knew very well. In fact, during the 1983 episode in Moscow the only undetected 'cry for help' I could think of was to send a postcard containing an 'obvious lie' home to my family, asking them to contact Bill Tucker, who might perhaps have been able to arrange some publicity to help get me out if I ended up languishing in the Lubyanka. The 'lie' was that the Bolshoi Opera was absolutely lousy: having raved about it the previous three years I had visited it, this was palpably untrue.

When I returned home I found that this method of communication was a total failure in that the children had said to my wife

'Oh, don't worry, Dad is just having a good time with the vodka'

and the card was thrown in the rubbish bin.

A decade later, in 1993, I happened to meet a now retired Bill Tucker at a Pugwash Conference on Chernobyl in London. The first thing he said to me was

'Do you remember the postcard?'

and then he burst into peals of laughter. It is quite something to provide a well known journalist with one of the most amusing memories of his career!

### Publicity Follow-Up

There were two effects of this publicity: the first in the sales figures for the book (which in the event paid for a family holiday in Tunisia, so at least I had something to be grateful for to Tommy and Yuri!!), and the second in some additional anecdotes related by other visitors to Moscow during the time of Brezhnev. The best of these are now described.

### The British Airways Pilot

You have several alternatives if you return to your Moscow hotel bedroom and find an unknown lady waiting in bed for you. Two of the more unusual are illustrated here.

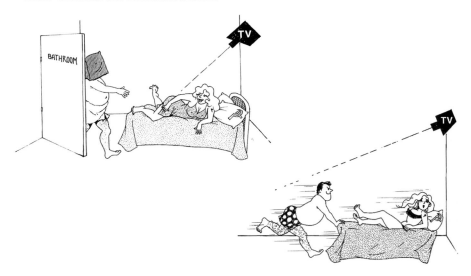

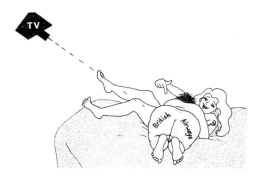

An airline pilot encountering this situation did neither of these: he went straight to bed. He was, of course, video-recorded, but—as he later told his superior back in London—he thought it was OK as his face could not be seen. What gave him away were two tattoos, one on each cheek of his bottom: one said *British* and the other said *Airways*. This was at the time when British Airways was first allowed to fly direct to Moscow from London. Previously, passengers had first to fly to Helsinki and then onward by Aeroflot or a Finnish airline to Moscow.

## Chandeliers

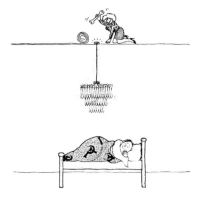

It was always considered a good idea to look for 'bugs' in your hotel room and I well remember Professor Sir David Smithers telling me how, while attending a medical conference in the 1950s, he had found a bug in the chandelier; this was somewhat unusual as most bugs were supposed to be in telephones or in the bed frame.

409

At one conference, however, a lady engineer even looked under the carpet, where she found a bug screwed to the floor. Being an engineer, she used the set of spanners she had with her to unscrew the bug, which she threw into the wardrobe, and went for a good night's sleep. The next morning at breakfast there was a buzz of conversation throughout the restaurant. The previous evening some poor chap had a chandelier fall on top of him whilst he was in bed!

## 1984

In March of 1984 I attended a workshop at the World Health Organization headquarters in Geneva on 'Computers in Radiotherapy and Oncology' organized by the European Association of Radiology. In the middle of the week I was invited to lunch at a Russian's flat and—for a cold March day—was given a very nice meal of beef Stroganoff, strawberries and cream, and Stolichnaya vodka. The purpose of the meeting turned out to be an opportunity for my lunch hosts to say of Tommy and Yuri:

'It was a mistake'.

'They were *cowboys*: agents provocateurs.'

That was all until the Chernobyl accident two years later.

## 1986

The Chernobyl accident occurred in April 1986 and in August of that year the first 'Post-Accident Chernobyl Conference' was held at the International Atomic Energy Agency in Vienna. The conference was attended by government delegations representing all countries in the United Nations and I was fortunate to be among the UK delegation headed by Lord Marshall, then Chairman of the UK Central Electricity Generating Board.

Most of the delegates were nuclear engineers and industrial radiation safety experts, as distinct from medical physicists and cancer statisticians. Indeed, only four months after the accident there was no information on which one could make any educated guess concerning the future incidence of and mortality from diseases such as thyroid cancer and leukaemia. However, it had been time enough for me to collect all the major published literature with a rather vague idea of possibly writing a book on the accident.

Imagine my surprise when I saw that the Deputy Chief of Protocol for the USSR delegation was none other than the short, squat Vladimir Zagourelko: *the* Vladimir Z of **Moscow Case History, 1983**. However, it was a greater surprise to him, because when I tapped him on the back of his shoulder and he turned round, he had a look of pure shock. I well remember that I felt very angry but all I asked was

410

'How are your wife and family?',
to which he replied
'What do you want?'
and I answered
'To see Academician Ilyin (who was the Joint Leader of the USSR
delegation and responsible for health matters as the Director of the
Institute of Biophysics in Moscow) and obtain his support for my
writing a book about the accident'.
 Zagourelko duly arranged the meeting with Leonid Ilyin, but for the rest
of the conference totally avoided me.  Ilyin asked
'Why should I support you as an author?'
and my off-the-wall reply obviously worked:
'If you have an American journalist writing the book you will not
like what is written. If you have a Western communist journalist
writing the book, nobody will believe him. So I am your best
chance.'
The rest, as one might say, is history, and I was provided with
unique assistance with photographs and visited Chernobyl in December
1987, meeting with Mikhail Umanets (Director of the Nuclear Power
Station), Professor Angelina Guskova who treated the injured (including
undertaking some bone marrow transplants with Robert Gale from the
USA), Major Leonid Telyatnikov who was the chief of the fire service at
Chernobyl at the time of the accident, and many others.

Collaborative work on the Chernobyl accident continues to this day,
mainly with the World Health Organization and the Russian Academy of
Medical Sciences Medical Radiological Research Centre in Obninsk some
100 km from Moscow. Some of the publications are listed below.

All without any KGB problems.

### 1995

In November 1995 I was invited to attend the WHO, Geneva,
international conference on 'Health Consequences of the Chernobyl and
Other Radiological Accidents'; this was, in effect, a 10-year post-
Chernobyl report.

It was a very successful conference and many were enjoying
themselves at a final drinks party when I was approached by one of the
WHO public relations staff who was a Russian. He was a rather large,
tall and drunk Igor R---v, who snatched a half-full champagne glass out of
my hand and insisted on substituting a glass of yellow vodka—'hunter's
vodka', which contains quantities of port, brandy, bison grass and other
constituents in addition to vodka. The conversation then proceeded along
the following lines:
'The KGB never forgives or forgets'.

'Don't tell me something that is obvious.'

'I've been told not to speak to you as you are a British spy.'

'Rubbish: you've been reading too many Le Carré novels.'

What an exercise in PR! It turned out after a bit of detective work that he had been at WHO in 1983, had left to work elsewhere and had only recently returned. He claimed that one of his colleagues was the son of a Colonel-General who had been chief of the KGB 5th Directorate, responsible for ideology. This was indeed correct but apparently well known throughout WHO.

A second claim was that he was a family friend of Academician Leonid Ilyin who had written a book on his personal experiences of the Chernobyl accident (this is also true); the book had been badly translated into English and I was to be approached as an English language editor. However, Igor impressed on me that I could not mention his name! This was after I tried to question him about his curriculum vitae, which included experience as a Moscow taxi driver and as one of the team of directors/producers of the Russian-to-English language video 'Olga & Viktor' made at the time of the 1980 Moscow Olympics.

Ilyin and many other Chernobyl accident workers whom I know were actually at the conference, and a couple of days previously Ilyin had mentioned to me that he had written this book; although there had been no talk of any involvement in a new English language edition, he had plenty of time to raise this topic.

Twelve years on from Moscow 1983, the meeting with Igor the so called PR expert was not a pleasant experience. Obviously not all of the *cowboys* are in Texas!

REFERENCES

Ilyin L A 1995 *Chernobyl: Myth and Reality* (Moscow: Megapolis)

Ivanov V K, Tsyb A F, Mould R F *et al* 1995 Cancer morbidity and mortality among Chernobyl accident emergency workers residing in the Russian Federation *Current Oncology* 2 102–110

Ivanov V K, Tsyb A F, Mould R F *et al* 1996 Study of the possible role of radiation in the induction of cancers in Russia after the Chernobyl accident *Current Oncology* 3 in press

Mould R F (Ed) 1984 *Computers in Radiotherapy and Oncology* (Bristol: Adam Hilger)

Mould R F 1988 *Chernobyl—The Real Story* (Oxford: Pergamon)

Souchkevitch G N, Tsyb A F, Repacholi M N and Mould R F (Eds) 1996 *Health Consequences of the Chernobyl Accident. Results of the IPHECA Pilot Projects and Related National Programmes. Scientific Report* (Geneva: World Health Organization)

# Hospital Department Noticeboards

Noticeboards are an excellent source of cartoons and two examples from Europe and North America are the *House of Fun* (above) from Munich in Germany and the *Stressed Zebra* (below) from Atlanta in the USA.

'I THINK IT'S STRESS'

# Science According to Mark Twain

'The great thing about science is that you can reap such a rich harvest of conjecture from such a minimal investment in fact.'

413

# Anecdote of William Beaumont

Today for any doctor visiting hospitals in Detroit and its surroundings the name William Beaumont is associated with the hospital of that name in Royal Oak, but there is no obvious link with earlier famous people of that region such as Henry Ford. Nevertheless, William Beaumont has a unique niche in medicine in the USA, and is described as America's first physiologist.

William Beaumont was born in 1785 and early in his career became the doctor at the US Army post at Fort Macinac on Macinac Island, Michigan. In 1822, when this story begins, the island was a large fur trading station, and it was in the local store that a 28-year-old fur trapper, Alexis St Martin, was badly wounded when a musket loaded with duck shot accidentally went off two feet from his, blowing a hole the size of a palm of a hand. The damage included the fracture of two ribs and the rupture of a portion of the left lower lobe of his lung.

The stomach had been perforated and nobody expected the unfortunate casualty to live. In the event, he did survive, and William Beaumont conceived the idea of 'an experiment of introducing food into the stomach through the orifice purposely kept open and healed with that object ... very soon after the first examination', according to a supposed eye-witness account written many years later.

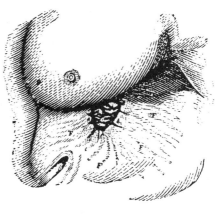

William Beaumont and an engraving of the wound beneath the
left breast, drawn with Alexis standing upright.

414

However, Beaumont was not as callous as this sounds and it is proven that he had repeatedly attempted to close the fistula before he ever thought of the experiments that made both himself and Alexis famous. The latter had made a remarkable recovery only three weeks after the accident and by then had a healthy appetite. One problem, though, was the fact that part of the sixth rib had been blown entirely off and the only solution was to amputate it halfway between sternum and spine. Nevertheless the major problem was that any measure other than using a lint plug for the stomach fistula resulted in all the contents of the stomach flowing out.

The gastric experiments on Alexis continued for several years and what was termed the second series began in 1829. William Beaumont's three important discoveries at this time were that gastric acid is indeed an acid, that various stimuli are important in regulating gastric secretion, and that gastric juice has the power to cause digestion, although not equally for all foods.

A typical experiment involved keeping Alexis in a fasting state for 12 hours and inserting a gum elastic tube. It was noted that there were only some 60 cc of gastric juice and no accumulation in the stomach. This and subsequent fasting experiments convinced Beaumont that there was no gastric juice in the fasting state.

He measured stomach temperature using a thermometer inserted through the fistula. Many experiments were performed, including, for example, Alexis being given in 1831, a 'robust meal' of roast beef, potatoes, beets and bread—albeit with a string attached to the food to guide it into the stomach via the fistula. Alexis, who was then living in Beaumont's home, was 'kept exercising about his usual employment as a house servant' and later the gastric contents were examined. The fluid was reported as 'tasting slightly acid, giving the flavour of dilute muriactic acid (another name for hydrochloric acid) and very slightly bitter'.

William Beaumont died in 1853 following a fall descending stairs in icy weather; by this time Alexis was 59 years of age and in poor financial circumstances. Until that date, although he had received many offers from various physicians to participate in further experiments he had always refused.

Now the situation had changed and he agreed to work with a 'Doctor' J G Bunting who, in the words of the editor of the *British American Journal of Medical & Physical Science* in 1850, had described Bunting as 'like the corn extirpator and itinerant dentists and oculists, this curer of stammering and stuttering remains just long enough in one place to dupe the people, pocket their dollars, and disappear before a sufficient time shall have elapsed to test his practice and expose his knavery'. Furthermore, in 1856 the editor of the *Providence Daily Post* published this damning statement: 'Dr Bunting is an impostor, swindler and villain altogether too well known in this city to need an extended notice from us'.

This was during the period when Bunting was exhibiting Alexis in a large number of cities including Boston, Washington, Cincinnati, Detroit, Louisville, Montreal, New York and Philadelphia, all in the space of four months. It was said that Bunting's curiosity and motivation were more like those of P T Barnum than of a true medical man.

The experiments Bunting put Alexis through included the insertion of a glass thermometer to show a temperature of 100 degrees and asking Alexis to swallow a tumbler full of water which he then ejected through the orifice by contracting his abdominal muscles. The spectators were then invited to personally inspect the gastric mucosa through the 'window'.

At the end of 1856 Alexis was still a poor man and completely disillusioned with the medical profession, so much so that he refused to undertake any more experiments; although he did negotiate several times, financial payments were not agreed.

Alexis died in 1880 at the age of 86 years, reputedly having fathered some 20 children, although only four were still alive at the time of his death. His family is said to have left his body in the sun for four days before burial in order to discourage those doctors who might still have had an interest in the stomach.

He was then buried in the churchyard in a purposely unmarked grave, at the uncommon depth of eight feet, and the coffin was covered with heavy rocks as a further measure of security. It was not until 1957 that his contributions to medicine were recognized: the site of his grave was found with the help of two of his grandchildren and in 1962 a plaque in French and English was placed proclaiming: 'Through his affliction he served all humanity'.

REFERENCE
Nelson R B 1990 *Beaumont, America's First Physiologist* (Geneva, IL: Grant House Press)

# An Expert as Defined by Nils Bohr

'An expert is a man who has made all the mistakes which can be made, in a very narrow field.'

416

# Cure for Crossed Eyes

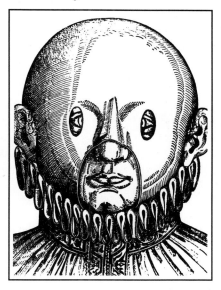

Headgear used in the 16th century to correct astigmatism.

# Baoding Iron Balls and Jingluo Medicine

For the princely sum of the equivalent of £4 I was able to acquire for myself some iron balls in Hong Kong's Stanley Market. They are hollow balls containing a 'rattle mechanism' which sounds when one rotates the pair around in either the left or right hand. Designs are available in various colours with different Chinese symbols, including dragons. Reproduced below is part of the text from the instruction booklet, which also contains a testimonial from the Light Industry Ministry of the People's Republic of China, together with the following quotation from one Pujie: 'The king of the treasures. Strengthen the ball players and make our country powerful'.

'Manufactured by the Iron Balls Factory, of Baoding City, Hebei Province, PR China, they are a traditional product dating back to the Ming Dynasty of 1368-1644. Originally the balls were solid but afterwards were designed hollow with a sounding plate. Of the pair, one sounds high and the other sounds low and they are for the elderly to build up physical strength and to remove disease.

Cover of the leaflet giving information on the iron balls.

According to traditional Chinese medical theory of 'Jingluo', the 10 fingers are connected with the heart, that is to say, by means of jingluo the 10 fingers are connected with the cranial nerve and vital organs of the human body, including the heart, liver, spleen, lungs, kidneys, gall bladder, stomach and intestines. On plucking the iron balls with fingers, the ball can stimulate the various acupuncture points on the hand, resulting in the unimpededness of circulation of vital energy and blood in the body.'

The description of the effects then continues, including the prevention and cure of hypertension and making sure that the mind is sober. After exercising for months or years, your brain will be in good health with a high intelligence 'drowning your worries and prolonging your life'.

# Fook–Luk–Shou

*Fook, Luk* and *Shou* are the famous trio of gods symbolizing the treasures sought by billions of Chinese both ancient and modern. *Fook* is the God of Happiness and is characterized by his contented smile and his big round stomach, signifying a well-to-do life. In ancient China happiness was closely associated with a large family tree symbolizing unity, stability and control.This thinking certainly has been well preserved

418

over many centuries; propagating the family tree remains one of the most practised and joyous events of the Chinese—a wisdom indeed shared by many others!

*Luk* is the God of Prosperity and is identified by his hat and special outfit worn in the past only by high-ranking government officers. Even today, the Chinese believe firmly that power equates to wealth, and politics is the door to the fulfilment of desire and ultimate personal satisfaction.

*Shou* is the God of Longevity and is the most impressive figure of the three; he has a bald head and bushy white beard and moustache that are meant to inform onlookers that he is over 150 years old. The 'shou' peach in his hand is said to be a God-given treasure for immortality. The shape of the peach complements very well its owner's head: big, round and probably of a single-cell structure!

Statuettes of the trio—from cheap pottery or plastic to very elaborate carvings in wood or ivory—can be purchased worldwide. Those of *Shou* (left) and *Luk* (right) shown here were found in 1985 on a wayside stall far north of Chiang Mai in Thailand (see also **Meo Hill Tribe Doctor in Northern Thailand**, pages 395–397). For some unknown reason there were no carvings of *Fook*, so perhaps I can only expect prosperity and longevity! However, using *Fook's* definition of happiness, I am not too worried as I now have two grandchildren as recorded on the dedication page of this book!

419

# Autopsy on Lenin's Brain

Vladimir Ilyich Lenin became seriously ill in the spring of 1922 and was partially paralysed and unable to speak. Although he recovered somewhat he was again incapacitated by semi-paralysis in December of that year. During 1923, although no longer the active leader of the state and party, he was still dictating papers; in March 1923 another stroke occurred and he was never to speak again. He died of a final stroke on 21 January 1924 aged 53 years.

A mosaic portrait of Lenin in the Central V I Lenin Museum at
2 Revolution Square on Marx Prospect between Red Square and
Sverdlov Square which contains the Bolshoi Theatre and the Lubyanka.
(From the *Museum Guidebook* 1979)

By 1925 a special laboratory was actively studying Lenin's brain with the aim of proving that he had an extraordinary mind. In 1927 this laboratory became the Institute of the Brain. Lenin's brain was preserved in formalin and alcohol, divided into blocks and set in paraffin wax for sectioning. In total 30 963 sections were sliced in this manner. Some even found their way to Germany during World War II where the Nazi brain study aim was the exact opposite of that in Moscow!

The Institute carried out many comparisons using brains from the general population, from Bolsheviks and from several academicians who were scientists, physicians and artists. It is not known if they had all given prior approval.

420

Lenin's brain weighed 1340 g and the average brain is in the range 1300-1400 g. His frontal lobe was found to have a higher proportion of furrows than others studied and this was taken as evidence that it displayed a 'high degree of organization'. It was recently reported [1] that in May 1936 a report to the Central Committee had completed its tasks in the previous 10 years and that its work consisted of 153 pages of typescript and 15 albums containing 750 microphotographs, tables and diagrams.

However, in the *Times* of 19 January 1994, Dr Oleg Adrianov, the then Director of the Institute, stated 'In the anatomical structure of Lenin's brain there is nothing sensational'.

REFERENCE
[1]   Volkogonov D 1995 *Lenin: Life and Legacy* (London:  HarperCollins)

# *Interview with Pierre Curie and 90 Years Later*

The earliest published interview with Pierre Curie was by Cleveland Moffat, a journalist for the *Strand Magazine*, in 1904.  Moffat's text and the accompanying artists' sketches are reproduced here.

Radium for illumination.

421

'M Curie took me into a darkened room where I saw quite plainly the light from the radium tube, a clear glow sufficient to read by if the tube were held near the printed page. And of course this was a very small quantity of radium, about 6 centigram. We estimate, said M Curie, that a decigram of radium will illuminate 15 square inches of surface sufficient to read by.'

Radium used at a reception in Lille.

'Radium offers a ready means of distinguishing real from imitation diamonds since it causes the real stones to burst into brilliant phosphorescence when brought near them in a darkened room, while it scarcely has any such effect upon false stones. M Curie made this demonstration recently at a reception in Lille to the great delight of the guests.'

Moffat also described the working conditions he found:

'M Curie was in one of the rambling sheds in the School of Physics (in the Rue Lhomond), bending over a small porcelain dish, where a colourless liquid was simmering, perhaps half a teacupful, and he was watching it with concern, always fearful of some accident. He had lost nearly a decigram of radium only a few weeks previously, he said. This was in an experiment when he heated a tube of radium to 2000 degrees Fahrenheit over an electric furnace and the tube exploded scattering the precious contents'.

This and similar experiments by Pierre Curie can perhaps be linked with information obtained 90 years later at the time when the bodies

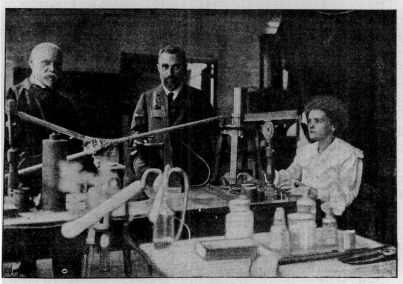

BIBLIOTHÈQUE
— DE —
RADIOLOGIE MÉDICALE
DU Dʳ BÉCLÈRE

1ʳᵉ ANNÉE. — Nº 1    Le Numéro : 50 Centimes    JANVIER 1904

# LE RADIUM

### PUBLICATION MENSUELLE

DIRECTEUR
Henri FARJAS

ADMINISTRATION : 36, Rue de l'Arcade, PARIS
TÉLÉPHONE : 124-03
Abonnements : Un An, 5 fr. — Union Postale, 7 fr.

## Le Laboratoire de la Rue Lhomond

M. & Mᵐᵉ CURIE

The front cover of the first issue of the journal *Le Radium*,
published in 1904. The man on the far left is Pierre Curie's laboratory
assistant; this photograph is invariably reproduced with the assistant omitted.

Reinterrment ceremony at the Pantheon in Paris, April 1995, showing the coffins of Pierre (1859–1906) and Marie Curie (1867–1934). The polystyrene foam symbols held by schoolchidren include β to signify beta-rays and Po to signify the element polonium, also discovered by the Curies. α and γ are hidden from view. The polystyrene cylinder seen far right next to β was devised by a historian without a background in science who informed me that it was meant to symbolize an atom.

of Pierre and Marie were reinterred in the Pantheon in Paris, in the presence of the then Presidents of France and Poland, François Mitterand and Lech Walewsa.

Prior to the ceremony, measurements were made of the radioactive content of the remains, those of Marie being remarkably well preserved, even to the rose in her hands which had been taken from a rambling rose bush (that still exists today) in the garden outside her laboratory balcony at the Institut du Radium, Paris. The remains of Pierre were found to be highly radioactive, which in the light of the type of experiment described above is not surprising. Marie's remains, however, were not so highly contaminated.

She died of aplastic anaemia and it is probable that she received a major radiation exposure during World War I when she was in charge of the French x-ray ambulance service at the Front, in places such as the Somme and Ypres, driving what were known as 'Little Curies'. She was also responsible for training the x-ray machine technicians (radiographers). Photographs clearly show that no x-ray tube shielding was used. Of course she also received radiation exposure from radium, but as primarily a chemist (rather than a physicist like Pierre) maybe she worked with smaller quantities.

424

The Curies are commemorated in a recent French banknote and as a final comment I quote the experience of an American radiation oncologist who was trying to obtain an illustration of this note for a paper he was writing: 'We found the note but then nearly went to jail trying to get it photocopied. There seems to be absolutely no trust left in this country, and we ran into a succession of individuals at various photographic facilities who felt that we represented the epitome of master counterfeiters. We were turned over to the FBI on two separate occasions'.

Front (top) of the 500 franc banknote showing a World War I 'Little Curie' ambulance and a schematic diagram showing the effect of a magnetic field on $\alpha$, $\beta$ and $\gamma$ rays. This is repeated on the reverse (bottom) together with a Bohr atom and the chemistry apparatus used by the Curies.

## Bizarre Claims for Radium

A rather strange use of radium for therapy is seen in **Early Medical Radium** (pages 54–59); devised by a Dr Saubermann of Berlin in 1912, this radium therapy enabled one to have a home supply of radium water as a panacea. However, even worse applications were to be found in the first two decades of this century.

# WEAK DISCOURAGED MEN!
### NOW BUBBLE OVER WITH JOYOUS VITALITY
### THROUGH THE USE OF

# GLANDS AND RADIUM

It is not clear what type of glands the 'weak men' have to take with radium: possibly monkey glands. For the ladies there was Atomic Perfume, given a number 58 which corresponds to an isotope of cobalt (Chanel Number 14 would correspond to carbon-14 used in radiocarbon dating!), and Tho-Radia cosmetic cream (also sold as a lipstick, a soap, and a baby powder!). The latter was a complete fraud: not only did it not contain any thorium or radium (when it was analysed by Dr Andrée Dutreix in the 1980s) but the supposed formula was based on the

426

recommendation of one Dr Alfred Curie. There have been Pierre, Marie, Irene, Eve and Pierre's brother Jacques, but never an Alfred.

In terms of drinks there was Zoe Atomic Soda which gave one infinite energy just like a nuclear reactor, and radon bulbs which could be used with scotch whisky instead of soda to provide 'scotch and

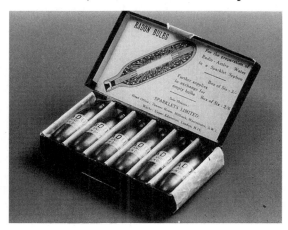

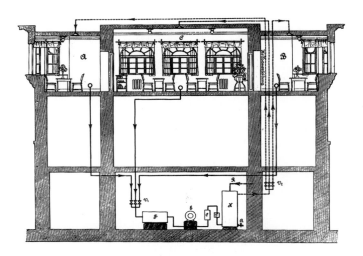

radioactivity'. The radon bulbs were made by the well known Sparklets Company which manufactures soda siphons and bulb refills.

To return to Dr Saubermann's radium therapy, which consisted of radon gas bubbled into water from a radium source: there was a remarkable design for a holiday hotel in Joachimsthal, Bohemia, the site of the pitchblende mines from which Marie Curie obtained the mineral ore from which she discovered radium. The architect's plan shows what might at first glance be thought to be an air-conditioning plant in the basement of the hotel. In fact this is a radon gas plant, the basis of which is a large quantity of radium. The gas was piped into the bedrooms as seen on the top floor, and in addition you could even breathe the radon gas through a contraption looking rather like a trumpet while relaxing in your bath.

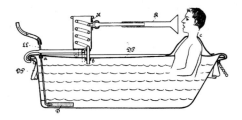

REFERENCE

Mould R F 1993 *A Century of X-Rays & Radioactivity with Emphasis on Photographic Records of the Early Years* (Bristol: Institute of Physics Publishing)

428

# HOSTETTER'S
# CELEBRATED STOMACH BITTERS,

PREPARED AND SOLD BY
## HOSTETTER & SMITH,
PROPRIETORS.

Nos. 58, 59 & 60 WATER, and 58, 59 & 60 FRONT STS,
PITTSBURGH, Pa., U.S.A.

A Swift and Certain Cure for Dyspepsia, Liver Complaint, and every species of Indigestion, an unfailing Remedy for Intermittent Fever, Fever and Ague, and all kinds of periodical disorders; a means of Immediate Relief in Flux, Cholics, and Choleraic maladies; a Cure for Costiveness; a mild and safe Invigorant and Corroborant for Delicate Females; a good Antibilious alterative and tonic preparation for ordinary family purposes; a powerful recuperant after the frame has been reduced and attenuated by sickness; an excellent appetizer as well as a strengthener of the digestive forces; a depurative of the blood and other fluids, desirable alike as a corrective and mild cathartic, and an agreeable and wholesome stimulant.

Sold at wholesale by all Druggists and Grocers in the principal Cities, and retailed by all Apothecaries throughout the United States, the British American Provinces, Mexico, Guatemala, the Independent Republics of South America, Cuba, Sandwich Islands, Australia, and portions of Europe.

## NEW YORK OFFICE, 59 CEDAR ST.
Hostetter, Smith & Dean, San Francisco, California.

# Father Murphy's Donkey

The following appeared in the *Baltimore Sun* newspaper in the 1980s.

'Father Murphy was a priest in a very poor parish. He asked for suggestions as to how he could raise money for his parish and he was told that a horse owner always has money. So, he went to a horse auction but he made a very poor buy, the horse turned out to be a donkey. However, he thought he might as well enter the donkey in a race. The donkey came in third and the next morning the headlines read "Father Murphy's Ass Shows". The Archbishop read the paper and he was greatly displeased. The next day the donkey came in first and the headlines read "Father Murphy's Ass Out In Front". The Archbishop was up in arms and decided something had to be done because Father Murphy had entered the donkey again and it came in second. The headlines read "Father Murphy's Ass Back In Place". The Archbishop thought this too much, so he forbade the priest to enter the donkey again the next day. The headlines read "Archbishop Scratches Father Murphy's Ass". Finally the Archbishop ordered Father Murphy to get rid of the donkey. He was unable to sell it, so he gave it to Sister Agatha as a pet. When the Archbishop heard of this he ordered Sister Agatha to dispose of it at once, so she sold it for $10.00. The next day the headlines read "Sister Agatha Peddles Her Ass for $10.00". They buried the Archbishop three days later.'

All statistics are soluble in alcohol

430

# *Organization Charts*

The 'Name a Fish Competition' (above) was found on the noticeboard of a World Health Organization department, but for legal reasons I have omitted the names of the director, secretaries, etc, allocated to the various fish! The international organograms (below) have also caused some amusement.

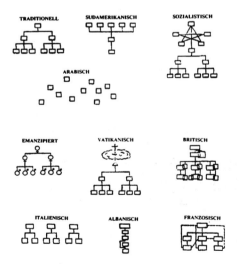

From top left to bottom right: Traditional; South American; Socialistic; Arabian; Emancipated; Vatican; Albanian; French. This organogram was originally given to me in German, which no doubt accounts for no German organogram being drawn!

431

# Vietnamese Cobra Tonic

This photograph shows a pharmacy with a difference. It was taken several kilometres from Ho Chi Minh City in 1994 when visiting a snake farm as a tourist attraction. The cure-all medicine is seen on the far right in eight bottles in which a small (dead!) snake is clearly seen. The liquid is the local 'firewater'. On the far left is a single jar of 'pickled snake in alcohol' containing a much larger specimen. Apparently when the cork is removed, the stench is one of the worst in the world: however, I was not brave enough to confirm this!

# Does Strategy Matter?

*(Contributed by Dr Michael Gürtler)*

Once upon a time, the Dutch and the Japanese decided to have a boat race on the River Rhine. Despite practising hard, the Dutch lost by a mile.

The Dutch team management decided to set up a project to find out why. Their conclusion: the Japanese had eight people rowing and one steering while the Dutch had one person rowing and eight steering. Consultants were hired to study the team structure. After millions of guilders and six months they concluded: too many people steering and not enough rowing.

432

The following year the team structure was changed to four 'steering managers', three 'senior steering managers' and one 'executive steering manager'. A performance appraisal system was also introduced to give the person rowing the boat an incentive to work harder.

This time the Japanese won by two miles.

The Dutch sacked the rower for poor performance, sold the paddles, cancelled the budget for a new boat, awarded high-performance bonuses to the consultants, and distributed the money saved to senior management.


# Mayan Artefacts and the Turin Shroud: Is Microbiology Better Than Carbon Dating?

Following radiocarbon dating studies of the Turin Shroud in 1988, it is generally held that that the Shroud dates back only to 1260–1390. However, a fascinating study by Leoncio Garza-Valdes and Stephen Mattingly of the University of Texas Health Sciences Center at San Antonio now casts doubt on these findings.

Their investigation began not with fibres from the Shroud, but with artefacts from the Mayan civilization which flourished in southern Mexico and parts of Central America around 300 AD.

### The Itzamna Tun

The Itzamna Tun is a phallic display from a large white and green veined celt of actinolite–albitite carved in the form of a penis, the ventral side of which has a reptilian head and human hands in carved relief. The centrally located reptilian mouth probably represents the cosmic portal. On the phallus side are three pairs of connecting holes as if for insertion of three cords or ribbons.

Three ribbons knotted together represent a Mayan and Olmec blood-letting symbol; when found in phallic depictions they indicate penile self-mutilation, the probable means by which the celt became anointed with blood. Three separate tests have confirmed that the blood is human, with human DNA, and have dated it to between 240 AD and 690 AD using radiocarbon dating techniques.

### The Ahaw Pectoral

The Ahaw pectoral is a carved gem made of emerald-green jasper and composed of five pieces; it is 19.5 cm high, 16.5 cm wide and weighs

The front (left) of the Itzamna Tun, which is 23.5 cm long and weighs 1415 g.
Blood found on the celt has been confirmed as human from some 1500 years
ago. The back view (right) shows clearly the three pairs of holes.
(Photograph courtesy Leoncio A Garza-Valdes, MD)

325.5 g. It was acquired without provenance in 1955, and in 1978 was studied in Texas by two mesoamerican archaeologists who concluded that it was one of the most important pre-Columbian artefacts known. It is similar to a pectoral worn by a shaman, one of the God-kings of Mexico.

In 1983 two pre-Columbian art connoisseurs from New York indicated that, based on their knowledge of mesoamerican iconography and style, the artefact was a fake and offered to buy it for their fake collection. However, it was not a fake.

While parts of the pectoral's surface have been repolished in modern

Frontal view of the pectoral (left) and (right) the individual parts which were joined together by fibre threads.
(Photograph courtesy Leoncio A Garza-Valdes, MD)

times the intact areas contain deposits of black varnish. This coating— composed of manganese and iron oxides, silica and clay minerals—is in general 10–30 μm thick but varies from a few micrometres to over 500 micrometres. A similar type of coating is also found on pottery in ancient tombs.

### Analysis of the Dark Varnish on the Pectoral

The deposits were studied by optical microscope, Fourier transform infrared spectrograph, energy dispersive spectrometry and scanning electron microscope, and wavelength dispersive spectrometer. Results showed the following biocultural remnants: charcoal fragments, cinnabar, copal, cordage traces, remnants of blue feather, root deposits and phytoliths.

Microcolonial fungi were found encapsulated in the dark varnish on the surfaces where the pieces of the pectoral had been bound together and where the strong grooves had not been repolished. These were removed for analysis by scanning electron microscopy.

### Distortion of Carbon Dating

The presence of varnish and of bacteria and fungi can distort the results of carbon dating; for example, in the Itzamna Tun 95% of the blood had been replaced by bacteria and fungi which were younger than the artefact itself, thus skewing the results of the carbon dating. The artefact actually dates from some time in the period 100 BC to 100 AD, a difference of some 500 years from the carbon date of about 460 AD.

435

Scanning electron microscopy on the Itzamna Tun, x20, clearly showing the
blood replaced by bacteria and fungi.
(Photograph courtesy Leoncio A Garza-Valdes, MD)

Scanning electron microscopy on the Ahaw pectoral, x80.
(Photograph courtesy Leoncio A Garza-Valdes, MD)

## The Turin Shroud

Given that deposits of bacteria and fungi on blood and on fibres can distort carbon dating, we now turn to the Turin Shroud.

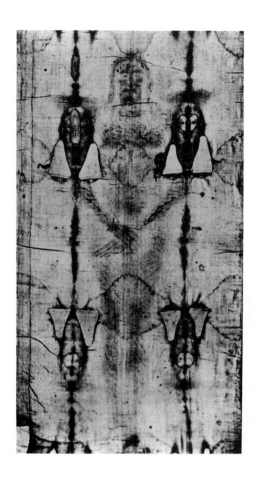

The Turin Shroud (as seen in a frontal view by the naked eye) is a negative image of a man with his hands folded. The linen is 14 feet 3 inches long and 3 feet 7 inches wide. The shroud is wrapped in red silk and kept in a silver chest in the Chapel of the Holy Shroud in the Cathedral of St John the Baptist in Turin, Italy.
(Copyright Holy Shroud Guild, Esopus, New York)

437

*Chronology*

| | |
|---|---|
| 33 | Death of Jesus |
| 476 | Last Emperor of the western Roman empire is deposed |
| 1095 | Crusades begin |
| 1215 | Magna Carta signed |
| 1357 | First recorded appearance of the Shroud: placed on display in a church in Lirey, France |
| 1532 | The Shroud is damaged by heat and water during a fire at Chambery, France |
| 1578 | The Duke of Savoy moves the Shroud to the cathedral of St John the Baptist, Turin, Italy |
| 1898 | Secondo Pia takes the first official photographs of the Shroud |
| 1988 | Radiocarbon methods date the Shroud to 1260–1390 |

In 1993 Dr Garza received a sample from Giovanni Riggi di Numana in Turin; he took the official Shroud samples for the carbon dating of the 1980s. It has been demonstrated that the Shroud's fibres are coated with bacteria and fungi that have grown for centuries and that the carbon dating method sampled the contaminants as well as the fibre's cellulose. From these analyses, the San Antonio team concluded that the Shroud is centuries older than its carbon date.

 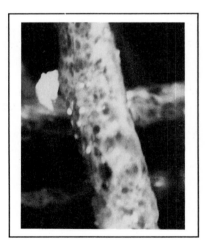

Under a scanning electron microscope, the Shroud's fibres (left) look crusted with colonies of bacteria and fungi, which form a bioplastic coating around the strands; a single fibre strand from the Shroud is shown on the right.
(Photograph courtesy Leoncio A Garza-Valdes, MD)

438

To date, the Catholic Church has declined to designate the fibres as an official sample: there have been so many claims of 'samples' that it is almost like claims for pieces of the true cross, which if placed together would make a forest (see also **St Blaise**, page 97). However, Dr Garza will soon be given a designated official sample and the San Antonio tests will be repeated.

Other relevant studies are being undertaken at the University of Arizona in Tucson, where they are preparing carbon dating procedures to test the hypothesis relating to ancient contaminated fibres on an ibis mummy bird that stylistically would date back to some time in the period 330–30 BC. Physicists will sample collagen from bone, which is relatively unaffected by bacteria and fungi, and compare its date with wrappings from the mummy. Textiles contain large quantities of bacteria and fungi because they have a much greater surface area by volume than a smooth object of similar size, and the wrappings are thus important for comparison.

Two samples of mummy wrappings will be tested, one cleaned of contaminants using conventional methods, and the other cleansed using a method developed by Dr Garza and Dr Mattingly, who claim that the conventional technique fails to remove the bacteria and fungi.

Dr Rosalie David, the Keeper of Egyptology at the Manchester Museum, some of whose work is described in **King Philip of Macedon and The Manchester Mummies** (pages 160–169), is quoted as stating 'This could be a great breakthrough in understanding the ancient world'. She has supplied wrapping samples from the Manchester mummy '1770', a 13-year-old Egyptian girl, for carbon testing using the Garza–Mattingly cleansing technique. This is of great interest because it has never been satisfactorily explained why the wrappings of '1770' indicate that they are 1000 years younger than the bones.

ACKNOWLEDGMENT
    I am most grateful to Dr Garza-Valdes for extremely interesting discussions in San Antonio, at very short notice, and also for copyright illustrations for this publication.

REFERENCES
Barrett J 1996 Science and the Shroud: microbiology meets archeology in a renewed quest for answers *The Mission Journal of the University of Texas Health Sciences Center at San Antonio* 23 6–11
Garza-Valdes L A and Stross B 1992 Rock varnish on a pre-Columbian green jasper from the tropical rain forest (the Ahaw pectoral) (San Antonio: Materials Research Society)
Stross B and Garza-Valdes L A 1993 Gems and the Maya shaman (San Antonio: Materials Research Society)

# A Mexican Worm

Snakes are not the only animals to be pickled in alcohol (see **Vietnamese Cobra Tonic**, page 432) and perhaps the most well known is the so called 'worm' in a bottle of Mexican tequila. This worm is in fact a larva which is found on the leaves of the agave plant from which tequila is made. It looks rather like a small black caterpillar about 2 cm in length. It is said that those who actually eat the worm are subject to hallucinations!

# Calf Fries in Texas

Forget about rhino horn and instead visit the annual NIOSA (Night in Old San Antonio) Fiesta which is one of the largest in the USA. Its international flavour includes Sauerkraut Bend, French Quarter, Main Street and Frontier Town. It is the latter which has a stall for Calf Fries, supposed to be one of the best aphrodisiacs.

They consist of thin sliced strips of bull's testicles which are breaded, fried and served hot (at NIOSA in a conical piece of paper, much as are French fries throughout the world). In Colorado they are known as Mountain Oysters and in Montana as Prairie Oysters. They taste like chicken.

## Wine and Women

Recommendation given by the Winemaster of the State of Bavaria at a dinner in the cellars of the Prince-Bishop's Palace in Würzburg, on the celebration of the centenary of the discovery of x-rays by Wilhelm Conrad Röntgen (shown below right) in November 1895. On posing the question 'What would you do if you had to give up either wine or women?', he then gave the answer that it 'Depends on the vintage'.

## Stalin and his Georgian Doctor: The Kipshidze File

Stalin's health deteriorated in his final years and he has been described as not looking like his public portraits but like 'an old small man with a face ravaged by the years' [1].   His death was announced on 6 March 1953; apparently he had died six days earlier of a brain haemorrhage and a stroke which caused paralysis although he had long previously lost speech and consciousness. There have been various novels, the most recent in 1995 [2], with the denouement being the poisoning and/or shooting of Stalin. However, no post mortem results were ever published, if indeed an autopsy was in fact performed. Thus the poisoning theory remains a hypothesis.

It would, though, be no surprise if he had been murdered because this was at the time of the infamous Doctors' Plot where the Kremlin doctors, described by Stalin as 'assassins in white blouses' and lampooned in a 1953 issue of the satirical magazine *Krokodil*, were mostly Jewish and were supposed to be acting on the orders of an international Jewish organization with headquarters in America.

Cartoon of the Doctors' Plot, *Krokodil* 1953.

The doctors were confronted by one Doctor Timashuk, a false witness who was rewarded with an Order of Lenin. It is also recorded that Stalin supervised the interrogation and torture of the prisoners. This

443

prelude was similar to the start of the 1934–1938 purge trials and no person in the Communist hierarchy could feel safe. However, Stalin died before the doctors could be put on trial, but less than a month after his death they were completely rehabilitated. The Deputy Minister of State Security, Mikhail Ryumin, who was blamed for constructing this plot, was executed in July 1954.

Dr Nicholas Kipshidze in military uniform in 1915.

Dr Nicholas A Kipshidze (1887–1954), although one of Stalin's doctors, escaped any involvement in this mythical plot for various reasons. He was born in Gori, Georgia, the same small town in which Stalin himself was born in 1879. Dr Kipshidze graduated from St Petersburg Medical Academy and then spent several years at the Pasteur Institute in Paris where, with Professor Dushan, he developed a special culture medium for use when researching different bacteria. For the rest of his life he worked in the Georgian capital Tbilisi, where he rose to become Professor of the Department of Internal Medicine in the Medical School.

Stalin and his cronies in the Politburo and the Central Committee usually spent summer on the Black Sea coast and in the mountains of Georgia: they always requested Dr Kipshidze to visit them and treat any ailments. In addition, for a number of years he also looked after the healthcare of Stalin's mother, who by then was also living in Tbilisi.

It was most unusual that Dr Kipshidze was not a member of the Communist Party, but what saved him from being a victim of Stalin, who hated doctors and never totally trusted them, was that he had always refused to move to Moscow when asked to do so by Stalin. Remaining in Tbilisi, far from Kremlin intrigue, most certainly ensured that he could continue with his work unmolested.

Biographical details of some of the Communist luminaries Kipshidze treated are given as footnotes. Most came to an untimely end and the Lubyanka was where Tukachevshy and many others breathed their last. However, with regard to Zhdanov's death in 1948, this is perhaps not too surprising if in 1946 his blood pressure was 235/100! For those interested in more details, a selected list of references is given [3–10]. The following extracts are from Dr Kipshidze's journal, translated from the Georgian language in which it was written, and use his own style of wording as much as possible.

Front cover of Dr Kipshidze's journal showing the author in the late 1940s.

### First Meeting with Stalin, Tbilisi, 17 October 1935

*[Stalin's Mother]*

This morning I went on a routine visit to Stalin's mother who for many years had diabetes. There was no need to prescribe insulin as it was only mild diabetes and I treated her using a strict diet and regimen. I had been seeing her for 15 years and she was a very intelligent lady who had a hard life.

She was very anxious this morning of the 17th and her pulse had risen to 120 from the usual 70, and I asked her why she was so anxious. Stalin's mother closed all the doors in the room, and then told me that her son Joseph was going to visit that morning and that she was very excited because she had not seen him for eight years. The reason she gave was that she did not like the Moscow climate where it was very cold, and also that she did not like to travel.

During this conversation I looked outside the window and saw a lot of people, the militia, the KGB and some others from the military, gathered in front of the house. I tried to excuse myself but his mother asked me to stay with her until her son arrived. I thought that it was not a good time to be here, even though I was her doctor. I gave his mother, Katerina is her name, my usual telephone number at the clinic and said that if there is any need for me to be called back then she should telephone.

I went to the clinic, undertaking my routine work with the patients, but after 50 minutes I was given the telephone message that I should quickly go to Stalin's mother. In 10 minutes I had arrived and when I entered, saw his mother sitting on the sofa and Stalin talking slowly in the middle of the room. She told her son that I am a good doctor and treating her very well. Stalin looked at me and smiled and said

'Oh, don't say so many good words about doctors. I also love this guy. I knew him when he was a young boy in the city of Gori'.

Although I had lived in Gori and attended the same school as Stalin, I was 10 years younger and could not remember him.

He mentioned insulin and I said that I had told his mother that with insulin, physicians are now obtaining miracles. When Stalin asked about complications I told him that there might be some with the pulmonary system: which at that time were very dangerous as there were no antibiotics.

His mother was 79 years old at the time, and we only spoke in Russian about the diabetes since she only understood Georgian.

*[Stalin's Childhood]*

Stalin then started asking his mother questions about his childhood:

446

'Why did you punish me very frequently when I was not doing good things?'

'Was I a good son during my childhood?'

His mother replied

'I will tell you why I physically punished you frequently: it was because I wanted you to be a very good sincere person'.

Stalin's comment was

'Now in the Communist country, schoolteachers do not physically punish children, and if children are physically abused at home, then the government and the state take measures against the parents'.

Stalin's mother smiled in answer:

'You are a good citizen now, a good person now, because I punished you all the time during childhood'.

Stalin in 1918.
(Courtesy: Mary Evans Picture Library)

*[Why his Mother Wanted Stalin to be a Priest]*

Stalin then continued

'Mother, why did you want me to become a priest after school graduation?'.

'Priests had a very good life at that time. They had a lot of food and

447

did not have to work hard. But now I want to admit that was a mistake'.

Everyone started laughing.

### Tbilisi, Morning, 18 October 1935

Today Stalin described to me how his father also used to punish him physically:

'Once, when he saw me reading a school book that was not the book for that year, he took the book away, lifted me up and threw me on the floor. After that I was in bed for a day with bad back pain, but in a few days I was OK'.

Stalin then told his mother

'I need to leave to meet some government officials of the Georgia Republic but will see you later this evening'.

### Tbilisi, Evening, 18 October 1935

When Stalin returned, his mother had prepared some sweets for him, but he did not eat them and only had one glass of Georgian wine and a few pieces of cheese and then left, promising to return: which he did not.

When I met him once again 11 years later he remembered the entire incident and asked me

'Do you know why I did not stay with my mother and why I left Tbilisi so quickly?'.

'Of course I don't know, but I think you had some important things to do.'

'No, it's not true. Some KGB people told me that people are going to kill me that night and they are organizing a court: but I recognized this and so I left and fooled them.'

Then he started smiling.

### Moscow, Kremlin, March 1936

It was the 15th anniversary of the Soviet State of Georgia and I was part of a small delegation from Georgia, who met with Stalin, Molotov and Ordzhonikidze. After the official greetings, Stalin looked at me, immediately recognized me, and we asked after each other's health. I was then praised by Ordzhonikidze at the meeting:

'Oh, you two are very healthy, but I want to tell you Mr Stalin, a few months ago I was dying in Tbilisi and this is the guy who helped me to survive'.

### [Voroshilov]

During this conversation Voroshilov came to listen to the exchange and I recognized that he had a severe cough. Stalin told me

'You see Mr Voroshilov is not very OK. You might see him and give him medication to treat him'.

After that, Voroshilov gave me his telephone number and later that week I saw him to prescribe medication.

*[Kalinin]*

At the end of the meeting I saw Kalinin whom I had seen in Tbilisi in 1935. He had been travelling through Tbilisi from Moscow to Erevan in Armenia for an important meeting. On the way from Moscow to Tbilisi he became ill with a very high fever and a temperature of 40 °C and I was asked to come to see him.

He had pneumonia and I advised him to stay in Tbilisi and not travel. However, he insisted on travelling, because that was the order from Stalin. At his insistence I therefore travelled with him on the train to Armenia and returned with him to Tbilisi by which time his health had deteriorated.

## Stalin's Dacha, Black Sea Coast, 7–12 December 1946

At 7 am I received a telephone call from the First Secretary of the Georgian Communist Party who told me that I was to fly to Gagara to see a patient on the Black Sea coast. I very much dislike flying and asked for permission to go by train but was told the patient could not wait.

Two planes were waiting at the airport, one of them a Douglas jet, and the chief of Aeroflot was also on the same flight.

When the plane landed at Adler airport, the nearest airport to Gagara, I was taken to Stalin's private residence in Gagara and the Politburo members with us were very nervous because Stalin was waiting for us at noon and there was no way they could have any excuse to be late.

In front of the dacha awaiting us was Stalin's chief secretary, Poskrebyshev, who told us

*[Stalin Treated by a Vet]*

'Stalin became ill last night and had a high fever, 39.6 °C, his abdomen is painful, but there is no vomiting or diarrhoea'.

Stalin usually treated himself with aspirin and also made his own diagnoses, which in this instance he claimed was flu. After I had examined him I realised that it was not flu and told Stalin that he had made a mistake and that he had taken his aspirin, and a pill of urotropin without asking a doctor. He looked at Poskrebyshev:

'This is the guy (Poskrebyshev) who told me to take these pills. Do you know, Dr Kipshidze, who was this guy before the war? Now he is a General but then he was a veterinary surgeon'

449

I replied
'Now I understand what is going on: vets can't even treat pets but
they always like to treat human beings'.

I then admitted to Stalin that he had coronary arteriosclerosis. The
two measurements of blood pressure taken that day were 135/75 and
145/80, the lungs were clear but the liver was enlarged by 4 cm. The
current problem, though, was that he was suffering from gastroenteritis
caused by food poisoning.

Stalin was worried about the seriousness and I predicted bouts of
diarrhoea, which indeed did occur. The fact that I recognized this two
hours before it occurred made Stalin consider me an expert physician and
to take all the medication which I prescribed. I gave him some general
recommendations and then we discussed alcohol.

'I never drink vodka or cognac, but I want to ask if I can have 150 g
of wine every day.'

'That's OK'

I replied and then finished with:

'Mr Stalin, the next thing I want to ask you is that you must live to
100 years and this is in my interest also'.

Stalin was very pleased and came and kissed me, gave me his hand and
said

'We are friends now'

which was very surprising because I knew that Stalin was not sentimental.

### [Zhdanov]

I stayed at his dacha a few more days and also examined Zhdanov, a
Politburo member, and when I recommended Zhdanov change his
lifestyle, Stalin interrupted

'Whoever is in the Politburo, they are not alcoholics. If they drink
they only drink red wine. They do not take any liquor'.

Zhdanov, although only 50 at the time, had very bad cardiovascular
problems and severe hypertension. Zhdanov was also with us when
Poskrebyshev had returned with Stalin's medication, but he could not
guarantee that it had not been tampered with in the pharmacy. Zhdanov
also offered, but I and Poskrebyshev took the pills to test them.
Everything was OK and three hours later we gave Stalin his medication
although we never told him about testing it first.

On Stalin's instructions I took Zhdanov's blood pressure and found it
was 235/100 and he told me that this was the same as in Moscow but that
when he was in Sochi a few days previously it was 180/100.

### [Ordzhonikidze]

One day Stalin asked me about Ordzhonikidze since he was very

Kirov (left) and Ordzhonikidze (right) in the early 1920s.
(Courtesy: Mary Evans Picture Library)

interested to learn if was due to poisoning when Ordzhonikidze was ill in 1934. I told him that there was no reason to think this, and that the problem in 1934 was bleeding from a gastric ulcer. I had treated Ordzhonikidze in Tbilisi in 1924 and this was remembered.

*[World War II]*
    Stalin then talked a lot about the Great Patriotic War and about the siege of Stalingrad, the battle for Berlin and other interesting events including the Tehran and Yalta conferences with Winston Churchill and Franklin Roosevelt.

*[Winston Churchill]*
    Stalin was talking about Tehran about Churchill being afraid that the United Kingdom might become Red.  Stalin told him
    'Oh don't worry, Mr Churchill, I think that the redness is not a bad
    colour at all.  Even more, red is healthy, a good sign for children'.
At that, Churchill laughed.
    Then in Yalta, on one occasion Churchill asked for vodka and caviar for his team and this was enjoyed so much that Churchill asked for a second order and then for a bill. Stalin answered
    'You don't need to pay Mr Churchill, this is our lend-lease'
and they both laughed.

451

*[Doctors' Salaries]*

Stalin was also interested in how much doctors were paid and I told him that the salaries were OK, but that while doctors who were working in medical schools were getting enough salary, general physicians were not. He went to his room and held out some money in his hands:

'Oh, this is your money'.

I have never been in such a difficult position and I did not know what to do.

'Mr Stalin, I have enough salary and treating you is a great honour
for me and I do not need any payment'

and I placed the money back on the table. Stalin then took the money and put it in my pocket and after a further refusal, shouted

'As the Prime Minister of the Soviet Union I order you to take this
money'.

I was really very frightened and told him

'Thank you very much Mr Stalin. I'm not a party member so I
cannot donate this money to the party, but I'll give this money to my
clinic to buy some more equipment'.

Stalin calmed down, but said

'Listen, don't do that, because if you do I'll get very angry'.

The amount was 5000 roubles (in 1980, 100 roubles were the equivalent of £101—RFM) and although I had a private practice I had never received such money before.

*[Stalin on Private Practice]*

I overheard him telling Zhdanov and Poskrebyshev that he liked physicians in the Soviet Union having private practice.

'It is not capitalist, but the way to choose a doctor.'

The final event I attended at the dacha was a dinner, and again during conversation, Stalin stated his approval of private practice. He also found out that I liked bananas and ordered some for me.

ACKNOWLEDGMENT

I am most grateful to Dr Nicholas Kipshidze, now a cardiologist in the USA and the grandson of the original Dr Nicholas Kipshidze, for providing me with all the information relating to his grandfather and as a gift presenting me with one of very few copies in existence of the original Georgian language journal (published in 1987 in a limited issue) containing the medical records of Stalin and his associates, and for taping for me the English translation of the journal.

REFERENCES

[1]  Deutscher I 1979 *Stalin* (Harmondsworth: Penguin Books) Reprint of revised edn, 1966
[2]  Meade G 1995 *Snow Wolf* (London: Hodder & Stoughton)

[3]    Andrew C and Gordievsky O 1990 *KGB: The Inside Story of Its Foreign Operations from Lenin to Gorbachev* (London: Hodder & Stoughton)
[4]    Pipes R 1990 *The Russian Revolution 1899–1919* (London: HarperCollins)
[5]    Pipes R 1995 *Russia under the Bolshevik Regime 1919–1924* (London: HarperCollins)
[6]    Volkogonov D 1995 *Lenin: Life and Legacy* (London: HarperCollins)
[7]    Rybakov A 1988 *Children of the Arbat* (London: Hutchinson)
[8]    Ustinov P 1983 *My Russia* (London: Macmillan)
[9]    McCauley J 1993 *The Soviet Union 1917–1991* 2nd edn (London: Longman)
[10]   Shukman H (Ed) 1993 *Stalin's Generals* (London: Weidenfeld & Nicholson)

FOOTNOTES

**Frunze, Mikhail Vasilyevich (1885–1925)**

Was a revolutionary as early as 1905 and a leading general in the Civil War after the 1917 revolution. He is regarded as the father of the modern Red Army and has a military academy, which is on the road between Moscow airport and Red Square, named after him. He succeeded Trotsky as Commissar of War. He fell ill in November 1925 and some of his doctors advised surgery, whereas others did not because they feared that he was too weak to survive the operation. However, the Politburo ordered Frunze to have surgery and he died in the operating theatre. Trotsky suggested that Stalin orchestrated the pro-surgery doctors' opinions as a means to condemn Frunze to death as he had sided with Zinoviev in an argument against Stalin. Zinoviev lasted longer but in 1936 was a defendant in one of the show trials and shot.

**Kalinin, Mikhail Ivanovich (1875–1946)**

Was one of Lenin's first supporters in Russia of the Bolshevik faction and was the formal Head of State from 1919 to 1946. Co-founder of *Pravda* in 1912. The city of Köningsburg was renamed Kaliningrad and in Moscow one of the main roads was the Kalinin Prospekt which went in the direction of the Arbat from the Kremlin walls.

**Kirov, Sergei Mironovich (1886–1934)**

Was a revolutionary first in the Transcaucasus, based in Azerbaijan, but in 1926 was transferred to be party boss in Leningrad and by the early 1930s his power almost rivalled that of Stalin. In 1934 he was assassinated at party headquarters. This was used by Stalin as an excuse to carry out a purge in Leningrad, where some tens of thousands were deported to Siberia or executed, or both. In 1956 Kruschev strongly implied that Stalin had engineered the assassination. The Kirov Ballet is named for this Bolshevik.

**Ordzhonikidze, Grigori Konstanovich (1886–1937)**

An early Bolshevik who was in prison in Baku in 1904 with Stalin and a member of the Central Committee in 1912. Was instrumental in subduing Georgia in 1922, which was then mainly Menshevik. In 1926 he was Chairman of the party's Central Control Commission, responsible for eliminating dissension among party members amongst other things. By 1932 he was

Commissar for Heavy Industry but later, with many other industrial leaders, he objected to the purges of his colleagues, in particular the trial of his deputy Piatakov. Ordzhonikidze suddenly died in unexplained circumstances in 1937 and in 1956 Kruschev stated that Stalin had driven him to suicide.

**Tukachevsky, Mikhail Nikolaievich (1893–1937)**

Fought in World War I in the Tsarist army and then entered the Red Army and was responsible for the defence of Moscow in 1918. Played a leading role in army reforms and served as deputy to Frunze in 1924, Chief-of-Staff 1925–1928 and Deputy Commissar for Defence after 1931. Was tried on a trumped-up charge of conspiracy with Nazi Germany, and was executed in the Lubyanka in 1937. The top military commanders who were shot during the purges were: 3/5 Marshals, 14/16 Class I and II Army Commanders, 8/8 Admirals, 60/67 Corps Commanders, 136/199 Divisional Commander and 221/397 Brigade Commanders. In 1988 Tukachevsky was cleared judicially and rehabilitated by official decree.

**Voroshilov, Kliment Yefremovich (1891–1969)**

Although a Marshal of the Soviet Union, Voroshilov was a mediocre personality and showed no leadership qualities. After the revolution he was in charge of the 5th Army on the Tsaritsyn front but refused to cooperate with the military experts and the front became a disaster. Nevertheless, after the death of Frunze he became Commissar for War and remained in this post from 1925 to 1940. By then, Voroshilov had been involved in so many military catastrophes that even Stalin recognized his faults and downgraded him. He also fully supported the purges of the 1930s. When attacked by Kruschev after Stalin's death he made an apologetic statement to the Party Congress and thus just managed to retain his party membership.

**Zhdanov, Andrei Alexandrovich (1896–1948)**

Succeeded Kirov as party boss in Leningrad and organized the deportations and executions which followed Kirov's assassination. He was still the party boss during the 1941–1944 siege in World War II. Was responsible for the severe control of ideological guidelines in art, literature, philosophy and medicine. Zhdanov died in 1948 in mysterious circumstances. However, immediately following his death there occurred the notorious *Leningrad affair* in which some 2000 people, many of whom were Zhdanov's associates and subordinates, were purged, probably due in part to Malenkov who had been his great rival at one time.

# Mice and a PhD Thesis

The mouse is an animal which, if killed in sufficient numbers under controlled conditions, will produce a PhD thesis.

*(Journal of Irreproducible Results)*

# Japanese Medical Anecdotes of the Late 18th and Early 19th Centuries

Until the last quarter of the 19th century Japan was virtually a closed country—Nagasaki was its only seaport of entry and trading took place only with China and with the Dutch. Illustrations found in the Hiroshima University Institute for the History of Medicine and in the Siebold Memorial Museum in Nagasaki provide interesting insights into these early years.

One of the earliest schools of obstetrics was based on the work of Gen-etsu Kagawa and a textbook entitled *San-ron-yoku* was written by his son Gen-teki in 1775. The illustrations of the normal (left) and abnormal (right) positions of twin foetuses shown below are taken from this book: the heads are represented by the solid black circles and the placenta by the shading.

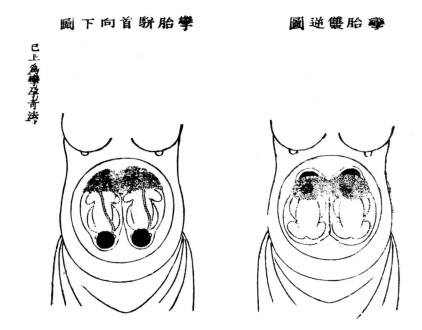

Another textbook of the late 18th century, by Kakuro-ryo Katkuro, was first published in 1793. Katakura was one of the experts of the Kagawa school of medicine, and was influenced by the medicine of Deventer in Holland. The illustrations shown overleaf are from a revised

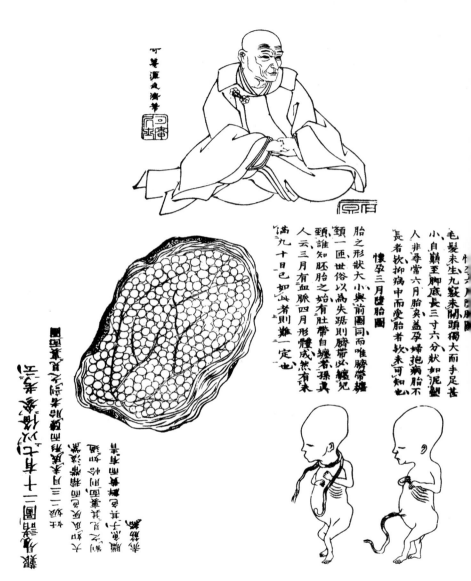

緣肖生先陵鶴

胎之形狀大小與前圖同而唯臍帶纏
頸一匝、世俗以為失跌則臍帶必纏兒
頸、誰知胚胎之始有肚帶自纏者孫真
人云、三月有血脈、四月形體成然有未
滿九十日己如此者則難一定也

懷孕三月墮胎圖

毛髮未生九竅未開頭偏大而手足甚
小、自巔至腳底長三寸六分狀如泥塑
人、非尋常六月胎矣蓋孕婦抱病胎不
長者欬抑病中而愛胎者欬未可知也

edition of 1799, and are of the author, the placenta at 2–3 months, a
strangulated foetus and a normal foetus.

456

The medical history of breast surgery in Japan began in 1805 when Seishu Hanaoka first operated on the breast under general anaesthesia: the illustration below shows the bandaging after surgery. His first patient was his mother and Hanaoka used an anaesthetic derived from herbal plants. This was the world's first use of general anaesthesia and predates by some 40 years its first use in the USA, as described in *The Birth of Anaesthesia* (pages 374–379) which describes the work of William morton at the Massachusetts General Hospital, Boston.

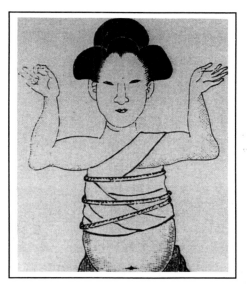

The introduction of western European medicine into Japan is credited to Philipp Franz von Siebold (1796–1866); he was born in Würzburg almost exactly a century before Röntgen discovered x-rays in the same Bavarian city. Siebold came from a medical family—both his grandfather and father were doctors. His father also published papers on obstetrics at the end of the 18th century.

Following graduation in surgery, obstetrics and internal medicine, in 1822 Siebold was appointed an army surgeon at Batavia in the Dutch East Indies (now Jakarta, Indonesia). However, he did not stay here long and in 1823 was sent to Japan to investigate trading possibilites and to study Japanese culture and products.

On arrival in Nagasaki he assumed responsibility for the health of the employees of the Dutch East India Company and also began lecturing in medicine and botany to Japanese students. By 1824 he had set up a clinic and medical school (on the site of the present Museum), and was issuing licences to practice medicine.

457

Siebold in 1826 painted by the Japanese artist Kawahara Keiga.

Siebold soon took a Japanese consort called Taki, who in 1827 gave birth to his daughter Ine. His contract with the Dutch was only for five years and he left Nagasaki in 1828. However, his ship ran aground in Nagasaki harbour and unauthorized export items—such as maps of Japan—were found amongst his belongings. He was interrogated as a suspected spy and was deported at the end of 1829. Fearing for the safety of Taki and Ine, he left money with trusted students but did not return to Japan until many years later in 1860, by which time he was 63 years old.

Siebold corresponded with Taki after his departure and they exchanged small gifts, including the pair of mother-of-pearl inlay portraits of Taki and Ine shown at the bottom of the facing page that was sent to him in Europe.

Taki married some time after Siebold's departure and in 1845 Siebold married a German noblewoman. His family in Germany comprised three sons and two daughters and the eldest son, Alexander, accompanied his father when he returned to Japan in 1860. It is recorded that Siebold and Taki had a nostalgic reunion. She is remembered to this day for a beautiful hydrangea: Siebold was a botanist as well as a doctor and named the flower which is now the city flower of Nagasaki— 'Hydrangea Otaksa'—for her (Taki's original name was Sonogi but she was also known as Otaki-san).

シーボルト記念館

Otaksa
長崎市鳴滝2丁目7-40
☎(0958)23-0707
たき

The story of Siebold and Taki has many parallels with that of Puccini's opera 'Madame Butterfly', although Lieutenant Pinkerton was in the US Navy and Siebold was a German in the service of the Dutch East India Company. The Siebold Museum denies that there is any association, but the background music for visitors nevertheless contains the famous aria from 'Madame Butterfly'!

The medical aspects of this story do not end here: Ine received medical training from Siebold's Japanese students and also from Dutch physicians and went on to become Japan's first female practitioner of Western medicine. She was also later appointed an Imperial Court physician, and as an obstetrician assisted in the birth of Emperor Meiji's

Chosabro Kusumoto.

child. Her son Chosabro Kusumoto was also a physician, and was one of the earliest medical professors of Osaka University Medical School. At the time when Osaka University consisted only of the faculties of medicine and science he became its second President. Together with the first President, Aihiko Sata, he is commemorated by a bronze bust (dating back more than 50 years) in front of the new Osaka University Medical School Buildings completed in 1993.

Taki's great-granddaughter Taka Mitsuse with her husband Morobuchi Mitsuse, who from 1869 was the Osaka Prefectural Medical School's interpreter. This photograph was taken in 1877.

460

# A God Who Cures Nightmares and a God Who Can Cure All Diseases

There has previously been a **Charm Against Nightmares** (page 299) but there is also a god who cures nightmares. The gilt–bronze statue (left) of Yumechigai-Kannon—believed to have the power to change nightmares into pleasant dreams—dates from the 7th century and can be seen in the oldest wooden temple in the world, the Höryüchi built more than 1300 years ago.

The photograph (top right) of the 18th century Buddhist god Binzuru (Pindola Bharadväja) was taken at the Todaiji temple (bottom right) in Nara. This is the largest wooden building in Japan: 155 feet high with a frontage of 187 feet and 166 feet from front to back, and housing a Buddha 49 feet high. Clothed in a red robe, Binzuru was a disciple of Buddha and was said to have been an expert in the occult. It is commonly believed that when a person rubs part of the image of Binzuru and then rubs the corresponding part of his own body, any ailment there will disappear.

# *Not the End?*

The latest (Imogen 1996)

# Index

With such a large number of anecdotes covering an extremely wide spectrum of medical topics, it is essential that easy access by subject is available to the reader. I have therefore produced this extensive cross-referenced index. Included in the major subject areas are diseases and illnesses, countries of origin of anecodotes, saints, alcohol, aphrodisiacs, famous and not-so-famous physicians and surgeons, x-rays, radium, statistics, advertisements, quacks, American anecdotes, cures, gravestone and memorial inscriptions, mummies, gods and charms, spy stories, and medical recipes and potions. Where dates are listed, they are given within ( ) brackets.

467

471

485